THE HUMAN BODY.

THE HUMAN BODY

AND ITS

CONNECTION WITH MAN,

ILLUSTRATED BY THE PRINCIPAL ORGANS.

BY

JAMES JOHN GARTH WILKINSON,

MEMBER OF THE ROYAL COLLEGE OF SURGEONS OF ENGLAND.

"HE cried with a loud voice, Lazarus, come forth. And he that was dead came forth, bound hand and foot with grave-clothes: and his face was bound about with a napkin. Jesus saith unto them, Loose him, and let him go."

PHILADELPHIA:
LIPPINCOTT, GRAMBO AND CO.,
SUCCESSORS TO GRIGG, ELLIOT AND CO.
1851.

PHILADELPHIA:

T. K. AND P. G. COLLINS, PRINTERS.

CONTENTS.

	PAGE
Dedication,	vii
Preface,	ix
The Human Brain,	25
The Human Lungs,	83
Assimilation and its Organs,	132
The Human Heart,	169
The Human Skin,	252
The Human Form,	289
Health,	343
Appendix,	407

TO

HENRY JAMES, Esq., of New York.

My Dear James—

This book is indebted to you for its appearance; for, without you, it would neither have been conceived nor executed. I dedicate it to you as a feeble tribute of friendship and gratitude that would gladly seek a better mode of expressing themselves. It may remind you of happy hours that we have spent together, and seem to continue some of the tones of our long correspondence. *Valeat quantum!* It could not lay its head upon the shelf without a last thought of affection directed to its foster parent. That prosperity may live with you and yours, and your great Commonwealth, is the prayer of,

My dear James,

Your faithful friend,

JAMES JOHN GARTH WILKINSON.

St. John's Wood, *London,*
May 25, 1851.

PREFACE.

At the end of a work, an author writes the preface or beginning, unable to neglect the law which urges that extremes shall meet. Prefaces need no apology, and a reader intending to study a book, does not do himself justice unless he peruses them. A preface gives the author's last conception of his aim—the most comprehensive eye with which he see it. And unless the reader looks through this eye, he cannot enter into the author's mind. The study of a book is the temporary putting on of the faculties and insights of another; and the sooner the assumption takes place, the sooner the reader begins to read aright.

To assist this we have a few remarks to make that may prepare in some measure for the following pages.

We labor under difficulty in procuring the right audience for the present discourse. The subject of which it treats has been so much narrowed to a class, that on the one hand that class, the medical profession, claims it as an exclusive knowledge; and on the other hand, the public mind is in abeyance with regard to it, and looks upon it as a property for ever alienated from its possession. We therefore run the risk of finding no readers, unless we can persuade the public that the knowledge of the human body belongs to every man, woman and child, and has no more necessary connection with physic, than with art, industry, philosophy, divinity, or any of the other occupations that we do in the body, and by the body. To write a treatise on the subject through which this persuasion shall run with vigorous consequences, has been a leading motive with us in the present work.

Persons for the most part have no idea that the sciences belong to the great world in the first place, and that the classes who are actively cultivating them, are but little bands of pioneers, that are contending with difficulties at the out posts, and slowly winning a new magnitude of knowledge, which as soon as it is settled, belongs afresh to the large country of popular common sense. On the contrary, they allow each party of settlers to hoist the flag of a petty kingdom of their own, without insisting, as ought to be done, that the adventurers shall at once become the colonies of the mother state. Thus it is that professors of all kinds have kidnapped the sciences, and the people fear to take so much as a walk under the walls of these bristling strongholds. But scientific feudalism is evidently about to pass away.

This desirable result will be accomplished by the growth of large towns, that is to say, popular doctrines, of the sciences, which will belong to the broad industry and insights of mankind, and will not contest, but swallow up, the castles of the present chiefs, and reverse the feodal direction of duties and fines. Already we have seen the process going on in the history of civilization, and we are about to witness the same thing in the progress of science and of thought.

It may take place somewhat as follows. The masters of science will pursue their own way, warm in their little senates, and cheered by their subjacent schools; satisfied with perfectionating the chairs into which they have been inducted. They will gather useful facts, and more and more comprehensive formulas, and appealing to rarer and rarer qualities in their scholars, be most out of sight when they are most at home, until at length, by extremity of cleverness, they will become invisible to all but adepts. Lords of the last ether of things, they will only exist as influences, and not as appreciable substances. In the mean time the people, happily unconscious, will listen to flesh and blood, which seems to talk to them about themselves. Abstractions will have sailed away to the flying island of the professors, who will exert a strong attraction upon the whole wealth of the world's inanitions. A clear stage and sward of common sense will then be at the popular disposal, and the problem of universal education may be conceived upon new grounds. The safeguard of the people will be this—that when the learned have done their best, it is no matter to *them;* their hearts will be unseduced, and their brows unterrified, by the lordliest and most captivating formulas. One or two plain Johnsonian stampings on the ground, will be sufficient to convince them that their ignorance and carelessness on these scores is most invulnerable and sublime.

But here comes a knot claiming *Deus intersit.* Nor can the difficulties of man be without a response from the mercy-seat: new thoughts and new persons will come speeding to make new things possible. Already we see that the whole of the sciences may reappear on the popular side. The waning moon of the schools gives place to the full-orbed Dian of a more generous light. All the common truths that have been neglected since the foundation of philosophy; the stones that the builders have rejected; that great orthodoxy that has bided its time while ages of conceit were cuffing against its serene face, will rise out of land and sea, and out of the graves of the hearts of many generations, and come in hosts such as no man can number to the people in their hour of need. The doctrine of final causes, which is God in the sciences, and which atheism hates, will ramble over the pleasant fields, and teach them to childhood as a book; and out of its mouth will come lessons of order and fitness, which will make the world as familiar as a father's and a mother's house.

We find it to be a law, when a branch of knowledge has been cultivated for ages, and still remains inaccessible to the world at large, that its principles are not high or broad enough, and that something radically deeper is demanded. If it does not interest universal man, that is sufficient to prove the point. This law is illustrated by many things, and particularly by the history of the arts.

Once upon a time all books were perpetuated by copying with the hand; whoever would possess a volume, must undergo the toil of transcribing it, or pay the price of that to another. This was the narrowness of the circle of the learned. The perfection of the copyist's art was soon attained, but the utmost rapidity and cheapness in this mode of multiplying books, could not render them to the mass of the public. How was the seeming impossibility to be surmounted? By some meaner process, which should deteriorate the appearance of books to a degree commensurate with the humble fortunes of the poor; so that if the rich man's Bible cost him £30, a copy of but one sixtieth the excellence should be produced for one sixtieth the sum? Far from it indeed! The means of making the poor man a proprietor of books, lay in a glorious new art that clothed all literature in a bodily frame of surpassing beauty and usefulness, and placed it in the hands of the common people in a form that before the invention of printing the greatest kings would have envied; and which even Virgil or Cicero would not have disdained as the material pedestal of their immortality. This art, simpler and more universal than writing, was not lower but immeasurably higher than its predecessor, whose services were for the noble and the learned.

Another illustration: The means of locomotion or material progress—what is their history? Up to a recent date the coaches and high-roads furnished nearly the only mode of land traveling. Journeys by them were restricted to a small portion of the community. The more the coaches were perfected, and the better horsed, the more expensive and select they became. How shall we popularize traveling? By a viler expedient—of canals, carts, and the like? This too existed, but it was used merely for necessity, and did not attract, or tend to make all men into travelers. To effect the latter result, an invention grander and cheaper than had then traversed space was required. To move the rich needed only a four-horse coach running in an agony of ten miles an hour; but to move the poor required cars before which those of the triumphing Cæsars must pale their ineffectual competition. Thus though the problem was the enfranchisement of the meaner classes from the fetters of pedestrianism, yet the only solution of it lay in the increased convenience of all ranks from the noble to the peasant, and not in the degradation, but the elevation of the locomotive art.

And so it must be, as we apprehend, with human knowledge; the arts of education that will summon the people to learn, are *toto cœlo* different from, and greater than, those which have been sufficient for the schools. A petty magnet is sufficient to take up a few hundreds of isolated persons; but when the nations are to be attracted, there is nothing less than the earth that will draw their feet.

Here we touch the gist of the matter; for it is in fact powers of attraction in knowledge that are demanded for the new education. There are three heads to this, which form one. In the first place, attractive knowledge gains the learner and keeps him. In the second place, it enlarges his genius, and out of that, his memory; whereas dry knowledge cultivates his memory at the expense of his mind. In the third place (or in the first again), such knowledge is coherent with itself, and tends to be all known whenever a part of it is known, giving the learner a constant sensation that he is developing it for himself, which lets him into the legitimate delight of mental power.

But only that is attractive which is allied to our business and bosoms, and seems to have a life that understands our life, and *vice versâ*. On the other hand, repulsion is the effect of death and unkindness. Hence, to limit ourselves now to the human body, no popular science of it can exist, but one that fills it with at least as much life as its pupils feel throbbing in their bodies. Knowledge draws them never until they are forced to cry out: "Ah! I see myself more than myself in that wonderful glass!" If to their curiosity about themselves any dead body near them mutters "germ-cells,"* they feel dusty, degraded, and abhorrent. They must be rendered better, bigger, and worthier for every look they give, or their eyes will be averted from their books.

Knowledge, however, is progressive, or its cars are of different sizes. It will only be by slow degrees that we can accommodate the world with seats in the trains of science. New inventions will be requisite for each new population that is to be drawn. In the meantime it is good to see this, and to place as an end the education of the universal people, because that education will require the largest and noblest principles of common sense. To educate a Mechanic's Institute demands far greater principles and more prolific love, than to do the like for a Royal Society: you have to bring things down and to incarnate them, to connect them with material and substantial uses, and to give them both souls and bodies, for the former; whereas in teaching them to the learned, you escape into laws and formulas, and shirk the problem of realizing the sub-

* The intellectual correspondence of the doctrine of cell-germs and convertibility of forces, is found in the doctrine of subjectivity, which implies in the first place that you are shut into yourself; and in the second, that whatever comes to you, puts on your state, and is nothing but your own walls vibrating. This is the philosophical, as the other view is the physical, jail. The way out of it is by walking through the walls, which look granite, but are impudent mist.

2

ject to the audience, by insisting that it is the fault of the audience where it does not comprehend. This is according to a vice common in schools, of neglecting the dullards, and petting the clever pupils. On the other hand, public education is for the publicans and sinners of the brain—the stupid and the quick alike, both of whom are constructed for the largeness of a common understanding.

But if to inform the Mechanic's Institutions be already so much vaster a task than to propagate learning to the learned (who are the people that can take it easiest), what a much ampler knowledge still is required for the educations of clowns and sempstresses, and those great classes of society who have almost no organs but hearts and hands. Knowledge must be like music and nursery songs before these clodhoppers can dance to it.

We have attempted, but most imperfectly, to begin a discoursing on the human body which goes somewhat in the direction of this great outlying population; being convinced that to win them to the truths of the creation is a mission enjoined upon the future time. In doing this, we have felt that there was no condescension implied, but the strain of all our faculties into the most universal sphere. And although, by the law of things, there is more to be done than when we commenced to do, yet shall we be satisfied if we have struck a single cord of that instrument whose earliest notes portend to us the grand Hallelujah choruses and symphonies that are to be.

The matter of universal education may well claim the serious regard of all classes of society. It is indeed a panacea, though not as it is at present conceived. To teach superior artisans and *savans* a few sciences, and by all means to stimulate the love of property, may keep things together by giving a large army of "specials" an interest in what is called order; but the great mass is not touched by such motives or informations. The facts prove that it cannot learn these lessons, or care for these objects; because, like the old coaches, they are not big enough to carry it on its voyage through time. There is nothing for it but a new method and kind of knowledge; an Orpheus who can thrill into dance not the present nimble figurantés, who can dance to any tune, but stocks, stones, birds, beasts, and fishes, who had no fantasy in them before. That method must take the best common sense of all these people, and show them that just that, when carried to its highest stages, is the very truth of all the arts and sciences. As soon as this can be done, all stories will move one degree higher; the kitchens will be parlors, and a new drawing room will flower out of every house for the great reception days. The order which will then conserve all things, will not be solely that of the policeman, but it will be like the order of creation, or like that of the notes in a good harmony, where each keeps the rest to their posts.

We have selected the human body to make a trial of the above method, on account of its central importance and significance, and because it is already the theatre of our common sense: nevertheless, any other object of science might have been chosen, and with the same results. For wherever we go, we find that common sense comes first; and when the subject is completed, again comes last. First glances are always charged with it, in a more or less latent form: the business of investigation is simply to eliminate it as pure as possible from its accidents.

In no science does the present state of knowledge appear so manifestly as in physiology: in none is the handwriting on the wall so plain. Great is the feast of professors here; but *Mene, Mene, Tekel, Upharsin,* is brighter than their chandeliers. Chemistry and cell-germs are the walls on which the lightning writes. Well may we call them walls; for it is impossible to conceive anything more limitaneous: prison stares us in the face while we are in that company. Who of woman born can go further than to distil himself into gas, or to pound himself into cells? Annihilation, which God forbids, must be the next stage of smallness. These respective doctrines are the last solid points which are possible, and by nature itself there is no passage beyond them. After these, the scientific men themselves must evanesce; for already their watchword to each other is, "Hail, Bubble Brother! Hail, Nucleated Cell!"

Before it came to this chaos, there was everything to show that physiology, studied with the present objects, had completed its rule. Its great outlines had long been traced, and its general problems had ceased to occupy attention. It had become more and more complicated and microscopic, and leaving the human frame, it was gradually slipping down into the brute creation. Meanwhile the leading questions that a child would ask respecting structure or function, were forgotten, or postponed indefinitely. The connection of the science with our living sympathies, which was never strong, had ceased altogether. The function of its hierophants, to enlighten the ignorant, had become impracticable, for want of vitality and attractiveness in their knowledge. These were signs that its kingdom was moved from the midst of it, and was about to be given to another. Be it observed, however, that we do not blame the existing sciences for what they are *not;* our task lies only in recording that they have had their three sufficient warnings. Their state is like that of us all—birth, growth, and decline; and like us in due time they are to be supplanted by their sons, when the world's business requires further workmen. For the sciences become immortal in their generations; solid in the general subordination of the past to the present; and living in the perpetual aspiration and appeal from the present to the future.

I know that even the chemistry is all right in its own way; and I cannot help admiring the thoroughness of the Liebigs, who after having analyzed the rest of things, put men and women into the retorts, and with pen and ink ready, write down so much dirty water and fœtid oil, and so many ounces of scientific dust, as the future state which comes over. This is positive fact, if not for the physiologists, at least for the candle makers. I ask nothing more than that it shall be in its place.

And who can quarrel with the microscopists? When Leeuwenhoek, in the seventeenth century, determined that everything that was too little to be seen, was his empire, he really laid his hands upon the whole of things, in a certain small sense. And whenever our great duties and rights are tolerably well fulfilled, little things will be intellectually and morally magnified into a new importance. Ultimate structure will then coincide with primary structure. But until then, the microscope is before its time in physiology, and must wait upon lower callings. While the reasons of minute forms are so totally unseen, it is their prettiness chiefly, their scenic charm and glow, which is of use to human eyes. I have often wondered why, in the difficulties of the artists who make patterns, they have not resorted to Mr. Queckett and the Microscopical

Society, for the exquisite traces which they can show. Solomon was not arrayed like the lilies of the field. The Manchester manufacturers would do well to dress out the ladies of this generation in the spoils of the colors and forms of these brilliant creatures. There would be something new as well as charming in rhinoceros-kidney mousseline-de-laines; in shawls according to the inner splendors of the burnished beetle's wing; in veils worked to represent the many-eyedness of the blue-bottle; or in a mantilla on the back of a professor's wife glorious with mimic cell-germs. Here would be a mission for the microscope, and a final cause for the corporation which represents it, which would then take rank as a sub-committee of Drapers' Hall. It would be good to be small, when it enabled you to dive into little cracks and holes whence real beauty could be fetched. And by gaining this practical object, of ornamenting outsides, the Lilliputian people would cease to infest physiology, and leave Gulliver's inside alone.

Such, we believe in our hearts, would be the art corresponding to the present state of this science. For we notice that every little discovery is so straitly held and claimed by the finder, and the label of "My Truth" is so decisively pinned to each fresh zig-zag of cell work—as though the man had not only seen that special quirk, but made it—that we cannot help thinking that Show is the fire which is heaving under this uneasy stratum of the Small Seers. Our proposition, which we hope to see adopted long before the next "World's Exhibition" in London, exactly jumps with this passion for display, but by carrying it over to our ladies renders it both beautiful and harmless. Let "my" family wear the blazon of "my truth" as they walk before me to church.

There is one part of our work which we think it right to mention, but for which we do not apologize. Throughout the following pages, we have taken for granted the Divinity of Christ and the truth of Christianity, and with this tacit assumption we have labored to connect the whole of our general views. There is no escape from some step of this kind: the atheist takes for granted his atheism, and works in its darkness; he sees no God because he looks from none, and for none: and in every sect, that most comprehensive aspect of nature which is its dependence upon Deity, colors and shapes the whole of the sciences which that sect elicits. We would undertake in many instances to assign to particular Christian denominations, the scientific views which have been contributed by each, and to pin their insights into matter, space, time and history to their prayer-books and hymn-books. In short, science is always impregnated with either religious, or irreligious life. By virtue of this it is, that the coldest treatises end in some kind of sermons, and the *nodus* of every theory of the world is a corresponding god.

In presence of this necessity, it occurred to us long ago, to assume openly our own creed as the supreme key of knowledge to which we could arrive, and to peril our faith so far as our readers are concerned upon the success of the experiment. And here we seemed to be truly scientific according to the common rule. For after some facts are found, the method of science consists in assuming a hypothesis to account for them; and if that hypothesis serves the purpose, it passes over to a theory, and in time is received as the truth of the case. And though the hypothesis might seem unlikely, and be liable to many ugly reasonings against it, yet if it answered to the facts, it was justified, and all ratiocinations to the contrary died. By pursuing this method, we have convinced ourselves, that our Lord is written down in the pages of nature herself as the truth of her whole creation.

This is the method of tolerance. For while we work from this point of view, it is our own fault if we meddle with others who are trying to settle the same problem from different grounds. There is a prize to be won by the religions of the earth: they are so many accounts of God: the order and laws of creation and spirit are the check which adopts or rejects these accounts at some certain point. The book is closed, and who shall open the seals and read it?

The same is the method of scientific persistence and unity of aim. For no one lifetime can be expected to measure the adjustments of the problem. To all argumentation we can scientifically reply, that we have not yet concluded our experiments. In so great a cause, ages of ages are allowable times for operation. Religions are a vast social matter; no man has a right to declare the trial done, until the success of creation-reading warrants him. Thus we gain indefinite time against the skeptics, allowing them also their own time for doubting. If God permits them leisure, why should we feel bound to hurry them? But then they must be equally patient with us.

The same also is the justification of causes and missions. For these are to action what theories and hypotheses are to science. And every man is not only justified, but bound, to plead his religion; and if it contains that force, to send it forth over the world. He is trying this scientific problem on the practical side.

The same again is a criticism upon criticism. For in a working science, we do not desire to know what *à priori* probabilities or improbabilities may attach to our hypothesis, excepting in so far as they determine our choice of it at first; after this we have simply to apply it, and to test it in all convenient ways. There is only one objection which not only slays but buries principles; which is, that they break down in fair practice, and are disowned by the nature of things: other objections are impertinences, which may, or may not, be true. For this reason, the modern criticisms upon Scripture have nothing to do with the question: whether miracles are likely or unlikely, possible or impossible; whether God would have dictated a Book which contradicts itself; whether the attribution of human passions to Deity is consistent; whether Christ as God could have died upon the cross; whether the Gospels were written in the first century or the third; whether Christendom be not a wide mistake of which Chapmandom is the correction:—all these are matters which we postpone, from our extreme inability to do two things at once. If it be found that Christianity is the theory of the world; that the Divine Man is Lord of the sciences; that the biblical Revelation is the truth of truths, which opens a Shekinah of light to the later races more than to the first; that the Gospel alone can rule the nations with a rod of iron; then the finding of this from age to age will sufficiently conserve the text against the stings of the Straussian school. The more so, because if their principle of criticism first, and faith afterwards, were admitted, the result must be atheistic confusion. For if, on account of what contradicts our notions of convenience in Scripture, the Bible be untrue, then for the same reason, nature, being full of contradictory essences, tigers and lambs, men and vermin, is no work of God; but a single flea is enough to trip over the nature-textuary into the abysses of denial. The armor of these greatest truths is, however, not so ill-jointed as to let in such lances. It demands that the critic shall try his criticism by not only accounting for, but ruling the world. If he cannot do these two things, his rack of texts proves as good as nothing.

This method also is a defence against books. For were we not entitled, as children of science, to take for granted our Revelation, and to make our lives, and if need be, our eternity, into its trial, we should be bound on a quest over the whole universe, to listen to what everybody said. The sandals of this terrible flight would soon reverse us, and take the place of our heads. By nature however we make up our minds very soon, *malgré* the possibility of meeting some one who shall one day upset us: there is a quick hour when every man burns his Alexandrian Library as heartily as the Kalif Omar. And thus we limit ourselves to a particular walk, and like cobblers, "stick to our last." The atheists do this just as much as other people; the vacancy of the air of their studies, suffices them for the induction that "God is nowhere." Why should not the Christian professionally accept this necessity of not roaming through all books, but work from the Best to his vocation?

We were forced upon this track of thought, by noticing that the Rationalists had got to nothing as punctually as if nothing had been their aim; and that their inductions were of no consequence, supposing them to be true; which however concluded against their truth. We found also that they were like the fellow who claimed Virgil's "*Sic vos non vobis*" to himself, but could not complete the line which the great poet had left half finished. In the whole company of them, and in all their promise of offspring, there was not a spark of revelation; though to hear them talk one might have imagined, that they knew the way of making myths, and that writing Bibles was their *forte.* For these and a thousand other reasons, we left them on one side, and took another tack.

Here also we quitted those two little parties who think that they are the only two, namely, the contenders for the principle of authority on the one side, and for that of reasoning on the other; and taking some gold and silver from both, we determined to choose the party of science, as that to which the Lord of science was about to commit the kingdoms of the earth. And as eyes are the great faculties of sciences, we determined thenceforth to pray for eyes above all other powers of that kind; that we might not have to appeal to either of those effete parties, authority, or criticism, but might terminate some of our perplexities by sight. The more we have proceeded, the more convinced we are, that this ground of science is the coming earth of a new and more glorious time, and that He whose feet burned like fine brass in a furnace, will pour His love through it, and give to it to conquer and to heal.

At the same time we have no faith in the present state of the sciences, excepting as the ministers of some industrial arts. For educational purposes they are almost worthless, because they terminate the first plain questions with unsatisfactory replies. For spiritual purposes they are equally negative. And hence we regard them, with all their seemingly large retinue of facts and colleges, as only provisional occupants of the mind. Indeed they are in so rapid a flux, that it is hard to say what they are from one year to another.

But when science becomes Christian, we may have some natural theology upon the face of the earth. And here again our views depart from those of many good men. If Christ be the God of the Christian, then natural Christology is the only theology of this kind which is possible in a Christian state. In Mohammedan countries, natural theology, or the culmination of the science of nature in Deism, or what is the same thing, Theism, must be permitted as an inevitable growth; but this is plainly not the case where Christ is worshiped. We feel it necessary to insist upon this, because even those who accept Christ's Godhead, strangely pass Him by when they are attempting to trace up all nature to their God. The consequence is, that it is only the truths of mere development and creation that occur in the sciences, and not those of love and redemption; whence moral and spiritual life is banished from the book of nature. The church must look to this; and bishops and clergy whose place it is to give prizes for natural theology, must consider whether as at present taught, it is not an active branch of Deism. We venture to express a hope, that in the distribution of laurels to the successful candidates for the great Glasgow prize, this Christian exigency will be remembered; and the award be made not to those treatises that so easily trace the parallelism between certain views of science and the God of fancy, but between the integral sciences themselves and the God of Revelation.

If the judges upon the important occasion referred to, adopt the standard which we suggest, the end of natural theology will be altered, and the effects upon science may be of the most considerable kind. At present, natural theology has undertaken the impossible task of "finding out God," Who can only be found in so far as He has been pleased to reveal Himself. The Deity thus elicited, or as Fichte rightly says, "constructed," is a scientific abstraction answering to the concrete figure of the Vulcan of the Greeks—that is to say, a universal Smith. The course of the natural theologians is as follows: they see in the human body and the world the principles and applications of the arts in a surpassing degree: the skull displays the virtues of the arch, and the hand embodies wondrous pulleys and levers; whence they infer that God is acquainted with mechanics. And from all the other parts of man, the clay patronizes the Potter in the same way, and the Deity which arises out of the whole is at best an infinite handicraftsman. This is anthropomorphism, or the distillation of God out of our own limits and thoughts, our own space and time. The Paleys, Broughams, and the authors of the Bridgewater Treatises, seem to have been satisfied with this vulgarity of heathenism.

We hold, on the other hand, that a scientific natural theology is different, and that it accepts the living Lord of Revelation, and consists in tracing the correspondency of His revealed attributes in the sciences; being in effect the synthesis of the knowledge of the real God with the sciences of real nature. In this case it leads the waters of life into science, and is the most indispensable of all the single studies of the natural world. As for the evidences which it affords of God's existence, these do not consist in demonstrations of artistry and carpentry, but in symbols of spirit and love pervading creation, and reconstructing the natural mind of man upon spiritual models. The proof that nature is full of Deity, lies in its power, when rightly seen, to soften the heart and moisten the eyes of the unbelieving world, and without a controversy to send the scoffer to his knees. In that affecting moment, genius and devotion are twin-born. But no demonstration of Vulcan, even though approved by bishops and clergy, has any effect of this kind.

Many of the thoughts in the present work may seem new : yet the larger portion of them is not our own. We believe implicitly that "the stone which the builders rejected" will be "the head of the corner." It has become therefore a scientific canon with us, to look out for this stone in all quarters. Hence we have been led to many discredited sources of information; and if borrowing good things be plagiarism, we are guilty of it to the utmost of our powers. Wherever it can be shown that we are not original, so much the better : we desire for ourselves and others to enter the circle of the great dependence of things, secure that there is no independence of heart or mind upon any other terms.

And now our task is done. The style of it differs somewhat from the common, because our creed differs; and we disbelieve for the most part in what is called the severity, strictness and dryness of science. We hold that nature is the drapery of spirit, and that analogy is the cement of things, and the high-road of influences. We are therefore afraid of nothing, however fanciful, on this ground : nature is more fanciful than any of her children. Besides which, we have found practically that metaphor is a sword of the spirit, and that whenever a great truth is fixed, it is by some happy metaphor that it is willing to inhabit for a time : and again, that whenever a long controversy is ended, it is by one of the parties getting hand on a metaphor whose blade burns with the runes of truth. For these reasons we dare speak in parables.

Yet is it fearful to disagree so widely with our friends and fellows. We love them, but crave space to breathe where they are not. Difference makes the world larger; and without a quarrel with the old modes, we have emigrated to another country, where we hope for peace; particularly as we trust not again to come forth with the pen.

THE HUMAN BODY.

CHAPTER I.

THE HUMAN BRAIN.

OUR first conception of the human body is that of a living subject; life is the dim personality which animates it, as well as the atmosphere in which it moves and breathes. Upon this lowest floor of our existence there rises an edifice of many stories; upon simple life, which is the vegetable in the animal, there is founded a life of life, which is mind. The mind is many-chambered and many-storied. Life dwells in the body, but the mind, or superior life, inhabits the head, or according to anatomy, the brain. The brain, then, is presumably the body of the mind, and whatever is wisdom or faculty in the mind, is furniture or machinery answering to faculty, in the brain. And as the mind is the man, the brain is his representative, or the man in another degree.

This is the solid voice of the head, or the most general truth of consciousness, which lies in the head, and speaks from the head. And our plan is, to accept as oracles, these gifts of our instinctive sense: to regard them as the elements of knowledge: and not to question until we have accepted them. It is then the consciousness, or instinctive natural history of the organs, from which we commence: not what man says of the brain, but what the brain says of itself in man. Thus we shall first endeavor to gain the impressions which the organ under discussion makes upon the mind, or begin from natural experience. We shall then briefly consider its anatomy, or pass to scientific experience. Next we shall attempt to give life to the part by considering it in motion and emotion.

Lastly, we shall pursue the analogical method, where it is not too difficult; and assuming that every principle runs quite through the world, we shall endeavor to show that each organ has kindred in every sphere; and thus, out of the consanguinity of things, we shall try to deduce the fact of a native coherence in the world, whose links are a real logic, and which, when transplanted into knowledge, will spontaneously constitute the association and unity of the sciences.

But this will be better understood in the sequel. We now proceed to the facts of the present case, to visit the mind which we have found, in its proper mansion, the brain.

At the outset, we would guard the general reader against an error which requires to be removed. It is commonly supposed, because anatomy has been cultivated by a class, that it is difficult to learn. On the contrary, any one with a common understanding, and of course industry and attention, may possess himself of the leading parts of anatomy. Ladies may learn them as well and as harmlessly as the other sex. Plates moreover are satisfactory means of acquiring a view of the human frame which is enough for public education; for although insufficient for the surgeon, the knowledge derived from plates will enable the public to enter upon the study of organization, both with cleaner hands and clearer heads than if they busied themselves with the ever-varying detail of dissections. It has indeed been usual with practical anatomists to decry anatomical plates; and yet they are a degree better and truer than the dead body; for they contain the mind of the artist, and are superior to any single subject, being not mere copies, but carefully collected from many subjects; and as they are generalizations, they are adapted to be vehicles of general knowledge.* They may be regarded as translations of the body, available for those who cannot read the original. And indeed those who can only see and not touch anatomy, as they have all power of destruction removed, are favorably situated for constructive truths—for the theories of bodily

* Many of the current anatomical plates have descended through books for centuries, improving on the way, and may be traced from the masters of the Italian and Dutch schools of anatomy to the present manuals: a plain sign of their truth and serviceableness.

motion, proportion and gravitation; just as the astronomers are indebted to the distance of their view, and the superficiality of their objects. Good science then, we repeat, will not refuse to attend upon anatomical plates, though not the science of the surgeon.

The brain is an oval mass, filling and fitting the interior of the skull, and consisting of two substances, a gray, ash-colored or cineritious portion, and a white, fibrous or medullary portion. These substances occupy different positions in different parts of the brain. The gray portion constitutes the circumference of the large and front division which is called the cerebrum; it also enters into the interior of the same in various parts, and forms both the centre and circumference of the cerebellum and the centre of the spinal marrow. The white portion makes up the greater share of the brain. Besides these divisions of substance the brain also presents divisions of form. It is parted into two great masses, viz., the cerebrum and cerebellum, and at its base there are two other portions, named the annular protuberance and medulla oblongata. These are the four primary divisions on the surface of the organ. The spinal marrow, which runs down through the vertebræ or back bones, is the continuation of the medulla oblongata. The above parts collectively are the bed and trunk of what is called the nervous system. From all those already mentioned there arise certain white cords, the nerves, which come out through holes in the skull, and through notches between the back bones, and run to all parts of the body, gradually splitting into filaments, which at length become invisible by reason of their fineness. There are generally reckoned to be eleven pairs of nerves arising from the brain, and thirty-one pairs from the spinal marrow. Besides this great nervous plant which continues the life of the head and its appendages, there is also another proper to the body, as it were a creeping or parasitical system, which weaves its meshes among the branches of the former: this is the system of the sympathetic nerves. It is not obviously referable to a centre, like the system just described, but it has many small centres scattered throughout the body, but especially near the important organs, the heart, the stomach, &c. These centres are called ganglions, and are said to consist principally of gray matter. Innumerable fine nerves radiate from them to the viscera, and to the great blood-vessels; and also fibres pass to communicate with the capital nervous system.

From the general tenor of these two systems it appears that the cerebral is the engine and representative of the mind, and of the body as constituted in the hierarchy of the mind, while the sympathetic system is the nervous engine and representative of the body considered independently, as possessing a life or mind in itself so far as this system can arouse it. Consciousness does not here come in question. Many of our impressions are unconscious, nay perhaps all through the longer part of their course, though travelling along the cerebral lines. And again, the bodily organs, as the liver or the kidneys, require to exercise processes of selection, and acts of composition and elimination, to which nothing less than a stupendous bodily judgment is adequate. Mental judgment would not be fine enough for the work; and it is only mental judgment and faculty which are conscious.

The brain and its parts, including the spinal marrow, are clothed with three membranes or skins. That next the brain is the pia mater, which not only encloses the brain, but dips down into its folds, and probably ramifies more and more finely through its minute divisions, acting as a framework to the nervous substance: it is full of delicate vessels which supply the brain with blood. The next membrane is the arachnoid, which is said to form a shut bag like a double nightcap, in the inside of which a lubricating fluid is given out, the whole constituting a kind of "water bed"* on which the brain may undulate. The next membrane is the dura mater, lining the movable arachnoid on the side towards the brain, and lining the bony skull on the other side, and being separable into two layers to suit these two situations or offices. It is the strongest of the membranes of the brain, and gives off membranous pipes which receive and envelop the nerves at their exit from the skull and spine: it also sends down tight sheets or processes between the two halves of the cerebrum, and between the cerebrum and cerebellum, whereby it

* Water beds of this kind, *i. e.*, serous membranes, are prepared for all the viscera; and bursæ or water cushions are frequently apposed in the joints; where these are met with they are the means and evidences of some motion performed.

supports the larger divisions of this soft and tender organism: and between its layers it contains the great channels of venous blood known as the sinuses of the brain.

We see then that the consciousness is physically nourished from without, as the brain is nourished by the blood of the pia mater; that it is physically movable, or can assume varieties of shape and state, as the floating brain can move its person upon its fluids, and specifically upon the arachnoid bed: we also observe that the consciousness, like the brain, is limited, the dura mater and the skull being the emblems of that limitation.

The brain is supplied with blood by the carotid and vertebral arteries. The carotids are the first great vessels which issue from the stem of the arterial tree, and they supply the cerebrum with the first blood of the heart. The vertebrals are the first streams from the subclavians, which supply the arms with blood. Thus the first blood of the body and the limbs is alienated to the brain. Its veins, which bring back towards the heart the blood that has passed through the before-mentioned arteries to the pia mater and substance of the brain, empty themselves into certain peculiar channels contained in the layers of the dura mater, and termed the sinuses of the brain; which sinuses communicate with each other at last, and pass out of the skull by a bend or curve where the carotid enters; there constituting the internal jugular vein, which carries the first considerable stream that is returned to the heart.

The cerebral globe is divided into two halves, and each of these into lobes; the lobes again being subdivided into convolutions, which have furrows between them that dip down into the brain, and are covered by the pia mater. By means of these foldings, the surface of the cerebrum is much increased, and space is economized; this folding of surfaces into solids being one of the principles of the body, whereby distances are brought together, and association or organization is promoted. In proportion to the number of these twists or convolutions is the power of the brain. The mind's revolvings are here represented in moving spirals, and the subtle insinuation of thought, whose path lies through all things, issues with power from the form of the cerebral screws. They print their shape and make themselves room on the hard inside of the skull, and are the most irresistible agents in the human world.

At a considerable depth the two hemispheres unite together, and below their union, if the cerebrum be opened, we come to certain cavities termed the ventricles of the brain. Of these there are four, all communicating with each other; and a fifth is enumerated, but of small size, and disconnected from the rest. The four cavities form a continuous chamber, and always contain more or less fluid. Thus the brain, far from being prepared for rest, is constituted internally upon the movable pivot of this fluid cushion of the ventricles.

The cerebellum lies behind and underneath the cerebrum, of which it is said to be one-eighth the size, and it is divided into lobes and lobules. It consists of gray and white substance, not disposed in convolutions but in thin plates. There is said to be no direct communication between the cerebral hemispheres and the cerebellum. The latter is evidently not a revolving, thinking, or spiral organ, but a battery of determination and power: thoughtful consciousness does not connect itself with the back of the head, or with the cerebellum. Its form too, double fisted, does not answer to the cerebral functions. Its visceral or hidden situation also brings it into analogy with the other viscera, in which there is no freedom of thought, but fixed acceptation and permanent action.

The nervous system, though apparently homogeneous, is constructed of distinct pieces, which are extraordinarily united, and extraordinarily capable of separation in their functions. The first piece is the proper spinal marrow, with all the nerves of the limbs, trunk and head which issue from it. This lowest pillar of the cerebral system is in a manner complete in itself, and receives impressions, and executes actions, on its own account. It consists of a running axis of gray matter of a peculiar form, and in front gives off the nerves that convey bodily motions, at the back receives those which carry bodily sensations. The circle of its operations is therefore as follows. When an impression appeals to it from the body through its quasi-sentient nerves, this mounts to the gray centre to which the nerve carrying the impression belongs: an instant organic determination then occurs in the centre, a decision takes place, and a motion is sent down through the corresponding motor nerve to the parts which the latter supplies. For example, a pinch applied to the leg lodges its complaint at the gray centre, which at once by its nerves sets the muscles and the limb in that motion which enables the part to escape the distress. This is now termed, *reflex action*, which means that an impression is communicated to a nervous centre by a set of quasi-sentient nerves, and a motion *reflected* from that centre by a set of motor nerves. It does not necessarily involve consciousness. A paralytic man, with no feeling in his legs, if the soles of his feet be tickled will move them away from the irritation, just as though he perceived it. In short, the spinal brain is unconscious, or let us rather say, we are unconscious of its feelings. We term its nerves therefore, not sentient, but *quasi-sentient*.

We come now to a second and distinct piece of nervous system, of whose operations we may still be unconscious, viz., the medulla oblongata, whose nerves are connected with the organs of respiration and the ingestion of food—with the functions of breathing and of eating, which although they may be permeated by sensation, and controlled by the will, may also occur independently of either; as during sleep, when breathing still proceeds, and in various cases when the movements of eating and deglutition are performed without cognizance. All that is necessary for continuing the actions of the parts of the body supplied from these centres, is, that a quasi-sensation be communicated to them, which the centres act upon through the motor apparatus of nerves. The conception of so mere a machine in man, is perhaps difficult to realize; the spinal brain, however, with its dependencies, represents an automaton man as the basement of the nervous system. It is the organization of routine or insect progress. By it we walk, work and eat when we are not thinking of those operations; and thus the inherent properties of this routine system save us from much expenditure of attention, and allow the brain and the senses to be emancipated as necessity requires from the lowest wants. Had we to perform our animal functions by direct volitions, we should have no time for anything better: if each breath were a distinct voluntary act, breathing alone would fill our lives: nor in this case would walking or any other external function be possible, for the will does but one thing at once. As it is, however, a number of bodily acts are momentaneously and harmonically performed, each through the separate vigilance of its own agent in the recesses of the spinal brains.

We have alluded already to the act of walking, which affords a good example of automatic function. When we are walking without attending to our steps, the foot coming down to the ground conveys the quasi-sensation of its contact to the spinal centres; these are roused to a corresponding motion; in other words, they command the muscles of the other leg to put it into a forward movement. No sooner is this executed, than at the end of the movement, another manifest quasi-sensation is afforded by the fresh contact with the earth, which contact reaching the centres, engenders a second motion: and so forth, throughout the walk. This is a simple circle, in which quasi-sensation excites motion at the centre, and motion produces quasi-sensation at its extremes. Thus the foot on the ground represents sensation, and that in progress, motion, and the two contemplated together represent the links in a chain of nervous fate.

The next piece of the nervous system consists of the nerves of the special senses, and of certain central parts at the base of the brain, to which those nerves run. The latter are the sensual* brain, from which fibres emanate that ultimately become the olfactory, optic, auditory nerves, &c., which run respectively to the nose, the eye, the ear, the mouth and the skin. The central endowment of this nervous piece is sensation. In itself it provides a simple circle from sensations to motions. Impressions which are perceived by the senses mount to the sensual centres, which dictate suitable actions. The instincts of some animals low down in the scale, are to be referred to this class of mobilized sensations.

Instances of purely sensual actions are comparatively rare in man, so rapid is the transit of feeling into desire, which is not a faculty of mere sense, but the gift of a mind in direct communication with

* We prefer the term sensual to sensory, which seems coming into use. The word sensory, if applied so low down, would exclude from sensation all above it, and the cerebral operations would fall into abstract forms: whereas by allocating the term sensual to the nervous centres of the external senses, we leave the term sensory for the inner senses, to which it has been appropriated from old times.

the senses. The instinctive laying out of the body for pleasures, and its spontaneous avoidance of pains, are however cases of this order. Moreover we feel that a strong vein of these actions runs through many of our habits, and executes them for us, contemporaneously with our desires. Habitual vices, in proportion as they become fixed, seem to roll upon this fatal wheel, by which low pleasure runs incontinently into motion. The bottle is hung round the neck of the drunkard by a simple sensual circle of the kind, as well as by a longer thong of which perverse desire is the neck-piece. By a similar yoke, of seeing and leering, is the voluptuary led by his objects. In short, whatever we do, good or bad, without being able to control it, appears to contain a sensual kernel. Sensuality, however, is a passive faculty, and we shall have to distinguish presently an active part belonging to it, and playing upon it, and which for the present we term animality. This then is the third of the nervous stones which construct us—sensation, which set a-moving is sensuality. It leads us to dance pleasantly but involuntarily to the music of the senses. Its seat is in the base of the brain.

The fourth and remaining piece of the nervous system comprises the cerebral hemispheres, which reasoning by the method of exclusion we regard as the seat of the mind, or the properly human faculties; and not only of these, but of all those powers which belong to consciousness, and are over and above the senses. These hemispheres are predominant in place and size in man, as well as in the higher animals.

We have now chronicled three divisions of the nervous system,*

* The nervous system, or its functions, are sometimes dissected for us by diseases, as well as by narcotic agents, which benumb one part and leave the others free. For instance, during the administration of chloroform, first thought and will, and the fixity of sense which depends upon attention, become confused and waver: and shortly afterwards active consciousness is abolished. The cerebrum is detached from the train, and life goes on at slackened pace without its traction. The sense of pain however still continues. By degrees this too is lost: the second and feebler engine of sensation, the sensual brain, is also detached, and nothing but breathing is prefixed to the cars. This now becomes slower and more slow, and if the experiment be continued, ceases. Life is then generally irrecallable. In natural and artificial trance the process of disconnection may proceed two steps further; and the nervous function may be detached

which may be grouped into two, an animal and an automatic; the lowest of these is a nerve machine, the higher in its twofold aspect is a nerve animal, with senses. We know nothing that expresses so physiologically the case between the automatic and the other part of this system as the ratio of the living creatures to the wheels in Ezekiel. "And when the living creatures went, the wheels went by them; and when the living creatures were lifted up from the earth, the wheels were lifted up. Whithersoever the spirit was to go, they went, thither was their spirit to go; and the wheels were lifted up over against them: for the spirit of the living creature was in the wheels. When those went, these went; and when those stood, these stood; and when those were lifted up from the earth, the wheels were lifted up over against them: for the spirit of the living creature was in the wheels."*

On arriving in our upward course at the cerebral hemispheres, we have come to a group of functions which represent the powers of the mind: consequently here we have no longer a simple centre to consider, but an organ with various essences. For the mind is evidently twofold; at the first analysis it is a rational spring on the one hand, and an animal spring on the other. It may be devoted to the truth of things, and proceed from thought to judgment, and from judgment to will, and so give the dictates of reasonable actions. Or it may be spent upon gratifications, and run from imagination

to a certain distance from the organism, and yet be capable of reassuming its place.

In the case of narcotism the separation takes place from above downwards: the mind is first taken away, then the senses, and lastly the breath: but sometimes the reverse process occurs, as at the hour of death, and first impressibility, and second sensibility are lost; whilst the mind retains its clearness, or even enjoys additional powers of circumvision and forethought. These states are opposite, although the beholder may mistake them for each other.

And not only disease, but the ordinary state, shows the separateness of the nervous pieces. Through the greater portion of life, thought and will are dormant, or the tops of the hemispheres are not in action, though the man is conscious. Many times also imagination and desire are completely at rest, although feeling is left; the eye immits rivers of objects without stirring any motions but its own as it rolls magnetically after the pictures. And then again we experience numerous twitches and convulsive movements for which we cannot account, or in other words, which are originated in the circle below the sensual brain.

* Chap. i. verses 19—21.

to choice, and from choice to desire, and thus stimulate animal actions. In short, there are two minds in man; the one which he possesses in common with animals; the other which is properly human. But where do we find these two in the cerebral hemispheres?

To this we reply that the whole cerebrum is the general sensorium; that is to say, the residence of the animal mind, or the mind of the senses. For the fibres coming up from beneath diverge to the entire cerebrum, and terminate in its cortex. This brain, then, as related to the incitements of sensation, comprises the various powers which sensation stimulates, and into which it passes; that is to say, it includes memory, imagination, desire, and again the corresponding series of animal determinations which give birth to their proper actions. To this general sensory and motory head we also refer many faculties which go under the name of instinct; its great tides of change are emotions, joy and sorrow, and the like; and its general states we term moods. The passions are the lords of this worldly brain, by which man sympathizes with all nature in its own way, being governed by the moon and the weather, the circumstances of his society and his age, and whatever influences come from without. The absence of moral freedom characterizes its actions, when these proceed from itself alone. Its faculties are vague and general, and move altogether in their mass. It gives the pervading temperament and tone to the animal body, and being the highest expression of animal life, and bulky and forcible with our whole nature, its actions animate the frame with prodigious vigor and universality, as we see in the case of the various passions and emotions. Contrivance, cunning, and a number of conjointly human and animal qualities belong to this general sensory, and put on the appearance of wisdom or reason, not only in animals but in men: so that whether such or such actions argue reason is an equal problem in both creatures.

On the other hand the human mind, as distinct from its own animal mind, appears to reside in the cortical circumferences alone, and to play upon none but the very centres of nervous power, and not upon these in the gross, but with skill and discrimination. Thus it does not consist in new materials, or fresh parts, but in a fresh use of parts, and a new architecture of materials. The supreme superficiality of the cortex of the brain, its partition into convolutions, the separate movableness of these, their brain-embracing wholeness, show that a freedom is possible, and a universality conceded, to these parts, which neither the fibres *per se*, nor the other groups of cineritious substances, can enjoy. These conditions agree with the requirements of thought and will, both of which are central and concentrative; are seated at the top of the mind; work by separation and decision, seizing upon main points and governing provinces therefrom; and both of which finally can claim any portion of the domains of our animal nature, and control it, until its animal essence is reduced to harmony with reason's will. These cortical substances, in their distinct use, are the factories of mental skill. It is their perfection to except neither temperaments, nor weathers, nor the criminal influences of the stars, nor the imaginations, sensations, or apparent relations of things, as their allotted sphere; but they shape and cultivate things and ideas afresh; they recast influences on their own principles, of truth or falsity: their reception of outward nature is the philosophy of the man; their works are the creative arts of his mind; and their judgment is his moral soul. All this, we remark, demands no second brain apart from the general sensorium, but the command of this in its highest parts, where it is free to obey, and the reconstitution of those parts by the usage and skill of the rational power. In this respect the brain may be likened to the hand, which is a coarse, general and animal tool in the savage, and for some operations, such as grubbing in the earth, disqualified by its division into fingers; whilst in the civilized man it is extemporized by his mind and education into a running power of convolutions from which tools proceed and arts radiate, until nature is subdued and home is built: the separate use of the fingers being the sign and cause of this new essence in the hand.

It is therefore in the mind and not in the brain, and in the rational mind and not in the animal mind, and not in mere ends but in moral ends, nor in mere determinations but in moral determinations, that man is different from animals. Or to keep within our present subject, man is human or hyper-animal because he has a mind *extra*, which uses, or can use, the top of his brains.

The separateness of the animal from the rational brain is *functionally* more distinct than the separateness between the other parts of the nervous system. Do we not all perform vivisection upon ourselves every day in cutting reason off, and thinking and acting from animal motives; in keeping the mind under, while vague imagination, desire and pleasure over-ride it? in merging it in a sea of natural emotions and commotions, which allow reason no beginning, and will, none of its distinctness? But as for the *structural* separateness, the rational brain is no other than the mind itself as a distinct organization, whose existence is demonstrated by its play upon the cortical organs. This mind terminates the brain, and begins a new subject with new expressions; we cannot see it unless in its own way of intellectual sight; nor can we now pursue it, because its attributes hold no commerce with anatomical terms.

The nerves or brain form a representative system, which does not itself come in contact with objects on the one hand, or with actions on the other, but deals in the one case with the images of things, in the other, with the commands of actions. It results from this, that whatever will produce the central impression, sensation, imagination, or intellectual vision, will cause the appearance of the object, whether it be present or not. Thus if the brain can radiate down to the spinal cord a vibration like that which the cord receives from any object as an impression from without, the same motions will be engendered as flow from the apposition of a real circumstance or object. So again, if the brain can shake the optical centres as light shakes them, or can extemporize the part of light within them, the man will have the sensation of light as though it were present from the sun or a candle. And so too if the soul or spirit, or any other spirit or influence, can make the imaginations or the thought-movements in the cerebral substance, these will seem as much our own thoughts as though no such influence had been exerted. But in both cases, be it remembered, there is an object out of the faculty excited; though in the one case the object is out of the organism externally, in the other case out of it internally.

Each of the centres then, namely, the automatic, the animal, and the rational, are susceptible of a twofold excitement; first, from a circumambient world, or from beneath, through their own proper circle: secondly, from the organism, or organisms, above them, and thus indirectly from a higher circumambient sphere. Each also has its memory; which in the spine is habit; in the animal brain is proper memory; in the distinct cortex is principles, which hold things together in the bonds of ratio or reason, as memory combines them in the bonds of a common sensation. Habit then is the spine of the nerve-man; memory is his world of sense, or his senses; and reason is his proper head.

Let us now pause for a moment to ask cursorily the use of the brain to the mind, as illustrated by the foregoing observations. Now, what is the use of the spinal cord to the senses and the brain? for this will give us a similitude of the answer to the previous question. Its use is, to carry the general cerebral principles into an automatic or mechanical sphere, and there to set them up in unconscious operations. Thus, the spinal cord makes motions which look as though they proceeded from emotions, when yet there is nothing felt. This dramatic mechanics, is the marrow of the nervous system, and consequently of the body. As a principle it reigns throughout it. The whole system is a *quasi* thing; a mental theatre or drama. The spinal cord moves as though it felt; the medulla oblongata breathes and eats, as if it were instinct with appetites: the senses feel, as if they were conscious: and the brain understands, as though it were a spirit. The cheek, too, blushes, as though it were ashamed; and so forth. But all is *quasi*, and depends upon a reality somewhere which is in none of the actors; and which reality, proximately, lies in a spiritual organism, or in the human mind. Take this away, and the mimicry is soon at an end.

Thus far, then, the use of the brain to the mind is, to enable the latter to personate itself in a dead world, which it could not do without a brain and body, really dead, and yet seemingly or dramatically alive.

It is not long since it was believed that the actions of the spinal marrow were evidences of consciousness, and that feeling was implied in its habitual regular movements. And it is still thought by many that sense or feeling is necessarily connected with consciousness: and by almost all, that consciousness is where it seems, namely, in the head, and not in the mind. But strike away the lower fallacy, that muscular or convulsive action has any necessary connection with feeling, and the other fallacies also totter. For if without feeling or motive we can be impressed and act through the spinal cord, we can, without *an inherent mind,* understand and will in the cerebral hemispheres. Or, in other words, as dead a structure in the brain may be in apposition with the mind, for mental purposes, as that which is added to the brain in the spine for impressional and motor ends. The corollary is, that life is not in the body at all, in the brain any more than in the nails; but that the body is essentially dramatic; can feel, *as it were,* think, *as it were,* and will, *as it were:* which, indeed, is the reason why the soul, desirous of doing all these things in a world which likewise is dramatic, adheres to a frame which is so perfect a medium of representations or mundane actions.

But let us also consider for a moment the relation of the mind to the brain, by an inverse analogy, namely, the relation of the brain to the spinal marrow. What, then, is the latter? The difference of place between the two, and the difference of calibre, are too obvious to mention. The one is the head of which the other is the foot, the one is the luminous globe of which the other is the ray.* Now, the brain performs or instigates on new grounds, with new efficacy, and in a thousand thousand new forms, the general automatic actions of the spinal marrow; for it extracts the secret and meaning from sensible impressions, and produces actions correspondent with the circumstances which that meaning shows to exist. All the actions of man are proximately brain-work; so also are all his perceptions: whereas a few automatic movements and convulsions are the whole of the actions that can be assigned to the spinal marrow. We have here a glimpse of the transcendent nature of the next higher term of the series; and if by the rule of three we may say, as the spinal cord is to the brain, so is the brain to the mind,

* Here we may remark that we agree with the ancients, that the spinal cord is a continuation of the brain; although it is also, as the moderns say, an axis of independent centres; but its dependence is a longer truth than its independence.

we shall admit that perfections, amplitude, and newness of function will occur in the latter which it baffles words to describe.

The same truth is presented by comparison of the parts of the brain within itself. Thus from the great cortical surface of the cerebral globes the white fibres radiate downwards and inwards, the calibre of the radiant mass becoming smaller as it travels. Presently the rays are arrested by a new bed of cortical substance, that in its turn sends down radiating fibres, which similarly converge and decrease. The same process, of encountering the cortical part, starting afresh, and always diminishing, is repeated, until both the cerebellum and the cerebrum terminate in the small medulla oblongata, and in the narrow spinal cord. Thus powers are stopped off and arrested as the brain descends; or reasoning backwards, as Gall has done, they are added on as the medulla oblongata ascends. What must be the addition that takes place in the mind? what the new breadth in its cortical spheres? what the acres of the sheaves and bundles of its intellectual light? and what, on the other hand, the gulf of loss and degradation that lies between it and even the highest portions of the brain?

Thus far we have attempted a slight sketch of what may be said with some certainty of the functions of the nervous system. We have found that it consists firstly of an automatic apparatus, the spinal brain, by which contacts are apprehended, and motions executed, without the intervention of our consciousness: secondly, of an animal brain, which is to all intents and purposes an animal, or imagines, desires, lusts, contrives, plans and acts from animal motives, though very imperfectly, from defect of instinct, which is the limiting perfection of the beasts; and thirdly, of a rational and voluntary function, playing in its revolving cortex, and evidencing the presence of an invisible mind, whose *action* reveals the human brain. Thus we have found that the brain *per se* is not human, but perpetually humanized; and that in its openness to that which is next above it, and its docility to the spirit, lies its grand endowment. In thus proceeding from below upwards, we have been separating parts whose perfection lies in their harmonious union. We must now make amends by declaring, that the influence of reason, permeating the animal brain, gives it powers super-eminent over instinct; and as man domesticates the animals, or chooses those which suit his purpose, and abolishes the rest, so does reason govern the moods of the brain, feeds upon its tranquil emotions and compresses those which are fierce, governs its imaginations, and, in a word, civilizes the savage countries of the original head. All this is no work of passion, or simple pleasure or pain, but of artist-like struggle and contest, whereby reason, or the true ratio between the mind and the brain, begins to be established, and the little spots already cultivated are extended until the rest is won. In this high state of the brain the human faculties permeate the cerebrum, and the animal faculties, prodigiously cultivated beyond the wild state, are everywhere parallel with the rational powers. And so, too, the mechanical faculties, all the manufactures of thought, or the mere motions of the body: these are surrounded by a consciousness which knows what they will be, before they appear; they are developed into mechanism after mechanism—all the inventions of reason and will in the mechanical sphere. This is a condition of the nervous system which is seldom witnessed, but we place it as a counterpart to the state of disunion considered before: the ordinary state is a kind of composition between the two, and can easily be constructed out of these extremes.

I do not know that we can escape generalities in treating of the functions of the brain. Certainly at present, when we go inwards, whether into the head or the mind, a few powers present themselves, very little colored, and with none but a general outline. The gray and white shadows of metaphysics seem to answer to the cineritious and medullary groups of the nervous system. But there is no doubt that the truths of this class are valuable, as the pure science of the body; and that they are the outlines of a multitude of moreinteresting sciences.

There is indeed a branch which it has been thought throws a broader light on the nature of the brain : we allude to phrenology. This office of phrenology we regard however as a misapprehension. As we understand phrenology, it is a science of independent observation, which is completed in tracing the correspondence between the surface of the living head and the character of the individual. It was as such that its edifice arose stone by stone in the hands of the illustrious Gall. He noticed that portions of the surface of the head stood out in those who were prominent in certain faculties, and putting the bodily and mental prominences together (for which may he be honored!), he arrived by repeated instances at the signs of the character as they are written upon the head. He completed the dark half of the globe of physiognomy, and letting his active observation shine upon it, he found the rest of the head representative of the whole character, as the face is expressive of the mind. Expression, we may remark, is living representation, and representation is dead expression. The representation of the man by his head had always been vaguely felt, and the best sculptors and poets had imaged their gods and heroes with phrenological truth. But Gall made their high intuitions so current that all could buy them. Now this department of physiognomy surely night be carried to the perfection peculiar to itself without the head being opened. Nay, it would be best learned without breaking the surface; for the beauty of expression and representation lies in their bringing what they signify to the surface, and depositing it there. But for this purpose the surface must be whole. There is no interval between life and its hieroglyphics, but the one is within the other, as a wheel within a wheel. The thing signified by the organ of form is *form*, and not a piece of cerebrum: *love* is meant by the protuberance of amativeness, and not the cerebellum: and so forth. It is superficiality, and not depth, that is excellence here. The deep ones had dug for ages in the brain, and found nothing but abstract truth: Gall came out of the cerebral well, and looking upon the surface found that it was a landscape, inhabited by human natures in a thousand tents, all dwelling according to passions, faculties and powers. So much was gained by the first man who came to the surface, where nature speaks by representations; but it is lost again at the point where cerebral anatomy begins. Gall himself was an instance of this, for he was one of the greatest and most successful of the anatomists of the brain. But when the skull is off, his phrenology deserts him, the human interest ceases, and his descriptions of the fibres and the gray matter are as purely physical as if they were of the ropes and pulleys of a ship.

It must however be supposed that the brain has a definite ratio to the head, but what that ratio may be, is an undecided question. It is difficult to prove that the risings and fallings of the skull correspond exactly with those of the brain. This is of no consequence to phrenology as a science of observation. And it does not follow that the *representation* of faculties is equivalent to physical correspondence or similar undulation of surface. The nose represents the sense of smell, although the olfactory nerve does not lie under it in the form of a nerve nose. And destructiveness may lie in its bony den without exactly fitting the bone. On the contrary we might suppose that where activity was involved, there would be room for exercise; and that the inner table of the skull would represent something more than the limits of the greatest exertions of the faculty—the arms-length, spring and hatchet-play, for instance, of the destroying organ.

Moreover, looking at the instance of the face, it does not appear certain that the ratio is between the surface and the parts immediately beneath it. Concealment and *projection* are elements of representation. The eyes are put forth far away from those cerebral origins which they signify, and with which they communicate. The parts that functionally underlie the eyes are not the structures nearest to them inwards. The superficial-making process is often slanting, as is seen in the ducts of many organs, which carry the produce by which they represent the organ to a spot remote from the surface above it.

We are inclined however to believe that there is a fitness between the parts of the phrenological head and the brain underneath them. And we suspect also that the ratio is one of quantity. For when we consider the whole frame as representative, the front half of the body stands for expression, or that which represents the mind actively, and in the face intelligibly; while the back half stands for representation or reaction. The front impresses spiritually, the back materially; or in other words, the front acts, and the back reacts. Now the reaction consists in the gross pushing of the frame, while the action is skillfully supported upon this, and busy in the front. This pushing for a rest necessarily is quantitative, and moulds to its shape what it comes against, if the latter can take an impression. The skill of action on the other hand similarly moulds, but at a distance from itself, and upon the models of quality or design, which issue from its creative fingers. Now it seems as if the brain, considered as made of organs, and as determined to without, leans the backs of those organs against the skull, their fronts being turned to the origins of the nerves of the face and body; and thus gives upon the skull the back or passive side of the character, and a portion of the front or active side in the face.

But still the push of the organs against the skull and their protrusion or apparent bulk, would not show that the protrusion upon the head was primarily due to the subjacent organ. In a soft and yielding mass like the brain, the strong and energetic parts would, it seems, hold their places, displace the others, and cause the weakest to go to the wall. Thus a stout eminence on the head might signify firmness, though the organ of firmness were far away, making room for himself in all directions, and ousting the feeblest parts of the brain into the poorest places, aside from the rest. In fact, power is always felt physically at a distance from itself, where reaction and resistance begin. Thus feeble races are pushed up into the mountains before their conquerors, where their condition indeed signifies power, but it is unfortunately the power of their enemies, and their own weakness. We might then expect that in the reactive skull the greatest prominences should have under them the most passive portions of the brain. And if this be the case, the phrenology of the brain would be the inverse of that of the head, and each depression or elevation on the skull would not be the result of a simple pressure, but of the varying balance of two or more faculties or organs, pressing each other up or down as the case may be. It does not seem impossible that such a phrenology of the brain should be constituted; which if it were, it would signify the strength and weakness of the character, both written on the surface. Undoubtedly many of the exaggerations of character proceed from the imperfect resistance offered by some of the faculties, and show their strength in the direction of our greatest weaknesses. Thus their forcible nature argues the compliance and not the active strength of the mind which immediately executes them. If this hold good of the skull also, then it contains no organs, but merely passive evidences of the faculties, and we are brought again to the point, that phrenology is more properly to be called cranioscopy, and to be regarded as the mute complement of physiognomy.

Moreover the correspondence between insides and outsides cannot be calculated upon with nicety. Circumstances not only environ essentials, but alter their shapes and seemings. The skull, as a circumstance surrounding the brain, may represent it badly, as a poor gift of language may choke the utterance of a rich heart. It by no means follows that there is harmonious co-development of the parts, but on the other hand, the instances of this perfection are rare. Brains may be born into inconvenient cases, just as good human minds, veritable immortal children, are born into idiot brains. Difference of form, also, and consequent difference of distribution of the constituent parts, may be expected between different races. If the climate and the wants of life are various, the shape of the life and its parts will be various also; the faculties will consent to the circumstances and grow in their training. Ideality will not fight with hope for any precise chair of bone, when the relations alter, and another piece is naturally offered. A faculty squeezed in by circumstances will rise up somewhere else, or cause the protrusion of some other part. It seems clear, then, that the brain will consent to be packed differently; that if its external world or climate, viz., the skull and membranes, are of a new type, its resources will not be overset, but developed in a new direction. But this does not disprove phrenology, though it may perhaps cause us to hope that there are phrenologies besides the European, and that this little science also is of a spherical richness.

Perhaps we have been too much accustomed to regard the exterior head as a mere wrappage of the brain, whereas, like other externals, it is independent like the brain itself, and has its own centres of structure and function. We have likened it to a country or climate that the brain inhabits, and pursuing the analogy we may say, that the inhabitant did not make his country, nor can he modify it, excepting in so far as he is modified by it. True, he is destined to mould it to his wants; but then the wants themselves draw their forms from the climate. Hence we see that in the greatest shell or skull which is built up around every man, namely, his vault of sky, and what it contains, a typical difference exists which cannot be reduced for different vaults to a single rule: the skull of heaven has many shapes, and societies or brains fill them differently: destructiveness in one wages itself upon lions and tigers, or is committed to the arms and hands; in another, it goes forth with the powers of the mouth or the sword of the spirit. Phrenology, following this order, must rather start from a type, and gradually depart from that, than attempt to make the type universal in the relations of shape and place. In short we must admit a comparative phrenology in the human race itself.

Phrenology, however, is one of a group of sciences different from anatomy, and its truths are of larger stature than those which we are considering. It belongs to the doctrine, not of the human body, but of man, and is one of the lesser departments of anthropology. It furnishes also a contribution to that which is the science of sciences, namely, the significance of forms. Considered as a branch of observation it has not been assailed successfully, because no one has paid so much attention to its facts as the phrenological student. We take his word for its truth, at least provisionally, since the oppugners have formed no contrary induction, which in destroying phrenology, might supplant it by a better practical system. And if we are not mistaken, the world will give it a long trial, were it only that it deals with the substances of character, and seems to create a solid play-ground away from the abstractions of the old metaphysics. Color and life, substance and shape, are dear to mankind, as homes against the wind of cold speculation. We cannot give them up for patches of sky a thousand miles from the earth, or for anything, in short, but still more substantial houses.

An important set of problems concerning the brain and nervous system remains to be noticed; namely, the doctrine on the one hand, and the denial on the other, of a nervous spirit or fluid. On this subject interest has ceased, in consequence partly of the uselessness of the controversy. Each party has wielded a sword to which the other was invulnerable; a sign that Providence did not intend the dispute to be settled upon the terms of either. Nor perhaps can we ourselves command a decisive victory for our opinions, which are affirmative of the existence of a nervous spirit. Nevertheless we think that our side has been increasing in strength of late, and that the contest must be renewed.

In the first place we remark, that no part of our frames is so little given up to sight and sense as the live brain and nervous system. No part is so much alarmed, or presumably so much altered, by exposure. Nay, there is no part in which there is less to see than in this supreme organization. Detail and structure and the broad lines of design begin to be inappreciable, as a preparation for the indwelling and shadowing forth of something which is absolutely invisible to the senses. *Per contra* there is no part in which there is so much left to fancy, to imagination, and to the inner eye of reason. Tempered thoughts seem to be the only steel that can open the viscera of thought. The method *à priori* appears to be as much prescribed by nature for investigating organs that work *à priori*, or from life downwards, as, they say, it is proscribed from those where the senses give immediate information. The truths that we can see, we do not discern by reason; but conversely reason must discern those which the eye cannot see. If all truth is *à posteriori*, then the physiologist can see the living brain in vision. But as he has lost this faculty, reason in the meantime must have place.

Now neither of the parties in the above controversy has ever seen, still less dissected, the living brain, meaning thereby the brain in the exercise of its functions, thinking and reasoning in its peaceful head. On both hands, then, imagination has been active in all that has been said. The question therefore is, as in several other cases, Which party has the best set of imaginations? measuring the excellence on either side by the fruits that it produces in its department. Which set of conceptions—the affirmative or the negative—comes nearest to life, and to those attributes which will fit the highest organ of man's body, and one which receives the influences of his mind? For reasons of analogy, necessity and experience, we adhere to those imaginations and thoughts that postulate a nervous spirit.

Be it observed, however, that we do not dogmatize respecting the physical properties of this which we term a fluid. In reasoning by analogy, we are forced to take with us the garb of the known sphere, and to talk of a fluid, because the body furnishes the word. We desire, however, to hold the term loosely, and no longer than until a better analogy, or the real name, arises.

It is well known that influences mount from the body to the head in sensation, and descend from the head to the body in voluntary motion; and anatomy demonstrates that the course of these influences is, in sensation, from the skin and sensories through the nerves to the brain; and in action or motion, from the brain through the nerves to the muscles, bones, skin: in short, to the body as set in motion. Furthermore investigation shows, that the course of the same is, from the nervous fibres and fibrils, through the medullary fibres of the brain, to the cortical or gray substances at its surface and in other parts. What do we deduce from this known transit of influence? Or what is the meaning of the ambiguous term, *influence?*

Two views have been current. On the one side it is maintained that the fibres are solid, and that the influence is a vibration which traverses the fibres in both directions. On the other hand, the influence is regarded as a real *influxion,* and the fibres are regarded as conduits, permeated by a fluid. There is a third hypothesis, of a gelatinous or other fluid in the nerves, which propagates forces by waves or undulations, but does not itself move forward: but this is only a subdivision of the doctrine of vibration. And there is a fourth view—that we can know anything about the matter. But we do not yet know what can be known. In thus classifying opinions we by no means intend to convey that they are held sharply by any one at present: on this subject the state of most inquiring minds is mixed; indeed the apathy felt of late respecting the controversy is too profound to admit of distinctness of parties.

I. If the fibres were solid, and traversed, not by a fluid, but by a vibration or undulation, such vibration would be dissipated into the surrounding parts, especially in the brain, where the fibres lie close together. The softness and non-elasticity of all parts of the nervous system seem unadapted to mere vibration being communicated, for vibration depends upon tension, and the only condition of tension in such a system lies in a movable fluid distending the fibres. And when a nerve is tied, the sensation or motion which is supplied becomes arrested at once, which would not be the case if these depended upon vibration. Certainly nothing can be conceived less adapted than the brain and nerves, to the ordinate propagation of electricity or any other imponderable, unless limited to a fluid vehicle. The hypothetical currents of vibration, however, must be ordinate and distinct beyond measure, for on the one hand they are the lines of thought and will, sensation and action; and on the other, the fibrillation of the brain is exquisitely minute and separate even to the naked eye. We gather from these reasons, that if the nervous current be in any analogy with electricity, galvanism, or other imponderable, it must itself still be regarded as an *incarnate* imponderable; as might be anticipated from its domestication in the human body. But an incarnate imponderable, preserving its velocity on the one hand, and being embodied on the other, is only conceivable as an animal fluid or a fluid animal, whose speed and other attributes are life-like, thought-like, will-like, and sense-like.

II. The analogies of the grosser parts point to the tubular structure of the nerves. In the body, where there is an organ as a bed, and trunks, branches and twigs proceed from it, the derivations are hollow, and carry a fluid. This is the case with the heart, the lungs, the stomach, the liver, the kidneys, the pancreas, &c. On this ground we understand their functions; in their produce we recognize their use in the economy. Again, what reason can we give for the quantity of blood which is applied to the brain, and which is more in proportion than is sent to other organs, if it be only required for its nutrition, and not for supplying a large quantity of fluid. How also can we account for the physical debility consequent upon certain effects, unless by a waste and spending of a fluid? And recurring to analogy, does it not seem to be in the order of things, that the living principle should act through the fluid upon the solid, when we find that the more living parts of the body, as the brain and nerves, are the softest; and the less living, as the tendons, ligaments, and bones, are the harder and the hardest: also that in the earliest stages of formation all things are fluid, and at the very beginning, a fluid, and that hardness grows up by degrees, and plasticity ceases; old age consisting physically in stiffness, unyieldingness and ossification. Further, that the change of pervious canals into solid ligaments, fibres or threads, which takes place often in the body, is always a change from vitality towards the contrary; showing that the solid form is a degradation of the previous fluid, and not *vice versâ*. It seems then a prepared belief, that the nearest thing to life is the most life-like, the most movable, the most quick, in short, the most

5

fluid; in a word, that a nervous fluid is itself the first organ in the body, and the physical handle of the spirit.

III. To come to experience, what are we to think of the pulpy, semi-liquid texture of the brain itself, and of the immense proportion in which fluid enters into its composition? That there is a *cerebral fluid* is past doubt, for the brain consists of almost nothing else. And if three-fourths of the mind's organ are fluid, certainly that portion is subject to mental arrangement, like the rest. When the mind is gone, the arrangement is gone, and hence the dead brain gives no response respecting the nature of the fluid. Nor indeed would the sight of the living brain reveal the mystery, unless the observer had some thoughts commensurate with the organization which prevails in these provinces of the soul. A fibre of truth, tubular from heaven downwards, and a fluid mind travelling in it—in short open faculties—would be the only conditions of discernment, even if the skull were a window through which the brain could be seen. But at least we can conclude, that the so-called serum of the dissecting-room is only the corpse of the cerebral spirit. It is true the dead body retains its shape when life has gone, and is therefore instructive still, but the animal fluids undergo alterations even in form, when they fall out of their native shapes, the vessels, as witnessed in the coagulation of the blood. And for this reason these liquid corpses teach absolutely nothing of the properties of the living fluids.

To illustrate these remarks, we will put an analogous case of a larger order. Suppose that London were visible from one of the stars, and were known to be a city of the living, but its inhabitants were not within the power of vision. And suppose further that the stellar people were of those who believe nothing but what they can see. It is clear that the outworks and great channels of London life falling under their ken, would be mistaken for the living things. And as life always brings motion to mind, the vibrations and shakings of the houses and walls of our metropolis, would be postulated as the life which was alleged. But suppose further, that on some coronation day vast crowds of the inhabitants assembled, and formed a dark mass before the eyes of the gazing star; this would of course be taken to be of the same substance and rank with the houses, and the identification of life with bricks and mortar would still stand. In this case the truth, or the living multitudes, though seen, would lend itself to the fallacy; the only escape from which would be by analogy, proclaiming that solid houses everywhere are dead, though inhabited by fluid or freely movable living beings; by imagination or hypothesis, guessing that towns and streets are for men, and not *vice versâ;* and by reason or theory, affirming mankind, and accounting for the appearances of the city upon the wants of its substantial, though separately invisible inhabitants.

But IV. the existence of an animal spirit has great historical probability attached to it. For the course of knowledge has consisted, not in confirming abstractions, but in merging them in some adequate reality, such as we are now claiming for the life and spirit of the brain. The concrete form of things, or the tracing them home, is the final victory of knowledge touching mere existence. So long as life is an indeterminate phrase, applied without distinction to the whole system, the study of that life has not commenced, for its presence has not been gained; but when its proper currents are found, and the mind traverses them, then the separate knowledge of their properties can begin, but not sooner. And, moreover, the triumphs of this age are peculiarly due to the introduction of the mind to the empire of the fluids. The steam engine and its nervous spirit, steam; the railway and its locomotive fluid, the train; the wire and its electric spirit, show the practical benefits of the subordination of the solid to the fluid. And in human progress, it is the fluid and the modifiable that give motion and impulse to the otherwise fixed. What are quickness, conception and imagination, but the fluids of the mind: regard them at work, and you can bring them under no other analogy. They stir the old, hard world, and permeate all things, and like nervous fluids are present in a moment where their mission is, with the power of arranging and quickening virtue that they have received in the fountains of thought. Indeed, I see not that there is any known sphere of things, whose analogies do not cry aloud for the existence of a fluid brain governing the solid, and like it, organized, though on a more living plan. Thus until a nerve-spirit be admitted, How can the science of the brain be in fraternity with the other arts and sciences?

V. If this fluid is *quasi*-life, or what is the same thing, physical life, it may well be conceived that the nervous tubes will close against it in the moment of dying; as the dying arteries contract, and shed away the arterial blood. Death has no hold upon life, but its chill grasp is its means of losing it. Hence, microscopic observations upon the nervous fibres never can give the negative to their tubular form or fluid contents. The problem does not come within the brains of necrology. Moreover, how if the tubes were spiral, and not straight, which might be the case in a system where velocities are such, that distance forms no element in the calculation? In that case, even supposing the tubes were hollow, they might never be seen as tubes, by reason of their insinuations and turnings. For in the realms of mind and thought, the shortest distance between any two points or ends, is that which leads through all the means, no matter by what length of course. The zigzag of the lightning is in the straight line of smiting.

VI. The doctrine of a nervous fluid seems further to arise out of the construction of the system from successive pieces, each higher and broader than that preceding it. For this ladder takes us up to regard the mind itself as supremely nervous. Now each part has its centre in itself; but also is traversed by the part above it, on its way to the surface; whither all the pieces, the high and the low, arrive alike immediately, or are represented. Thus the mind comes down through everything and its spirit glitters in the face. And thus all the actions of man—automatic, sensual and animal—may be shot and pierced from the quiver of life, until they are nothing but rational and spiritual actions. It is to be remarked, however, that the visible solids terminate with the brain. If, then, the mind has fibres representing it in the brain, as the brain has fibres representing it in the spinal cord, the former fibres must lie in the fluids, for the solids belong to the brain itself. Thus, while the brain or organism terminates in its own centres, the cortical substances, or supreme solids, the mind enters into these by a series of corresponding fluid organisms, which represent the living or active portion, as the solids represent the recipient or passive. This is but imagination; yet imagination is the youthful eye of science, and provided it owns to its name, it is an innocent as well as a suggestive power.

Therefore, we proceed to imagine, that the mind broods above the brain, as the cosmical ether sits above the planetary air, and further, that the mind, or spiritual reservoir, fills all the interstices of the brain, where, however, it is determined by suitable fluid envelops, which accommodate, temper and envelop it, until it is brought into ratio with the fixed mechanism of the cerebrum. Hence, its solar vibrations are felt in the bodily organism, as light is felt on the earth through the splendid shiver of a medium which the air includes between its parts. And hence, there are as many kinds of nervous or cerebral spirit as there are nerves and brains; for it is the openness or intervalling of the latter, that admits and gives quality or fibrillation to the former. The brain, doubtless, is made with an express view to this recipiency: the fluids, which when entered by the soul become nerve spirits, are also predetermined; and hence there is the same ratio again between the mind and the nerve spirit, as between the brain and the spinal cord, in that the parts of both are alike, but differing in breadth or degree.

But indeed, the doctrine of the nerve-spirit forms but one of several problems, which have experienced the like treatment. The reality of whatever is above the senses, is questioned by many, and consequently the presence of the supernal in the sensual is denied; and, if the supernal be visible, still it is degraded by some ordinary name, and the spirit and message which it carries are sensualized and set aside. The thing is killed by skepticism, and then its spirit is easily called serum. It is so with Revelation. A divine mind there may be, but then man, say some, is solid at the top, and lets God in by a self-vibration: there is no open fibre between Him and man. We, however, affirm a nerve spirit of the human race, which is not man's, but God's in man—a genuine influx, a word, a revelation. It is just as visible as other books, though often it is not known what it is, because the mind is not known with which it communicates, nor consequently the peculiar animation of the book. We see on this mighty scale, what is the use of a nerve spirit, namely, to be immediately present with the directing mind throughout the organism; to speak in messages unwarped by the finite faculties; to be the First and the Last, instead of "sitting above the creation, and seeing it go." Were it otherwise, the government of the world would be virtually committed to man's reason, which would exclude that of God; for man would have no chart but his own faculties. And so, were the brain a solid, and shut at top, the government of the body must depend upon it, and not upon the soul, or only upon the soul, in so far as the brain chose to coincide with its messages. Government is incompatible with such an idea. But on the other showing, the brain cannot exclude the next higher series, which permeates it in its own right, and exercises a providential and healing virtue upon what is so apt to go wrong of itself. The truth, however, is, that the doctrine we are combating, is not in the sphere of the nervous or open system at all: it belongs to the muscular department of truth, or that which has both ends closed, and is a solid body. Its spiritual correspondent is "earnestness," spasmodic vigor, upon which so many and such famous men rely for the salvations of the time. We might collect other problems, besides that of the presence of the divine light in the world or Revelation; and that of the presence of the soul in the body, or spiritual existence, to show that the doctrine of the animal spirits suffer in the company of great truths; but enough has been said to show that deism has dependencies everywhere; and that besides the vast irreligion of excluding God from the universe, there is a series of lesser impieties and disloyalties of a similar kind, which rob the body of the soul, clip the world close to the limits of air, make man and nature truth-tight, and reduce all things to petty selves, which can choose whether they will have a God above them, a world around them, and a soul within them, or the contrary.

We postulate, then, that everything, according to its openness to the sphere above it, has a spirit, or a quasi-spirit; and when the organ is so constructed for opening as the brains—whose first great faculty is *openness*—when it can take in so much, the capacious influx or gift naturally takes the form of a nerve spirit, which is thenceforth the tutelar genius of the system. Further, that everything, whether a brain or a science, which is so roofed over that nothing but itself can come into it—so thick-skulled and self-opinionated—is dead at the top, however it may work reflexly or spasmodically, according to the exigencies of the lower life.

So much for the spirit within the brain. There is a spirit beyond it, of which we do not treat in this place. Only we remark that the spirit within is the physiological window to the spirit beyond, and that they who do not look through it, cannot admit the soul of man as having any ratio with the sciences of the body. Certainly, the soul is not their object; but woe to them if it have not its witness within their field. Even granting what we have said, the nerve spirit will be only on a level with the other fluids, unless it be interpreted by those higher powers which it serves. But what its function is can be told analogically, but not, as yet, otherwise. Thus, we may say that it exercises quasi-mind and influence in every part of the frame: that it is an atomic intellect everywhere, capable of representing the state of the body, and a will, capable of setting the organism in imitative movement according to its decisions and forms. By it alone can we account for the unity and coherence of the system. For every point of it is a sense or intelligence that shows the frame a model that commands instant imitation. And thus it may be likened to the truths of society, which command obedience by their bare showing, according to the health of the social frame. Let it be borne in mind, however, that the nerve spirit is different according to the divisions of the nervous system, and according to the nerves of every organ, and indeed, every particle, because each is open to the wisdom that it needs. Thus, in the nervous system proper, it is mind and will: in the liver, again, it is the mirror and model of the hepatic truths and operations. And so of the other organs. Dropped from above into their lives, revealed to their blood and races, it sets them in the fermentation and discussion of their problems, and on the *veritas prevalebit* principle, health cannot fail to obey it. It is the posture-master of the solids, and the charioteer of the fluids. We see, then, how full the body is of eyes—how instinct with spirit-like forces, and by the analogies of mental and social life, how irresistible are the causes of harmony, design, and co-operation in the bodily fabric. But were there not such a spirit, there would be no reason, but an immediate dictate of God's will, for the stupendous system which the human machine discloses; that is to say, there would be no wisdom *in* the body, answering to that supreme wisdom which exists above it.

And here, to assist our conceptions, let us revert to the current views of the solid pieces of the nervous system. These, as we have shown before, are all of them of spiritual machinery; in other words, they have the power of dramatizing mental functions. Even the spinal cord acts as if it were a sensible animal, guiding the fingers, for example, to the seats of pain by its automatic endowments. How much more must the supreme fluid, whose essence is motion and plastic virtue, be capable of dramatizing or representing the powers of the mind and soul, and their ever-varying relations to the microcosm. *A fortiori*, it will act like wisdom and like will everywhere, when its mere channels, wherever they extend, have these quasi-living properties. It is here to be remarked, that we are in the strictest keeping with experience; that we are simply postulating for the fluid brain the same thing, though in the fluid degree, which physiology claims already for the solid brain. But there is this difference, that the solid brain, or the nerves, are of course limited to the nervous system, whereas the fluid brain, or the nerve spirit, does not end with these conduits; but when it is shed from the fingers' ends, heart's ends and viscus' ends of the nerves, it can make use of every other structure as its vehicle, and thus import the drama of life, in all its degrees, into the flesh and entrails of the universal body. Agreeably to a common fact, that where a form ceases, its essence, if of true virtue, still passes on, and applies itself to utilities that it could not execute when confined to its bodily sheath.

We may perchance be accused of materialism for demanding such a physical wisdom. So an insect possessing only a spinal cord might accuse of materialism a being possessing the sensual brain, because the latter exhibited this as still a nervous system; and a creature terminated by the sensual brain might reproach materialism against a being who carried the nervous firmaments one step higher in the cerebral hemispheres, because those hemispheres were still nervous system. The insect conception of sense (if the phrase can be pardoned) is doubtless null; and could the creature evolve it, the nullity would appear in the want of body and structure, in the impression that pure spirit or abstraction comes on at the top of the spinal column. The brain, however, is a spirit pure enough to the spinal column. And so the mind, organic and in ratio with the rest though it be, is, nevertheless, the pure spirit of the human brain. The materialism lies with those who make the brain solid, the skull thick, and the mind an abstraction.

VII. In conclusion, we may be allowed to say that the doctrine of a nerve spirit is no new creed, nor ever was unorthodox until now. The greatest names in physiology are its adherents. But neither did these men see the tubes or the fluid, such as we have conceived it, in the dead subject: a fact which, as Haller says, "shows the weakness of the senses, but has no validity against the existence of a juice or spirit in the nerves." This reason had its proper effect with the great anatomists of the seventeenth and eighteenth centuries: it showed them the weakness of their senses, and stimulated them the more to use their understandings. They were the original geniuses who laid the foundation of a general knowledge of anatomy that will not pass away, but to which we must recur when we desire to refresh our minds with the first vivid impressions that the wonders of the body created upon the finest intellects that ever studied organization. Would it were in our power to bring before the reader the characteristics of these brave pioneers of anatomical knowledge. Would we could reintroduce him to the venerable Eustachius; to Malpighi, the father of visceral anatomy; to Ruysch and Morgagni, the purifiers of the anatomy of the schools; to Leeuwenhoek, who first seized the microscope as an exclusive field, and devoted himself to it for fifty years with an eagerness which has not been surpassed; to Vieussens, Lancisi and Baglivi, eminent alike for systematic knowledge and philosophical genius; to Bartholin, Verheyen, Heister and Winslow, whose methodical text-books kept their ground in the European schools for more than a hundred years, and who supplied their successors with much of both the matter and the form that exists in the manuals now in use; to Boerhaave, "the common preceptor of Europe" in the last century, and the consulting physician of the world, who gathered up the experience and deductions of ages in anatomy and physiology, and gave it a new and compact form in that wonderful little book, the *Institutiones Medicæ:* also to Boerhaave's pupil, Haller, who stands as a mountain between the present and the past, and reflects from his summit the departed learning of seventeen centuries. But their monuments are the living parts of their science. And we must be content with alleging, that these men, with no exception, believed in the animal spirits, and the tubular construction of the nerves; which, as the illustrious Glisson, our countryman, remarks, was in his time "accredited by nearly all physicians, and by all philosophers."

We now therefore assume that the nerves are tubes, and that there is a special fluid, cerebral, spinal, sympathetic, *i. e.*, following in its degrees and divisions the solid pieces of its nervous frame: and the question occurs, Where is this fluid engendered? We reply, that its matrix everywhere is the cineritious substance; that as the vehicle of what comes down from above, it begins everywhere with the beginning of the bodily order, which lies in the cineritious spheres. The latter show luxuriant provision for the purpose in the arterial meshes which supply them. But to pursue this subject would require a treatise on the life of the blood. We shall, however, recur to it in the sequel.

But another question presents itself—Is there a circulation of this living fluid? Do its centres impel it, as the heart propels the blood? Is there a cerebral *force* or a motion of the nervous system? Or is the brain, besides being solid and exclusive, also stationary and paralytic? Or on the other hand, if there be any natural force in brains, what is it?

First for the facts. Two motions are already admitted to have place in the head, one corresponding to the beating of the arteries or heart; the other, to the breathing of the lungs. By laying the fingers upon the open fontanelle of a young infant, the reader will feel the first-named motion. It is the stroke of the heart communicated to the arteries of the dura mater. This is no cerebral force, or proper motion of the brain. As for the second or lung movement, it occurs as follows. When the lungs expand to draw in the air, a tendency to vacuum is created in the chest, which causes the fluids all over the body to rush or to incline thither, to fill the threatened void. In short the lungs, besides breathing in the air, quaff down the venous blood from the brain, and suck up that from the body. The consequence is, that in inspiration the skull is more empty of blood than at other times. Hence when the brain is exposed, it is seen to subside during inspiration. This is the second movement observed, or the effect upon the brain of the respiratory pulse. But neither is this a cerebral force, but a physical subsidence, just as the corresponding rise is merely physical, and results from the act of expiration, which retarding the return of blood to the chest, forces up the brain upon this fluid cushion. This, however, is to be noticed, that when the blood leaves the skull during inspiration, it creates a vacant space which has a function, or which is not vacant in point of use: in short, it leaves room for the brain to expand. And if its real expansion be less in volume than its apparent subsidence, then the brain may be automatically rising at the moment when it is physically falling, in which case the latter movement will mask the former. When the lungs expand, then, there is room for the brain also to expand, and when the lungs contract, there is a physical reason why the brain also must contract.

We do not know that experience has gone further hitherto than to show the possibility of the movement of the brain, in showing that it has room or liberty to move. And indeed, if we consider the problem, we shall perhaps find reason to think that this motion, like the motions of other bodies from which we cannot separate ourselves, but with which we ourselves move, must come as a theory whose main case will lie in explaining all the facts.

Yet further as to the possibility of the movement, one of the most important truths of physiology lies in the insulation of the different systems that make up the body. In point of function, indeed, we have seen that the nervous system itself consists of individual pieces, which can act either separately or in combination. But in structure the whole nervous system is isolated from the rest of the body, though plunged or let down into it as into a well of flesh. The nerves end in loops or otherwise, but are not solidly continuous with the other tissues. In short, the nervous body floats in the fleshly body as the central yelk of the microcosm. We may liken it to a sword of lightning in an elastic scabbard, which scabbard is perpetually elongating and gathering itself up, but is always full of the subtle and separate fire. Movement is compatible with this state, nay, is perpetually stimulated by it where the whole subject is moving; just as the non-continuity of the planets with each other, and the interstice between these cosmical atoms, allows and necessitates their courses and revolutions. It used to be thought that nerve joined and became continuous with muscle and vessel, in which case thorough motion of the nervous system would not have been possible. We now find that it only rests upon these lower parts, of course with an interval, which allows of thorough or loco-motion; and we know, therefore, that thanks to the breathing movements, the nerve animal requires to plant its foot afresh with every inspiration, or it would slip down, and to raise it up again with each expiration, or the body would shock it in its ascent.

If the office of the brain lies in the distribution of the cerebral fluids, then the orderly administration of these requires a regular motion, or in other words, a rhythmical action. The heart and the lungs are evidences of this in the lower compartment. Their action is not an incorporeal vibration but a measured expansion and contraction. The blood cannot run to its destinations without the physical heart, nor the nerve spirit gain its ends without a propulsive power in the brain and the other centres. And more than in the case of the other organs must this motion belong to the brain itself, or be automatic, or the supreme organ would have no function of its own. It cannot, then, be derived from the heart or the lungs, as we said before, although no doubt these organs, and all the rest, are measured by its liberties, and concur to make its play and playground secure.

In the body, moreover, we find that stated motion is a condition and a sign of life. Does he breathe? or expand and contract, is synonymous with, Does he live? Now, as we shall have occasion to show in the next chapter, the universal movement of the body produced by the lungs, is the condition of all particular movements, muscular and visceral. For a body already in constant motion is easily guided about, deflected and managed for particular motions; but a body at rest absorbs a large amount of force before it can be moved. Thus if the whole body were not a perpetually-moving, breathing or living thing, there would lie so much dead weight on the feet of every action, and life would be clogged by matter. But as it is, the incarnate iron is hot for the strokes of the volitions.

Nothing can be more familiar than this, that motion is easy, and rest uneasy, to those who are already on the move. Now we may reason by a great syllogism from the man to the body, from the body to the lungs, and from the lungs to the brain: nay, and also from the brain to the mind. For example, if the brain were stationary, no thought could enter it without lifting it first: there would be no preparedness for thought, which is essential motion. Whereas, by a constant rhythm taking place in the whole mass, the mind requires to create no motion, but simply to act as a modifying, directing, or exemplary power, in order to produce all or any motions as states or details of the general swell. Thus, the rapid admission of thoughts requires a moving or active brain, just as does the distribution of the nerve spirit, which is a river of bodily thoughts.

It is indeed admitted that the brain is an active organ, so far as it undergoes the fleeting states of thought. These are the modifications of its activity, the ripples of consciousness. But what we are contending for is nothing less than a tidal cerebral sea. For as the thoughts, emotions and bodily actions fall upon the lungs, and produce the varieties of the breathing, because the breathing itself is there first as an impressible atmosphere, so do the thoughts and volitions produce the varieties of the *animation* upon the basis of an already-moving or animating brain. The temporary waves of thought, are but the surface which comes into our light: there is a deeper heaving besides, organic as the body, which has its own currents and shores, and is constant like nature, obeying the progress not of the moment, but of the lifetime. Its spirits are not transitory like ours, but night and day they do not sleep until death overtakes them.

The argument we are discussing is, however, theoretical; a condition which is common to the truths of the exact sciences. The solution of the problem in this light will depend upon whether the brain or the mind is accepted as the central truth. If the mind be taken as the fixed point, the principles of thought will have their just power, and the brain will be seen to revolve around the mind, and by its revolutions, or expansions and contractions, by its moving up to, or away from, the mind, to produce the times and seasons, the warmth and cold, of human thought and will; and as the variety of states will in this case depend upon the brain, and the brain is a physical organ, the motion will not be a mental or ideal, but a physical motion. In other words, if the brain is physical and not mental, then what seem to us to be mental motions in the brain are really physical motions; and in this case an open mind signifies an open brain, an active mind a moving brain; and human language, or the voice of common sense, contributes abundantly to the illustration of the organ. For the predicates of mind can then be assigned in a bodily, but not in a purely mental sense to the brain. If, however, the mind be regarded, as by the materialists or cerebral Ptolemaics, as the meteor and wandering lamp of the brain, then a mental, or in other words an inexplicable influence or movement will account for or rather accompany the cerebral states. But for ourselves we cannot think of the mind as an *ignus fatuus* in the swamps of the cerebrum, but as a distinct and superior organ, which has the cerebrum and nervous system between it and the body. But now if the brain has thought movements, there must not only be a law of succession in these, a beginning, a middle and an end of each; in other words, an expansion and contraction; but also, as we showed before, a ground swell of motion on which they depend, and which if it were not given by nature, no thought could lift the sluggish mass in time enough to give a thought-like action.

But after all, the question of motion is subordinate to the question of what the motion is? To show by rational arguments that the planets move, without demonstrating their courses, would leave the theory not only short, but curtailed of its strongest proofs. And to raise the problem of the cerebral motion without showing its times and rules, would be to rest in an embryo law, and to fail of the support which the body itself ought to proffer of so important a truth; if a truth it indeed be.

In turning to this new aspect of the question, we find in the body that there are already two movements, which we will designate the systemic and the sub-systemic; the movement of the respiration is the systemic, that of the pulse the sub-systemic. The breathing of the lungs is the largest revolution of organic life that the body executes; the beating of the heart is but a satellitial motion freely included within the former. And if organic life or motion be concentric, a strong presumption already arises, that the *animations* of the brain, according to the statement of Swedenborg, are coincident with the respirations of the lungs. Moreover we have already seen, that when the lungs *inspire*, the brain has room and invitation to expand, and that when they *expire*, it receives an admonition and pressure to contract. If the brain be impressible at all, and if its motion be physical, it can hardly fail to comply with these opportune times.

But again the motion of the brain is, or may be, voluntary, and the volitions as we know can play also upon the respiration. Nay every muscular movement however eccentric, is based upon the fixity of the respiration as a central stoma. Whence voluntary breathing becomes analogous to willing, and we are said to breathe the actions which we strongly intend. This points, in the next term of the reasoning, to the conclusion that animation, or the common function of the brain, is analogous to common breathing. For the argument stands thus—the mind of the brain, in falling upon the lungs, controls their states, and makes them voluntary; and in descending into the muscles, it always enters the lungs at the same time, raising up in the breath a central air or tendency, corresponding to the limbed tendency in the muscles. The motion of the mind of the brain, therefore, is voluntary, and the breathing can be voluntary; and if the brain and the lungs coincide in their extraordinary motions, is it not feasible that they coincide in their ordinary motions? This is the more conclusive and exclusive, inasmuch as there is no other viscus but the lungs manifesting a peculiar motion, upon which thought and will play. Now these are, physically speaking, the brain playing upon the lungs, or in other words, the brain moving to move the lungs.

But secondly, we affirm, that the correspondence is not merely general, but precise. It will be remembered that we are not now arguing the question of the motion of the brain, which we consider for the present established; but the rhythm which the motion follows. And we proceed to remark that there can be no doubt of its coincidence with breathing. The cerebral motion, the movement of the mind of the brain, is represented in the movements of the lungs. We all infer the manner in which a man's brains are moving, from the way in which we see his lungs moving. If we see deep breathing going on during consciousness, we know that it means profound cogitation; slow breathing concurs with deliberation, meditation, &c. And so on through a thousand states of thought and emotion, in which we reason instinctively on the principle of a correspondence between the motions of the brains and lungs. And diseases augment these phenomena. Stertorous breath signifies an oppressed brain; hurried breath in fevers, a hurried state of mind. And so forth. We take no similar indications from the heart, or any other organ but the lungs, which represent all visceral locomotion, as the muscles represent the locomotion of the outward body. The heart, indeed, is felt by its owner to have a correspondence with the emotions particularly, but it is not influenced perceptibly by the will or the understanding.

In the chest the condition of the nervous system confirms what we are advocating. For the great nerves run through a space governed by the pumping of the lungs, and their external coats, of necessity, are drawn outwards when we inspire the air, and are pressed inwards when expiration occurs. So also in the spinal cord. As these parts are continuous with the brain, the effect also is continuous; *i. e.*, the brain is subject to the reciprocal invitations and pressures of the lungs. The nervous motion is, notwithstanding, automatic; and if its expansion be invited, or its contraction promoted, it is only that these happen as the true circumstances of the freedom of the brain.

To carry the thread into another sphere, but one which is included in our plan, would it not seem that one of the first desiderata in brains is movement, and in all further progress, a history of the movements of brains? not only a history of fitful, but of organic and providential thought. In the realm of science this translation of our position is indispensable; the pistons of aspiration and practice go up and down, the brain opens for life, and opens the body for work, as truth after truth is brought in and converted for the moving intelligence of man. In the sphere of conception and philosophy the same strokes of the mental engine are perceived; and the more we contemplate them from the point of a Providence or a plan, the more regular they seem; the more rhythmical thought is found to be; the more its elements are measured; and the more the stirrings of the great brain concur with the tune of the stars, which measure the ages in their vortical tread. In fact the idea of a plan or a relative Providence, cannot subsist, without insinuating to us the oneness of all the motions of the brain, and their combination into a rhyme coördinate with the poetry of every universal law. Thus to look at them from above and beyond the organ, shows them all as one motion, coincident with the wants and aspirations, or in other words, with the breathings of their subject, man. And so again the outward and inward wants, the thoughts and the breaths, are married to each other. In isolated thoughts we cannot recognize so much; but the thought of epochs suggests a fate of thought, a movement involuntary as the respiration of sleep, in which the parts succeed each other as breaths, though full of the special will and intellect which are the life of the brain and the race.

For the last part of the subject, or the function of a brain moving or animating in this wise, it will lie in the distribution of the nerve life according to the principles and forces of the soul, and in the mind, of the understanding and will, considered both as powers and organs. If the fibres and nerves are the roads of life, and if the globular heads of the nerves are its stations and reservoirs, then the expansion and contraction of these in the times of the breathing, amounts to the constant injection of life into every portion of the body, at the moments when the body itself gasps or opens or wants to receive them. And this takes place in successive moments. Thus organic thought and will are present everywhere with a breathing or animating motion: a stimulus superior to nutrition is poured into the frame; and all this, with no thought on our parts. But on the other hand, during our intervals, and from our states, of consciousness, the distribution of the cerebral fluid is varied, just as the same consciousness varies the supply of air, or the evenness of the pulmonary respirations. And by the coincidence or synchronism of the latter with the brain-movements, the unity of life is maintained: the lungs dilate the frame to receive, while the nerves dilate to give. Thus the respirations of the brains and lungs are the beginning and end of the animal system, which therefore is poised in freedom like the planet, not supported upon a tortoise of dead matter, but swimming in double tides of motion.

The quantity, colors and kinds of the animal spirit thus inflowing, measuring what it is by what it does, are greater than those of all the other fluids; for it not only fills the brain and nerves, but occupies the interstices and the posts of difficulty throughout the body. The good things which seem to be so scarce that they are almost invisible, are yet at the last the only things—are all in all; and this which is the least and most hidden element, is in its volume necessarily the greatest, and the all-embracing. Whatever is more than gravity and rest comes forth from its active lightness. It is the life of the blood, the strength of the arm, the fire of the eye, and the bloom of the skin. Each intellectual city, each convolution of the brain, nay every spherule, sends forth its characteristic emissaries, possessing the body with the varied spirits of the head. Dryads and naiads, muses and furies, gods, goddesses, and heroes in endless populations, throng the columns of this old pantheon, whose last mythology is yet to come. The starry dance, the music of the spheres, the astrologic influences, the experiences of the supernatural, are but aims to express the perceptions and properties of this immortal nature which lives on the seeds of the sun. By this liquid flesh it is that the soul sees its face in the rushing river of creations, and feels the issuing universe, and the finest tremble of the stars. In this wisest vest of nature, it sits at the feast of nature's wisdom. This is the panic element of man in unison with the panic of the world. Could we see an apparition of the nervous spirit, waving and sweeping with luminous shoots into the curves of the body, we should behold a form complete in its details; a design exceeding the mortal building; solid as flesh to the eye of the mind; perpetually springing into life; yet though plastic, stable to its ends, and quicker than thought to execute them: shadowy, or terrible, to the senses, but safe reality to the soul. Then should we see, but according to our insight, that there are motions and mechanics which are the likeness and habitation of life, sense, passion, understanding; and we should know by solemn experiment, that our organization is an imperishable truth which derides the grave of the body.

But if the brain is not shut but open, not empty but full, and if the organ of thought and will is not stationary, but moving; if this is the first nature by which it answers to those ever-moving creatures: and if it operates, not by physical nothings, but by tides of animal life: further if the motion be most regular, and the down-rush constant—then there must be a definite channel laid down, or a nervous circulation. Of this circulation there are three elements, 1. The influence or influxions of the immaterial mind and soul, which come down as rays from a solar brain above the body, and are the order and supreme agent in organization. 2. The array of accordant agents or imponderable spirits of physical nature, which are the contribution of the world on its bended knees to the soul. And, 3. The highest animal juices, the cream of the body, proffered in the vehicle of the first blood of the heart. Thus the brain confers on the organization, first, thoughts and plans, wise as the soul, loving or unitary, and irresistibly organic: secondly, the cosmical and physical kinsmen of these, which are dramatically what the first are really, or which confer mundane efficacy upon the principles of thought. And thirdly, the body of the body, or the primal incarnation. These three, or life, nature, and body, are one in the nervous spirit. In a word, the nervous circulation, with every stroke of its spirit pulse, distributes the essential principles of thought, force and organization: and the body, therefore, is full of eyes or rational light; full of understandings and judgments; full of stupendously reasonable deeds; and materially an incarnation of the soul. This is what the brain is empowered to give; the brain being the common centre of gravitation of the three powers of mind, body, and universe.

The circulation then of the brain would be threefold; or that of all nature into mind and soul, and *vice versâ;* that of the Kosmos into the brain, and *vice versâ;* and that of the body into the brain, and *vice versâ.* But as we are treating of the body, we can only dwell on the lowest of these circulations, though the brain modifies the other spheres in the same way as the body modifies the brain. Now as to the bodily nervous circulation, it comes, like the rest of the secretions, from the blood, namely, that of the carotid arteries, whose fine twigs, inserted into the cortex, are the venæ cavæ of the cortical hearts. The secretion is there received, and meets the thoughts which build, and the magnetisms that clasp and cement it: by the contraction of these cortical hearts it is propelled all over the body into the terminal loops, from whose fingers' ends it flies with its ministering spirits, and is again received into the blood, whose life it constitutes, and which it incites and forces to construct the body on the principles of the wisdom of the soul and brain. Thus the nervous circulation has solid channels only in the highest or proper sphere; in the rest it runs through the fluid blood. Moreover this system has its excretions just like the vascular; in the ventricles of the brain; in every interstice of the nerves: so that when it comes into its loops, it has put off the body it assumed, and is again received into the red circulation. Thus the aged spirit of the nerves becomes the youngest life of the blood. This is necessary to the continuity of existence: for every end is a new beginning in the vortex of our providential universe. We now then see that without this openness of the brain, this animal spirit, this motion of the brain, and this nervous circulation, the soul could not be incarnate, nor the body animate; nor could the latter for a moment preserve that unanimity which gives it coherence, and constitutes it the ideal not only of physical, but of all social unity.

If, however, there be other principles of explaining these things, then we ardently desire to know them. But explanation must be attempted, for the people is weary of hearing that they cannot be investigated; and from men, too, whose powers have no preëminence to entitle them to set limits to the faculties of any new child who may be born into the world. We know of but One whose rights went to this extent, but His voice was, "Seek, and ye shall find; knock, and it shall be opened unto you."

It will not be difficult now, on these principles of motion, to discuss the problem of the respective uses of the cerebrum and cerebellum. And in order to approach the subject we would make the following remarks. The nervous system is a casket of stimuli of various orders superadded to the body: the bodily parts or animals are constructed mechanically without it, and only wait to be ordered into action by its spirits or voices. These spirits give different instructions according to the chamber whence they issue. The voice of the spinal brain to the body produces mimetic perceptions and motions; states that seem to be alive, for they act life perfectly. The spinal cord then endows the body up to the point of representing life by accordant motions. The voice of the sensual brain fills the body with sensual perceptions and sensual motions: it raises it up to the enjoyment of the cunning of the senses. Whatever is required for their gratification comes from this source in the shape of emotions and organic instincts. Under its influence the organs and viscera are like the animals hunting for their prey; and each takes what it wants from the common stock, as the living creatures take their pabulum from the earth. The organs would indeed do this without these sensual brains, but only as vegetables in a kind of wooden representation. Finally, the cortical surfaces, or the ganglia of thought and will, by their spirits diffuse their own light through the body, so that all and singular the organic acts and processes are done with the tincture of a higher than animal wisdom. All these would be done by the senses, but then their essence would be different; as the essence of bee architecture is different from that of human, though not different in its mathematics.

Thus the human body is all obedience to these three degrees of nerve-spirit, and if it stopped with the reflex actions, it would be but a dramatic mask, involving no wisdom beyond that of supreme mimicry: and in the same way, if it stopped with the animal brain, it would still involve no wisdom beyond the perfect adaptation of sensual means to sensual ends. It does not however stop here; but even its theatricism and animality become instinct with reason and will. And the like process which impregnates sense and motion, as we know them, with reasonable thought and voluntary action, strikes the same through the secretest parts of the organization, and makes the blood rational, and the bile rational, and in short makes the whole body human by the radiation of that which alone is human, from above its summit.

But now, in the human being, these upper states are not only fitful, but also intermitting or regularly periodical. Sleep comes to all, and takes away impression, sense and understanding, as well as motion, impulse and will. And in this respect waking too is full of somnolency, or abrogation of our superior powers. If then there were not some provision, sleepless and permanent, to keep us up to the human level, the answerableness of the body to the soul, and consequently the animation of the former, would perish many times every day, and certainly with the first slumber. For if all that is animal really died down to the surface of the earth in the seasons of sleep, the body, heavy mass as it is, and belonging of right to the ground, would be in the clutches of the grave, irrecallable from its congenial gravitation. To prevent this, there are two brains, a constant and an inconstant, but each corresponding to the other. The cerebellum does unconsciously and permanently whatever the cerebrum performs rationally and by fits. The cerebellum follows and adopts the states induced by the cerebrum on the organization, and holds the notes of the ruling mind. Thus immediately after sleep, the motions of thought may begin at once, for they have not been organically, but only consciously suspended. We see this in an image in the lungs. If the latter were voluntary organs, the man would cease breathing so soon as he fell asleep. But they are both voluntary and involuntary, the latter when not the former; and the movement is always proceeding, night and day, so that it has not to be created, but what is an easy matter, merely directed into the voluntary channels. Similarly so with the organic motions of thought and will: these are always going on, and merely require direction, not creation, by the cerebrum. Concordantly with this we can explain sleep, and much that occurs in sleep: *e.g.*, the fact that our thoughts and judgments are marvellously cleared and arranged during that state; as though a reason more perfect than reason, and uninfluenced by its partialities, had been at work when we were in our beds. This also—that our first waking thoughts are often our finest and truest; and that dreams are sometimes eminent and wise; which phenomena are incompatible with the idea that we die down like grass into our organic roots at night, and are resuscitated as from a winter in the morning. And it must again be adverted to, that this would not suit the Grand Economist; for after nature has ascended to one plateau of life, represented by a day, she will surely not tumble down into the valley because rest is needed, but will pitch her tent, and make her couch upon that elevation. We conclude then that the cerebrum is the brain of the mind, and the cerebellum the corresponding brain of the body; and as during sleep the cerebrum is a body, the cerebellum at such time is the brain of the cerebrum also. It may be added that the cerebellum, in adopting the mental states as her standards, is also a register of the mind, and converts actions into habits, or in other words, into structures. And thus the whole man, genius, intelligence, logic, sense, skill, action, may be called instinctive or given in nature, if we look at him from this providential side, whose organ neither slumbers nor sleeps.

Man then, as we see, is captured in sleep, not by death, but by his better nature: to-day runs in through a deeper day to become the parent of to-morrow; and the man issues every morning, bright as the morning and of life size, from the peaceful womb of the cerebellum.

If these views be carefully weighed, it will be found that they harmonize with the facts of the case, and not less so with experiments and vivisections. For if the harmonies of the frame, and the undercurrents which make its fitful states normal, depend upon the cerebellum, so also do the harmonies of motions, the production of which has been supposed to be the office of the lesser brain. If we are right, the removal of the cerebellum ought to leave for a time the conscious machinery and powers, but fitful as thought, sense or fancy: the body ought to change like a mind, now acting in part, now ceasing, and in short in everything capricious and partial, as might be expected when the man or animal was given up to his own wisdom, and had lost his organic providence. On the other hand the removal of the cerebrum ought to leave everything, and the power of everything, *minus* consciousness, and therefore *minus* the stimuli and beginnings of action which come through consciousness. This of course where no other parts were injured. We may also infer on these principles, that where the condition of animals approaches sleep or inaction, there is less need of a cerebellum, for these creatures are already on the ground, and comparatively secure. But where the least absence of mind might cause a fall, there a permanent organic provision is required, in the shape of a larger cerebellum. For we may lay it down as a formula resulting from all now said, that the perpetuation of high motion, composite motion, or harmony, is the function of this organ. On the same principles again it becomes probable, that the genital function assigned by the phrenologists to the cerebellum is founded upon truth. And now we leave it as a problem, whether there be, or not, an analogous provision for the sensual ganglia, to preserve the level of sense when our senses are annulled; and what the organ is? And the question may be repeated for the spinal cord also, with the insinuation, whether the sympathetic nerve be not its cerebellum? For we regard it as certain that the naturalness and economy of force, and the accumulation, are secured everywhere in the bodily system.

The above function of the cerebellum has its analogues in every sphere. We see it in thought, which has two elements, viz., that of consciousness and personal energy, and that of natural growth, the first corresponding to cerebrum, the latter to cerebellum. And these are often disparted in individuals. In some there is a preponderance of cerebral mind; their thoughts move quickly, but flightily; as we say, there is a want of balance; a want of body or nature in their minds; a defect of organic or cerebellar faculty. Their mental movements are random and inharmonious; they do not retain or accumulate wisdom; even repetition does them no good; but they strike out afresh in the vagueness of discourse, with no nature to back them. They have all the senses but common sense, which is the spring, incarnation and harmony of them all. In philosophy or collective thought the same division is visible. Philosophies are made, and also they grow; they are both cerebral and cerebellar. Universal tradition, the largest pressure of common sense, is the philosophical cerebellum. And here we see what complete experiments of vivisection have been performed; and what the result has been in philosophies that cut away the nature, accumulation, force and body of preceding thought; which extirpate the fixed organon of human growth, or the traditionary cerebellum. Dr. Carpenter, speaking of smaller things, describes to the letter the effects which follow: "It does not seem," says he, "that the animal has in any degree lost the *voluntary* power over its individual muscles: but it cannot *combine* their actions for any general movement. . . . . The reflex movements, such as those of respiration, remain unimpaired. When an animal thus mutilated is laid on its back, it cannot recover its former posture; but it moves its limbs, or flutters its wings, and evidently is not in a state of stupor. *When placed in the erect position, it staggers and falls like a drunken man; not, however,* *without making efforts to maintain its balance."** Such is the want of health or wholeness that comes from rescinding the natural brain that lies behind us, and beginning the intellect afresh with each passing day: for where there is no *vis a tergo,* there is no direction either in physics or metaphysics. And the sphere may be changed again with the same result. Law-making, which is the political cerebrum, stands in a similar ratio to the public morality, which is the political cerebellum; and where the latter is ignored, political vivisection is performed, and constitution-makers repeat Dr. Carpenter's phenomena on the scale of nations. And even in the highest sphere, where the cerebrum is termed prudence, and sometimes wisdom, and the cerebellum is providence, we see the same thing. Here the vivisection is frequent, and the results very confirmatory. We see the quirks of men whose actions, vigorous enough, are all tumbling to pieces; spiritual staggering and drunkenness; a positive sense that there is no God, but that man is the manager; and the like aberrations. All these instances are in the series of the cerebellum, and prove the universality of its functions. Man is in the leading-strings of God and nature, and what is greater than himself, to the end of his career; he is as a little child, whether he benefit by it or not; and the sovereignty of the things above him is represented by an organ or envoy from the everlasting, planted in his own head; and which, as we have now sufficiently said, is the cerebellum. But it lies unobtrusively underneath the cerebrum, because its guidance is nocturnal and unseen; and where it concurs with reason and will, it delights to seem to be their servant and not their master. When therefore the cerebellum or its similars are abstracted, the result is, motion and no harmony, progress without balance, thought without common sense, art without nature, philosophy without humanity, freedom without a career or destiny, and prudence alien to providence. In constituting this little series, we seem to hear the whisper of a reason why our own age has no revelation concerning the function of the cerebellum.

A word with regard to the *usage* of the brain. This grows with our years. At first all is impression and convulsion; very little

* *Manual of Physiology,* n. 912.

cerebral exercise concurs with this state. Then all is sensation, and action gratifying sense; and the cerebral hemispheres are called into play in a more important degree. Lastly comes the rational period mixed or marred with the others, and the brain movements are directed more finely to suit the new demands. The mind moreover handles the brain as the brain handles the body. At first the thoughts are little better that confused impressions, and the rational actions like sensual or spasmodic states. But by degrees the will gets into the cortex, and the understanding also, and mental continence is established. The brain sits and notices, and stands and runs, as its limbs are directed by the voluntary powers. It becomes to the full locomotive through a host of fears, trials and falls. Every nerve muscle now comes under control, and the brain enters on the command of its position. Learning to think appears to be analogous to learning to observe; and learning to will, to learning to act. Hence a good education would enable us to use all parts of the brain, as a fine gymnastic system calls into play all and unwonted parts of the body. But let the education be as good as it will, there is still a difference in brains, and one can do feats which another cannot attempt. The brain, however, is not the limit but the vessel of the mind, nor do the faculties of one sphere necessarily argue those of the other. This world's idiots may only be taking a long natural rest, to fit them for the activities of another station; where the brain cannot be a chariot to ride in, it is still a bed to sleep upon.

We have thus taken notice of some leading problems respecting the brain, dismissing many others; but still there are matters of importance on which a few words may be said. And first as to the use of the two symmetrical halves of the nervous system. This, we believe, is a deep problem, and capable of a large superficial reply. Have not all things two sides, and must not the brain and nerves image and use the doubleness of universal nature? Man has two halves, respectively called the male and female, and the body has two, in its right and left sides. Action is the function of the one, and passion, answering to action, is the function of the other. The left hand steadies the object upon which the right hand works. The woman constructs the material groundwork, the home, from which the man sallies, and to which he reverts, as his centre of operations. The one brain holds the object of thought, while the other brain works upon it with active power. This function of the two sides is a peculiar repetition of that of the cerebrum and cerebellum, without which consciousness would not hold, just as the male sex without the female would be a power without a fulcrum, a wanderer without the strength of home. Principles cannot be applied without a special envoy of their own, and the cerebellum must have the weaker sex or better half of the cerebrum in its interest in order to manage the cerebrum aright. If the reader consults his mind, he will find that in one and the same operation the process of steadying the grasp of thought concurs with that of exercising its active movements; that his soul has two hands; and these two processes or hands, widely different but united, are suited in the twofoldness of the cerebral lobes. But what is more, the two brains decussate or copulate, and the right head is married to the left body, and the left head to the right body; and this crossing of powers, whereby active and passive are not only collateral, but embrace, not only symmetrical, but one superposed upon the other, is a necessity for action and thought, considered as mental fruitfulness. The delights of harmony would not be felt if the two brains did not thus combine, nor would the brain have a circle in itself, unless each half had the support of contact with its partner before going forth into the body.

Furthermore, we have noticed in the nervous system a reference to something beyond and above, which soon lands us in the mind, as the first permanent station. Thus the feeling of the fingers conducts us anatomically to the spinal cord, in the centres of which, and not in the fingers, the sensation lodges. The cord at once refers us to the base of the brain, where sensation has its proper home. We know, however, that it is in the cortical centres that that attention lives, which is the inner sense, or the owner of sensation. These cortices, however, are dead and material *per se*, and thence the reference is straight to the mind, as the first organism that appropriates sensation, and calls it really its own. We stop here, not because the journey is done, but because the day of thought is spent, and our science wants a rest. Thus one lesson of this system is the unselfishness of our organization, and its recognition of a hierarchy which has no top save in that which is more than body, and of supernatural endowments. And in the downward series, running from wills to motions, the prime movers are as much above us as the original feelers in sensation: a superincumbent organism to be called spiritual, and which makes itself away to God with treble velocity of unselfishness, must be supplicated as at once the original arbitrator of will, and the last receiver of the thoughts which come from the world under the dresses of sensations.

This system, however, whilst we think of the body, is a land of mountains, from which the whole microcosm is visible. It rises white, with its pillars of alabaster, from the blood-lands of limb, heart and liver, and comprehends wider scenes as steep after steep is surmounted. Its elevation is measured by function, which gives the altitude over the level of the rolling juices and their shores. The representation of life by the spinal chain is already a summit towering above the flesh. The streams from this, falling upon the machinery of organization, give powers corresponding to the height and force of the descent. The rivers of sensation and impulse, again, descending from loftier tables, have the force of their greater fall, and press sense into parts and functions, into which no weaker torrents could gain way. And finally, the mental streams, each instinct with the view of the whole commanded from the top, have a power of descent and penetration which belongs to such intellectual rivers, whose sources are as the beads that the sun engenders on the needles of the supreme hills, heaven-kissing. We say nothing of the cloudy ranges beyond, or the spiritual temple, excepting that it too is of a new mountain order, though built out of the firmament, whose waters are above us as below us, and all the rest but currents in their eternal sea.

We may here assign two reasons why the brain is at the top; in the first place, because nature herself is the worship of rank and station, under their other names of excellence and power; and secondly, because the soul is the top of the top, and the brain by its place meets the soul. And as the arts and industries flow down from the brain, and the spiritual waters tend to find their level, so here a force is provided which carries them back, of their nature, when the channels are provided, to the altitude whence they came, or to the feet of the soul: that the life of man may not be wasted underground among his viscera, but may circulate from his head to his head, without a drop being spilt from the high nervousness of the body.

It is not surprising in this eminent region, that physiologists, loyal to their heads, have assigned to the brain the perpetual government of the body, or have regarded every function and every structure, as but a procedure or emanation from these commanding parts. Indeed, on the spiritual or nocturnal side, such a view becomes just, for the brain absorbs all the light and power of the system, and is the only noticeable part, or the body of the body. So when the sun ceases to shine, our domestic and motherly ground disappears, and contemplation, facing the vault, sees no longer the opaque earth, but only the ignited planets and the suns of systems, as it were the galaxies of the starry brain. And so with the body also: when its grosser senses cease to overwhelm us, and thought kindles in the teeming night, the body drops away, and organ after organ ceases to show, until we are left in the presence of a man, standing with luminous feet upon the darkness, and we see the ghostly human form, all nerve, feeling and volition; the brain as head and eye, body and limbs, founded not upon matter, but like the organic stars on its own sufficient form. We awaken, however, from these altitudes, and find that the planet too, and the material body, are things in themselves, nay, are our mothers, and deserve our best consideration in the homely way.

But the organic relation of the brain to the body generally, has not yet been well made out. It is clear, that the brain is the engine of the mind, and that the other viscera make up the body, which seems to be nourished on its own account. This, however, does not explain the immersion of the one within the other, or the subjection of each to the necessities of each. But surely, if the body gives the brain substance, the brain gives the body what it has to give, namely, brains. In this case, all the processes of nutrition, circulation, secretion, &c., must be controllable by the brain according to its perfection; and that this is the case, we know from the emotions which always and involuntarily transfer the state of the brain into the visceral lives. Convivial joy, the brain's joy, makes the stomach do double digestion with no harm to itself. Energy, brain tension, fires the muscles to feats that muscles never meant. Other passions scourge the liver or the kidneys into speed of manufacture, quite beyond their proper powers. The function of the brain then to the body is one of forcing or culture. The brain makes no clod of the body, no drop of the secretions: it makes no seed that the body grows. But it is the husbandman of the corporeal farm. The farm may go wild, and it is still something, though then called a desert. And organization may subsist without brains, but it becomes more and more tangled, lower and lower, until you cannot say that it is alive. The ratio of the brain to the body, is that of man to the planet. The planet is ready-made; every stone, plant and animal, night and day, the greater and the lesser light, are all there, and could not be created by us if they were not. But, man comes, the brain comes, as the cultivator. He is set there to have dominion over all; to be the image of the wisdom who made all; to spread himself as a head over all; and to modify all, as the last result, or the secondary soul of all. Therefore, until the brain has penetrated every viscus and function, it has not cultivated the body, as until man has grasped the climates, and forced them through their products and exotics, he has not cultivated the earth. The relation of the brain then to the body is, as the cultivator of the fields, originally wild, of nutrition, secretion, excretion, and the like. The cultivation begins as soon as the emotions of the brain begin, and every state of the brain plays in good or bad husbandry upon the brute or visceral powers.

But the brain could do nothing of this, if it were not itself among the natures that it commands. For what commander can speak to his soldiers, unless he be a common man like themselves, though an officer to boot? And, if man be not the supreme animal, the lion and the lamb *par excellence*, how can he wield the animal tribes? And again, if spirit have not all that matter has, how can the soul govern the body? Now, we have shown in detail, that the brain, as commanding organ, possesses the attributes of the lower organs in a superlative degree. We have shown that it is the heart of hearts, for it receives from the body and the universe a spiritual blood, which its cortices pulse forth in infinite streams throughout the frame. We have shown that it is the lung of lungs, for its animation is the breathing of the soul in the all-communicative ether. We have shown that it is the stomach of stomachs, because of its bold chymistry in the preparation of the food of food, which is the nerve-spirit. It is also the gland of glands, and the muscle of muscles, for it secretes the purest of juices, and obeys the beginnings of the motor force. Aye, and it is the primal womb of life and thought. In short, it is the body over again, piece by piece, with a truth befitting the brain. Hence, again, comes corroboration of our views, for we now perceive that we have assigned functions to the brain, of opening, breathing, moving, circulating, and the like, which are indispensable to its maintaining relations with the similar functions of the body. In short, we find that our deductions are but the claim of a common nature, as it were a common humanity, between the brain and the body. The brain, however, we must remember, is unmeasured by the body, and its attributes are peculiar, and not to be named by low names, excepting for the sake of illustration.

With all these advantages, however, of a community of nature and aims with the body, the brain could still do nothing, if there were not a physical motion in the body corresponding to the mental motion in the brain. If the body did not conspire or breathe with the brain, the metaphysical force which alone the brain *per se* possesses, could not be carried out. So, if the force of the seasons did not concur with the force of the cultivator, husbandry would be impossible. And so, if the moving disciplines of an army were not unanimous with the commander's voice, military operations could not exist. But we have said enough on this subject, which concerns the seconding of the fine brain thoughts by the powerful physical lungs.

The result of our observations hitherto is, that the brain opens the body to new influences, or gives it *animation*, and weighs upon it with the pressure of numerous changes or reforms; causing it to follow the mind, so far as the latter consists in the brain, through its vicissitudes. A similar animation, as we have seen, is introduced by man upon the earth, which he is born to subdue, and to reconstruct upon his own wants and ideals. By this means, for example, the inanimate ground is covered with waving vegetations; the vegetable kingdom is compressed by the animal, which browses down its increase, and serves as a partial end to arrest its exaggerations; and all together are braced round by man in girths and limbs of muscular arts, upon which sciences and volitions directly play for tightening the world to human aims, and carrying it through those revolutions of culture which are its aspects towards our wants. In this respect the trees are not inanimate, nor the beasts without progress; but they breathe and walk after man down the line of ages as after Orpheus in the days of old. Their proper brain, the *genus homo*, takes them along with him, and they become what he makes them, or are as he leaves them; as God has ordained.

The last part of our theme has yet to be written, or the comparatives and affinities of the brain. And here we may state, that we extend the province of comparative reasoning, and if the reader pleases, of comparative anatomy, above the human brain as well as beneath it. And we hold that the brains of the creatures larger than individual man are truly illustrative of his little brain, whereas animal cerebra are but falsely or negatively illustrative. By the creatures larger than man we designate societies, nations, races, or the individuals who cultivate the globe of history. The bodies of these are definite, fibrous, and individual, like animal bodies: they are mechanical also, though in a higher range of mechanics.

In the second consideration of the individual man, the brain is his genius—that which fills him with spirit, makes a truth or aim into his virtual intellect and will, and pours luminous rivers of these over his works. This *punctum vivens* of his mind animates the rest, and radiating its ideals far and near, irritates his apathies to death under hot arrows of zeal. This genius creates and then concurs with his wants; and the two together, or his life and his necessity, animate up to the shape and point at which determination can have actions done. These brain attributes, absent in none, are brilliant in some men, who take the name of geniuses on that account, and their deeds, by a fated fortuity, are treasured by their fellows as a common interest, though of no more than individual growth. These are the open men of their time, who hinder God the least: more rays shine through them than through the rest: you cannot say what their genius is, apart from what it shows and does, unless it be a natural road from heaven to earth: influx and the fluid kingdoms are their substances, and they know that the solid world is fuel laid up against the day of heat; also that truths and ideals are kings and priests, whose mortal namesakes, visceral and vegetating, are clay as in the potter's hands when that day comes. Their private thoughts seem the wants of the time and the schemes of societies; they are said to be sent and to have their mission; for the Maker has set them in the rhythm of his plan, and this world and that world heave to help them to shoot their lightnings to their destined ends. And still they are only the first brains that the epoch touches, and which, therefore, it publishes; and being the highest they are the longest visible as we pass away: but, as we said, every man is a genius or an end, a space crowded with ideals, and these ideals are the brain of the soul, or the personal life.

This word, genius, reminds us also of what we may call the Socratic brain, which attends upon the mortal organ. In this sense the brain-principle is an organization of guardian spirits, who live with our minds, fight our battles over our heads, whisper wisdom more than belongs to us, make our lights and resources exceed our days, and extend our debts into the unseen land to which we are adjourning. This vicarious function of souls is the result of their concatenation on the cortical plan. For here, where we are, our purer minds are infant, not yet detached from the matrix of the brain, and they sorely need guardians on the other side of time—as it were, parent hands and instructions to see them fairly through this big nursery, the world. We know not how little our lives would be, or how inanimate, if the gaps of power and the passingness of the day were not filled and compensated from another source where power is incessant and wisdom eternal. It would be as though each nervous fibril had but one cortical dot prefixed to it, and not the whole brain; or as though each mind stood alone, and were not environed and kept upright by an array of minds as long as the ages and as high as the heavens.

But epochs have brains as well as individuals, and these are the ruling spirits and ideas which are enshrined in their institutions. For epochs are the duration of the social life, and what is death and birth for the individual is but the exchange of old atoms for new in the marching epoch. These cerebral ideas are at first the private ends common to the race in a given period, which appear on a new morning in the field of individualities, and are the germ and birth-day of a social state. The principle assumed, the grouping of the cell, governs the composition from the least to the greatest, from the single family to the epoch of twenty centuries. As the grouping proceeds the action becomes grander, and the scale of operations is transferred from homesteads to continents, but the same cause is carried on in both extremes. Like fate it revolves through senates and priesthoods, whose maddest strifes it builds into its plan. The truths of the time enter the epoch through its own convenient sciences or eyes. The acts and charities of life are interpreted according to its familiar pattern, and committed to the spirit of its nerves. In short, the collective brain has an *animus* like the individual. And like the individual it has its decline; and also its better successors, which are the appointed angels of time.

These, however, are the animal brains of societies, receiving and transmitting the rush of destiny as it tramples through the chaos of the worlds. But another brain, with power over fate, is set above us as our social sun. For a firmamental organism of Prophecy and Revelation overarches the weltering centuries, and sends down spirit and divine light to the nations and races of the intellectual clime. Openness and circulation here are religions and adorations: the pressure of the life comes manifestly from above: as there is a God and Lord, it comes punctually to our wants, and the clank of our dire necessity is His mercy heard in terrestrial echoes; the sighs and heavings of mankind coincide with the birth of larger souls and societies, and with the advent of fresh dispensations.

CHAPTER II.

THE HUMAN LUNGS.

HUMAN LIFE is illustrated by every organ of the body. Each contributes a share to the general vitality. The brains are as the tranquil inward respiring of existence elevated into mind; a life which seems immaterial and motionless, until from the opened head the capacities of organization come to light, and the brain demonstrates that our noblest powers are incarnate, real and progressive. That which is the secret of the brains is the open lesson of the lungs. They live physically and largely the same life which the brains live metaphysically and most minutely. In the running wheel of life the imperceptible motion of the axle is thought; the sweep at the periphery is respiration. The brains give us the free principles of life, and the lungs, its free play in nature.

It is this idea of *the play of life* which is the principal point in our first knowledge of the lungs: it is in the completion of this idea that we must endeavor to bring out their functions. Of all the internal organs, not excepting the heart, the lungs move the most evidently. And as they are the plainest engines in our frames, we must, in that inevitable way from the known to the unknown, reason perforce from them to other parts, which also are engines, though more difficult to exhibit at work.

The nose and mouth are the two doors which open inwards towards the lungs; the nose being the special entrance to the chest, and the mouth, common to the chest and abdomen. The inner door leading to the lungs is the fissure of the glottis, which opens directly into the larynx, a cartilaginous box fitted up with muscles, membranes, and other appliances requisite for the articulation of sound. The larynx terminates in the windpipe or trachea, a pipe extending from below the middle of the neck to opposite the third vertebra of the back, where it divides into two tubes termed the bronchi. The trachea, the trunk of the windpipe, consists of from fifteen to twenty fibro-cartilaginous rings, which, however, do not form complete circles, being deficient at the back part, where the tube is completed by a strong membrane. These rings, like little ribs, are separated from, or connected with, each other by strong elastic membrane, so that first there is the membrane, then a ring; then again the membrane, then a ring; and so forth. The trachea is lined on the inside by a soft membrane continued from that of the mouth. It is the great stem which bears the ramifications of the lungs.

The two large branches of the trachea, the first bronchial tubes, run on each side to the lungs. On arriving thither, each divides into two smaller branches, and the subdivision continues, of each little branch into two twigs, and of each twig into lesser twigs, until at the last division the air cells terminate the tubes. These air cells are minute hollow chambers or vesicles, which hang like globules or grapes from the ends of the bronchia, and the air passes into them with every breath we draw, and is expelled from them more or less completely during each expiration. They form the characteristic element of the lungs, which are themselves nothing more than a vast, manifold and corroborated air cell. The amount of surface exposed by the cells is very great.

The whole of the constituents of the trachea exist virtually, in function and principle, in the smallest elements of the lungs, and the trachea with the lungs is a goodly diagram of the minutest bronchial twig with its delicate air cell. In the grand consistency of nature, the parts belong to the whole, and *vice versâ*, the mass being a spontaneous association of myriads of equally integral and so far independent structures.

We have now drawn an outline of the pulmonic tree. Its roots are the nose and mouth extending into the atmospheres; its boss is the larynx; its shaft the trachea; its first two branches are the bronchia; its other branches, twigs, and fruits are collectively the lungs; the fruits or air cells, however, are the lungs especially and essentially.

All the organs of the body are supplied by arteries carrying vivid blood, and the lungs are nourished by the bronchial arteries, which, running alongside the bronchi, form an arterial tree corresponding, in some measure, with their ramifications; the blood of the bronchial arteries being brought back out of the lungs into the circulation by the bronchial veins, which again form an inverse system of twigs, branches, and trunks, answering to that of the bronchial arteries. The bronchial vessels are of small calibre.

Besides these there are the pulmonary artery and veins, the former a very large vessel, which, coming direct by a single trunk from the venous side of the heart, accompanies the bronchia, and splitting into finer and finer ramifications, forms at last a "wonderful network" of blood-vessels around the air cells, the blood in which is separated from the air only by a membrane of extreme thinness. From the air cells this network reunites from twigs into branches, and from branches into the four trunks of the pulmonary veins, which pour the arterialized blood direct into the left side of the heart. Thus, the lungs are like forests of blood trees, the air cells being open spaces between, whereby the atmosphere is admitted to nourish and ventilate them; one set of trees, dull and venous, representing the blood before the ventilation, the other set blooming and arterial, representing the beauty and flower which succeeds where the vernal air has blown. This turn from autumn to organic spring is momentaneous in the lungs, which may not inaptly be compared to trees, inasmuch as leaves are the lungs of plants, and the vegetable kingdom, transmuting the earth and the atmosphere, belongs to the lung-department of material nature.

The lungs, like the other important organs, have a plentiful supply of nerves, which coming from both the cerebral and sympathetic systems, pursue the bronchia to the air cells.

The parts enumerated make up the active constituents of the lungs. In recapitulation, they are the tree of the air tubes, four other arterial and venous trees, and a nervous tree, terminating around and within the air tubes. All these are compacted by a system of membranes or skins, which make of the lungs not a five-fold or sixfold system of trunks, boughs, branches and twigs, but one solid though distinctly lobulated organ.

These membranes may be generalized under the name of the

pleura, including under that title all the cellular tissue which is directly continuous with the pleura. This pleura is a skin enveloping each lung; the cellular tissue is a web of skins that dips down into the substance of the lungs, and separates stem from stem, bough from bough, branch from branch, and twig from twig; at once dividing the parts from each other, and uniting them into a common body. The cellular tissue is therefore the bed in which the several parts of the lungs are planted. As it runs between the parts, and makes them into aggregate portions, or lobes and little lobes, it is sometimes called the interlobular tissue.

The lungs which we have thus endeavored to construct, are two conical organs, filling, with the heart, the cavity of the chest. They correspond in shape with the inside of the chest, and press below upon the diaphragm. The pleura which covers them, contracts and dilates with every respiration, and maintains its spring during life. The elasticity of this serous membrane is an indication that where serous membranes are present, as for instance over the brain, and over the abdominal viscera, a similar elasticity is intended, or an expansile and contractile motion like breathing is performed.

The chest, in which the lungs are placed, is a conical box, moveable in its parts, and capable not only of dilatation and contraction, but of infinite variations of shape. It harmonizes with the lungs in their movements, forming with them but one machine, so that it is indifferent whether we say that we breathe with the lungs or with the chest. In good health, when consent between the two is perfect, the bones and muscles of the ribs are of no heaviness in function, but rock and swim upon the lungs. Thus when we speak of the lungs in the sequel, we imply the whole engine of breathing even to the skin, and regard the chest itself as a dress or membrane inseparable from the lung-principle.

Inspiration, or the drawing in of the breath, is caused by certain muscles drawing out the walls of the chest, and enlarging its inward cavity, in which case the pressure of the external column of atmosphere causes the air to rush down into the windpipe and fill the lungs, which then enlarge to fill the cavity of the chest. Expiration or breathing out depends upon the relaxation of the muscles and the resiliency of the parts of the chest, as well as upon the elasticity and contractility of the lungs themselves. Thus the force of the diaphragm and of the muscles between the ribs engenders inspiration, and overcomes the elasticity of the lungs; the elastic power of the lungs produces expiration. The prolonged alternation of these two forces is "a contest in which victory declaring on one side, or the other, is [under ordinary circumstances] the instant death of the fabric."

In the act of breathing we notice four divisions, each of importance to our sequel. First, inspiration; secondly, a pause which ensues when the inspiration is completed; thirdly, expiration; and fourthly, a pause when the expiration ends; after which inspiration again occurs, and the same course is measured. Inspiration rises to a certain level, and there rests for a time; expiration descends also to its level, and registers it by a pause. Further, the inspiration may either take place continuously in one long breath, or by several smaller inhalations and pauses; the expiration likewise may either proceed without a stop, or it may be divided into several levels of exhalation, each with its own proper pause.

But let us come to the use of respiration, or the benefits of breathing. These are twofold: 1. The use of the air drawn in, towards the renovation of the blood, and of the air emitted, towards its purification. 2. The mechanical effect of the breathing upon the circulation and the body generally. We speak first of the first of these uses, because of the exclusive importance usually attached to it.

The result of investigations on this subject need detain us but a short time. It is in substance this—that when the air is breathed in, the expanded network of capillaries besetting the air cells absorb from it oxygen into the blood, and at the same time pour forth carbonic acid gas from the blood, which is carried out of the system by the air leaving the lungs. In consequence of these two changes, the reception of oxygen and the expulsion of carbonic acid, the alteration of the blood from dark venous to florid arterial in the air cells, is accounted for.

These changes, authors tell us, are purely chemical, and the same as happen to venous blood out of the body. We rejoin, however, that this theory is out of place, being chemical and not organic. It deals only with what is external to the body. The air of inspiration is on the way in, and the air of expiration is on the way out, but neither the one nor the other is a part of the living frame; however deep either may be in the passages of the body, still it is not *indoors*. The lining membrane or wall of the cell is the partition between life and death; inside the cell is vitality, outside of it, dead nature: within it, the man lives; without it, the universe environs him: the oxygen which is lost is missed from the outside, and the carbonic acid which is found is on the outside also; but on the inside no corresponding observation has been made, or can be made; and it is a questionable inference whether carbonic acid, as such, exists in the blood, or whether oxygen, as such, is there either. As an account of the food of the blood, and of the excrements of the blood, both external to the man, the chemical report is valid, but is it a satisfactory statement of any changes in the blood itself circulating in the system? Assuredly not; and let us here remark that the terms of every subject should be in keeping with the subject; the things which are life's should be rendered to life, and those of chemistry to chemistry. We cannot judge of the living, either by the raw material which they pasture from the world, or by the refuse which they leave behind them; or even by both together. Organization is the one fact in organization; chemistry disappears into it, and is seen no more as chemistry. The blood, as an organic creature, into which all things stream, and from which the body issues as the work of works, is the sole reality in our veins; it is as blood alone that its elements come before us. As well regard all heroic actions as instances of muscular exertion, and these as powers of lever and fulcrum, and other produce of mechanics, and not connect them with the fountains of human nature, as make the living union of all things in the blood into a congeries of chemical substances. Chemical indeed they are when they die and are experimented on, but chemical is not their name while they are part and parcel of our human blood. They must be addressed in the language proper to organic life, or the keeping of their science will be violated irretrievably.

We the more insist upon this, because by a very natural insurgency, chemistry has of late years been pushing what are called its conquests into the domain of physiology. But physiology has to consider the organic characters of things, whether animal or vegetable; chemistry, their inorganic or mineral elements. This is a broad distinction, and easy to apply. We would not rescind one chemical experiment, or deny the value of one fact therefrom resulting; but we protest against the logic of reasoning from chemistry to physiology; from bricks to architecture; from neutral matter to forms shapen for particular purposes, and with qualities that constitute their point, and their very existence.

At the same time there is no objection to regard life as a fire, and the chest as one of its principal grates; the fuel being the blood, and the draught, the fresh air of the lungs. In proportion as the outward world is cold, this lung-fire must be kept up by more inflammable materials, just as larger coals and logs are in use in December than in May. And moreover in proportion to the size of the fire is the quantity of smoke or carbonic acid disengaged from the system. All this is safe analogy and good experiment. But let us not be deceived into regarding this as *animal heat:* it is as purely mineral heat—this presumed "combustion" of carbon and fat—as the heat in an ordinary stove. Animal heat is that which warms life, or inflames the animal, as such. Its burnings are desires, the flames of animal existence. These are kindled by their appropriate objects. The universal animal heat of the body is the organic zeal or love of self-preservation hotly present in every part. This is the origin of hunger and thirst, which tend to continue bodily life, and lay the world under contribution, not despising even its mineral fire. There are different orders of fire; even in nature there is a substratum of heat of which we make all our fires. And so in the body there is an animal heat which lies at the basis of human warmth, even when the temperature can be fully accounted for by the "combustion" of the food. Take the sun out of nature, and the numbed flints will have lost their sparks; and take the soul out of the body, and you may indeed roast it or boil it, but cannot warm it with one ray of "animal heat."

Moreover there are several kinds of chemistry. The present chemical sciences are of the mineral degree, although their higher branches are indeed mineral-vegetable and mineral-animal. But there is no such science yet as either vegetable or animal chemistry.

Mineral chemistry teaches the composition of mineral substances by analysis and synthesis. It makes minerals by mixing together their components under favorable conditions. In like manner vegetable chemistry makes plants, and animal chemistry makes animals. Nature does this, and so far nature is the only vegetable and animal chemist. But though we cannot produce in her laboratory scientifically, but only blindly, we can observe her processes and learn the results. For example, the mixture of sexes produces animals, and in a certain sense plants also. The mixture of breeds varies these substances, namely, animals, and creates new compounds or animal varieties. The mixture of opinions produces ideas, and then we have intellectual chemistry. The present is distilled out of the past by the same law. Chemistry then is the mineral term. Raise it a step into the vegetable, and in plants it becomes propagation; into the animal, and it becomes generation: and so forth. This of synthetical or creative chemistry. The terms alter as the theatre changes. But to run one term through all the stages, is to miss the essence of all but the lowest. When science does this, it not only finds "sermons in stones," but is petrified by their discourses.

Dwelling therefore briefly, and under physiological protest, upon the oxygen and carbonic acid disengaged, or absorbed, during respiration, we proceed to remark, that the whole of the venous blood of the body, which is comparatively exhausted by its circulation, and also the whole of the new chyle or realized essence of the food, passes by the pulmonary artery to the networks of minute blood-vessels in their air cells, and so through the lungs, and in those fine vessels is counted out and thoroughly sifted, and its purification takes place. Whatever disabled portions it contains, are there taken to pieces, their broken elements thrown away, and the sound reconstituted. Whatever injuries the blood may have received from the passions of the mind, which as we know have all power to bless or to hurt it, are palliated by the removal of clouds of exhalations, as witness the odor of the breath. When the fire of life burns dark and fuliginous, the windpipe is as the chimney that relieves the body of its noxious smoke. Moreover, whatever crudities or superfluities the new chyle, or the milky produce of our food, may contain, are expelled by the lungs through the same channel. In short, the lungs are the general strainers and cleansers of the blood. Globule by globule they discuss its problems, separate its truths from its errors and its dead from its living, and hold it to its brief but energetic trials for purification and the consequences which follow.

Let us now turn from the act of expiration, from the air "laden like a mule," as has been aptly said, "with a burden and baggage of adulterations, and forced to carry them out," to inspiration—to the newly-arriving air, the provision for supporting, and the blast for rekindling, the blood. And let us spend a moment upon the admirable means of nature for managing this very air, and presenting it to the blood in the last place, clear, genial, and warm. First the nose, as the administrator of the sense of smell, takes cognizance of any odorous or stimulating properties in the atmosphere, and acts accordingly; if pure, sweet, and fragrant, it draws in the air by volumes, inspiring confidence and openness down to the very air cells: if manifestly noxious or impure, the nose closes in proportion, extemporizes a thousand valves that keep out the baser particles, and the air is driven against the sides of the passage, all the way to the same cells, its uncleanly accompaniments being caught in a viscid snarework all down the tube: it also gives notice to the mouth, whose mucus catches its share of effluvia, which are rejected by the shortest way. The mucus of these passages performs an important use, detaining clouds of particles which are unworthy to pass inwards, and this operation increases in strictness the further the air advances in the narrowing tubes. Consequently when it arrives in the cells, it is clothed with kindly vapors issuing from the body, has caught the tincture of the living heat, and is in fine unison with the blood. And the blood has no sooner sniffed it well, than it again becomes auroral and arterial. Immediately that this is accomplished, the air, exhausted for this primary use, is spent, as we saw before, upon the secondary and servile use, of undertaking and carrying forth the dead exhalations.

The blood that comes up to meet the air is all the blood in the body, for after circulating, it all requires refreshment or purification. It is carried into the lungs through the trunk of the pulmonary artery. But we mentioned another artery, the bronchial, which performs an office in the lungs. The old, exhausted and impure blood is conveyed in the pulmonary vessel; the blood which is pure and young as the heart can make it, runs parallel with the former in the bronchial. The pulmonary blood is to become regenerate in the lungs; the bronchial is the running model of its future state, and exercises a contagion of youth upon its pulmonary associate. For nature never prescribes an end, without showing a present example of it.

The air ministers to the blood an infinity of fine endowments which chemistry does not appreciate. How full it is of odors and influences that other animals, if not man, discern, and which in certain states of disease and over-susceptibility, become sensible to all: moreover at particular seasons all fertile countries are bathed in the fragrance shaken from their vegetable robes. Is it conceivable that this aroma of four continents emanating from the life of plants has no communication with our impressible blood? Is it reasonable to regard it as an accidental portion of the atmosphere? Is it not certain that each spring and season is a force which is propagated onwards; that the orderly supply, according to the months, of these subtlest dainties of the sense, corresponds to fixed conditions of the atmospheric and imponderable world adequate to receive and contain them; that the skies are the medium and market of the kingdoms, whither life resorts with its lungs, to buy; that therefore the winds are cases of odors; and that distinct aromas, obeying the laws of time and place, conform also to other laws, and are not lost, but are drawn and appreciated by our blood. Nay more, that there is an incessant economy of the breath and emanations of men and animals, and that these are a permanent company and animal kingdom in the air. It is indeed no matter of doubt, that the air is a product elaborated from all the kingdoms; that the seasons are its education; that spring begins and sows it; that summer puts in the airy flowers and autumn the airy fruits, which close-fisted winter shuts up ripe in wind granaries for the use of lungs and their dependent forms. Thus it is passed through the fingers of every herb and growing thing, and each enriches its clear shining tissue with a division of labor, and a succession of touches, at least as great as goes to the manufacture of a pin. Whosoever then looks upon air as one unvaried thing, is like the infant to whom all animals are a repetition of the fireside cat; or like a dreamer playing with the words animal kingdom, vegetable kingdom, atmosphere, and so forth; and forgetting that each comprises many genera, innumerable species, and individuals many times innumerable. From such a vague idea, we form no estimate of the harmony of the air with the blood in its myriad-fold constitution. The earth might as well be bare granite, and the atmosphere, untinctured gas, if the vegetable kingdom has no organic products to bestow through the medium of the air, upon the lungs of animal tribes. Failing all analysis, we are bound to believe, that the atmosphere varies by a fixed order parallel with that of the seasons and climates; that aromas themselves are abiding continents and kingdoms; and that the air is a cellarage of aërial wines, the heaven of the spirits of the plants and flowers, which are safely kept in it, without destruction or random mixture, until they are called for by the lungs and skin of the animate tribes. Fact shows this past all destructive analysis. It is also evident that accumulation goes on in this kind, and that the atmosphere like the soil alters in its vegetable depth, and grows richer or poorer from age to age in proportion to cultivation. The progress of mankind would be impossible, if the winds did not go with them. Therefore not rejecting the oxygen formula, we subordinate it to the broad fact of the reception by the atmosphere of the choicest produce of the year, and we regard the oxygen more as the *minimum* which is provided even in the sandy wilderness, or rather as the crockery upon which the dinner is eaten, than as the repast that hospitable nature intends for the living blood in the lungs. The assumption that the oxygen is the all, would be tolerable only in some Esquimaux philosopher, in the place and time of thick-ribbed ice; there is something too ungrateful in it for the inhabitant of any land whose fields are fresh services of fragrance from county to county, and from year to year. Chemistry itself wants a change of air, a breath of the liberal landscape, when it would limit us to such prison diet.

Here, however, is a science to be undertaken; the study of the atmosphere by the earth which it repeats; of the mosaic pillars of the landscape and climate in the crystal sky: of the map of the scented and tinted winds; and the tracing of the virtues of the ground, through exhalation and aroma, property by property, into the lungs and the circulating blood. For the physical man himself is the builded aroma of the world. This, then, at least, is the office of the lungs—to drink the atmosphere with the planet dissolved in it. And a physiological chemistry with no crucible but brains must arise, and be pushed to the ends of the air, before we can know what we take when we breathe, or what is the import of *change of air*, and how each pair of lungs has a *native air* under some one dome of the sky; for these phrases are old and consequently new truths.

We notice, indeed, a great difference in the manner of the lungs to the different seasons, for the genial times of the year cause the lungs to open to an unwonted depth. The breaths that we draw in the summer fields, rich with the sweets of verdure and bloom, are deeper than those that we take perforce on our hard wintry walks. Far more emotion animates the lungs at these pleasant tides. Nor is this to be wondered at, any more than that we open more freely at a table loaded with delicacies, than at a poorly furnished board. The endowments of the vegetable kingdom in the atmosphere not only feed us better with aërial food, but also keep us more open and more *deeply moved;* and we shall see presently that the movement of the lungs is the wheel on which the chariot of life runs, with more or less intensity according as the revolution is great or small. Now in summer it is great, and in winter it is small, for manifest motives.* Furthermore, our noses themselves, the features of the lungs, are in evidence that there is more to be met with permanently in the air than inodorous gases. For we cannot suppose that scent ends organically where we fail to perceive it with the sense. But enough has been said already on the flavorless world and noseless doctrine of the chemists.

* In a regular treatise on the chemistry of the lungs, the atmosphere would be separately considered in its mineral, vegetable, animal and human constituents, and in the effects of these, as introduced through the lungs, upon the body and the mind. In this work, however, we make no pretensions to treat the subject according to this larger order, though other considerations following out the above series will be presented in the sequel.

This extension of the subject has a practical bearing. The chemical view blinds us to the seeds of health and disease contained in the atmosphere. We pound it into oxygen, hydrogen and carbon, and find its ruins pretty invariable in all places under all circumstances. Plagues and fevers give a different analysis, and tell another tale. They prove that the air is haunted by forcible elements that resist segregation and distillation. The strokes of these airy legions are seen, though the destroyers themselves are invisible. In the atmosphere as a place of retribution, the cleanness or uncleanness of the ground and the people is animated by ever wandering powers, which raise cleanliness into health, and filth into pestilence, and dispense them downwards according to desert with an unerring award. But who could guess this from the destructive analysis into oxygen, hydrogen and carbon; which misses out the great shapes that stalk through the air, and laugh at our bottles and retorts often with a diabolic laugh? But we shall recur to this subject when we treat of Public Health.

To conclude this part of our subject, we have seen that the lungs raise the blood into its principles, and discuss them on a higher arena; that they continually refresh and enlarge it by bringing it into contact with the outward world in the shape of the atmosphere, where at once it gives up its antiquities as the free breath unlocks it; that the lungs also humanize the air as it enters, and fill it with the organic warmth and movements of the nose and the head. But further, the blood, in passing into the lungs, is held as it were above the body by virtue of the lightness of the sphere. And not only in the lungs but everywhere in the system, the pulmonic levity, or the rise of the surface, operates statically upon the fluids; so that each breath amounts to a posture or rather a hover of the entire capillary blood and nervous spirit of the body. This levity-giving is an intermediate function between the aërial and motor offices of the lungs. We shall speak of it again at the end of this Chapter.

We have now considered the chemical and physical functions of the lungs, and glanced at their statical office; it remains to treat the second part of the subject, namely, the respiratory movements, or the mechanics and dynamics of the lungs.

That the effects of the lung movements are not small, a short description will convince us. Every time that we draw in the air, our brains fall, from the venous blood being sucked out of the head to fill the threatened vacuum in the chest; and when we breathe out the air, the brains rise, from the return of the blood from the head to the chest being impeded. The same also takes place with the heart. During inspiration, the lungs breathe up the venous blood into that organ, and retard the passage of the arterial blood from it; during expiration they keep the venous blood away, and increase the onward impulse given to the arterial blood. So likewise in the belly. Largely emptied, as it is of venous blood in inspiration, and subject to the movements of the superincumbent lungs, it necessarily undergoes great motion during breathing.

Now, that movements like these have a universal part to play, it would be idle to deny. They actuate the body some fourteen times, or more, every minute of our lives. If we watch our neighbor as he sits upon his chair, we see, not his wind chest only, but the man himself, expand and contract each time he breathes. If we watch the face, we see a corresponding change: if we lay the hand upon the stomach, we feel it rise and fall as plainly as the chest. And common sense ordains, that in a machine divinely economic, the use of any motion is co-extensive with the motion itself, and if the motion be universal, the end it serves is likewise universal.

As we have seen, the pressure of the atmosphere distends the lungs and produces inspiration: the living contraction of the lungs causes expiration. The result is, to engender a power which alternately stirs the frame. As a familiar instance of what this power is, I appeal to those who have sat to the sun for a daguerreotype portrait. You know that the greatest stillness is needed to bind down that quick artist to the execution of single portraits; to make his successive ideas or touches fall in identical lines; otherwise, he will paint you not one, but a chaos of likenesses, equal in number to your variations of position. In the painfulness of your anxiety to sit still, to suppress the breath, you find that you are a frame which verily exists in motion; you ascertain what a struggle it is to combat your life's progress, to wrestle with the moving lungs. To the fingers' ends, to the toes' ends you pant and swell, and sink again, with irrepressible heavings; and the voice which emancipates you from the effort, and bids you breathe as you please, unobservant of the fact, is release from a straitness which could not be long endured. The same thing is experienced more or less, whenever it is necessary to use great stillness, to control the breath. The consciousness which is then awakened comes into collision with a power whose resources we never estimate practically but at such times of struggle. Ask any of those who have been engaged in *poses plastiques*, and they will tell you that of all hard work, standing still is among the hardest.

What becomes of the power created by the air falling with the whole weight of its column upon the movable lungs, and displacing or expanding them, and by the subsequent living contraction of the lungs? Can we imagine that its use is confined to the outside of the system? to the admission of fresh and the expulsion of contaminated air? This would be as reasonable as to suppose that the main office of a water wheel, connected with an extensive and complicated machinery, was confined to the water which falls upon it, and that the mechanical power engendered was not communicated inwards to the plant. At this rate nature would be less thrifty than our engineers, who know that power is precious, to be husbanded to the last degree, conducted where it is required, and never expended without a result. Suppose that a portion of the water be needed inside the mill, as is generally the case, this is easily supplied by some sideward allowance of the machinery, or by the pressure of the water itself, which contains the power in an unmechanized state; but the main action is never spent upon that which comes of its own accord. The blood is aërated in some animals, and the juice in plants, without any motion of the lungs, which may suggest that the aëration of the blood is not the grand office of these movements. But in machineries, any motion which is superabundant, or not turned to use, is hurtful to the object sought, precisely because motion always has effects, which in the latter case mix with the intended result, and confuse or disarrange it. This applies more strongly to the human frame than to anything of man's making.

Thus, we observe that there are really two questions which have been confounded with each other: Firstly, What is *the use of air* to us? and secondly, What is *the use of breathing?* And with respect to this second inquiry, we now see that it will be puerile to say that we breathe in order to breathe. Let us grapple with the problem, and solve it otherwise than by a verbal retort. We answer, then, that the use of breathing is to communicate motion to the body, to distribute it to the different machineries or viscera, to enable them each to go to work according to their powers.

Our position is, that the blood and blood-vessels make and repair the organization, and keep it in working trim; while on the other hand the lungs and the brains use and work it: like as an engine is made in the factory by one set of artisans, but is taken elsewhere to enlarged conditions of liberty or motion, to be worked by another class of persons. Thus, the heart's fabricative strokes are the lesser motion: the experimental play and employment of the lungs is the greater motion. By the one the hammer is plied upon the engine; by the other, the completed engine is made to use its qualities, and to work according to its construction. Every body contains two bodies—the one which is forming, and the other which is finished and working: the heart is the spring and centre of the first, and the lungs of the second; the one represents matter, and the other spirit; and the ratio between the pulses and the breaths gives the constant equation which subsists between these inseparable two.

It needs but little consideration to show that the organs and viscera of the body require a supply of motion to enable them to perform their functions. These functions consist, *firstly*, in the reception of a peculiar quantity as well as quality of blood, from which they, *secondly*, are to separate certain materials, or upon which they are to produce some change: the quantity and quality required varies also at different times. What is it that supplies them with this peculiar blood? Not the impulse of the heart and arteries, for this could cause no discrimination in the supply to different parts; on the contrary, its action is uniform all over the body. Each organ then requires an individuality to enable it to choose and take what it wants from the common system. How can this individual power be given to the organs, except by their exercising a motion of expansion and contraction, whereby they draw in or shut away the blood, as they find it necessary?

But again, if the organs are alive, the operations that they perform upon the blood demand a general motion on their part. If we are to regard them as dead sieves or filtering stones, then it may be sufficient for fluid to run into them, and the various secretions, the bile, the saliva, &c., may drip through them without any action on their part. The humanities and industries of the inner man may sit down deadly still, like mesmerized Turks. But is such a conception proper in a body overrun by spirited nerves, which in proportion as they are impressed or passive on the one hand, rise up in activity on the other? And if the parts of the organs are not only passively but actively engaged in elaborating the various juices, must not their activities combine by a law into one general action or motion common to the whole organ? Does not all function in a living body imply motion, and is not the sum of particular motions necessarily represented by an aggregate motion equal to all its parts? Though his blood may be circulating, yet a motionless man is a man doing nothing; and a motionless organ is just as ineffectual. To exist is one thing; to do is another and a further. In the whole man, the management of his motions constitutes his skill; in the partial man or the organ a corresponding management performs its functions. Without precise means set in real motion, you have no art and manufacture, no saliva and no bile. These latter are the most marvellous of fabrics; the body is the most stupendous of factories. Our commonest thoughts upon such subjects are the way to the best. Destruction, then, is in a manner compatible with rest, but construction never.

The sap is indeed distributed in plants without any apparent expansion and contraction of their organs, as it were by a magnetic or elective affinity between the parts of the plant, and the fluids they require. And this election is, doubtless, involved in the animal also. But then the meaning of animal as contradistinguished from vegetable, is motion as distinct from growth, or local as different from and superior to molecular movement. And the several organs of animals are animal like the whole. No vegetable tissue could associate in the body of life, but it would be the sport of activities which it could not share or reciprocate. A liver that was merely vegetating would be pressed to death in a body that is ceaselessly animating.

Therefore the motion of the organs is indispensable to make them parts of the whole, or to raise them into the animal sphere.

To return to facts, we find in the motion of the lungs communicated to the system, the very power which the organs require. For the body is a chain of substances and organs, whose connections are so disposed, that motions communicated from within, vibrate from end to end, and from side to side, and extend to the extremities of the limbs before they are absorbed. And in the intimate fellowship pervading it, and which is brought about by the skin and the membranes, we see the condition whereby a general motion, like that of the lungs, amounts to an attraction exerted by the frame and its parts upon the world without and the world within; by which, in each different voluntary expansion, it draws in as it pleases the fluid contained in its own cavities, as well as whatever it requires from the great ocean of the atmosphere. Such is the value of the movements of the lungs. They not only breathe themselves, but make the body breathe similarly with them; and in this consists its life, whereby it becomes an individual, and takes what it wants for itself, suffering no intrusion from within or without, whether from the blood of the heart, or from the pressure of the universe.

To follow the *gear* by which the motion of the lungs is communicated to other organs, belongs to anatomy and experiment, but the general fact belongs to common sense, and science has only to confirm it. We do not now enter upon the anatomy, but will content ourselves with observing the effect of the pulmonary engine upon the great departments of the system. And first for the effect upon the nerves and the brain.

First, with respect to the nerves, the motions of the lungs, occurring fourteen times *per* minute, act upon them more than upon any other part, because they are the most impressible of the organs. Now a large portion of the nerves runs through the chest, a space subject to threatened vacuum during every breath; and more than a third part of the spinal marrow virtually lies open into the same exhausting receiver. The plain consequence is, that the nerves and the spinal marrow are expanded with each inspiration. Either *that* —or they resist the inspiration, and in this case the unity of the body is at an end. But we cannot make the latter supposition. If they are expanded or enlarged when the lungs draw them out, of course a physical fluid enters them to fill the space created, and tends to fill the organs to which they are distributed. In this way the nervous system, the focus of life, opens the frame at the same intervals as the lungs, the circumference of life; the lungs being simply the want of living fluid, and the nerves the corresponding supply. This is an organic coöperation between effect and cause, whereby the highest purposes of the organization are seconded most absolutely, and yet most freely, by the lowest.

The nerves then breathe their atmosphere, the nervous fluid, at the same intervals as the lungs breathe theirs, which is the proper atmospheric fluid, and the breath of the nerves is the life of the lungs, as the breath of the lungs is the bodily action of the nerves. The nerves, however, are continuous with the *brain*, and *secondly*, we observe that *their* expansion is *its* expansion. It opens, for motion's purposes, into the chest, by the nerves, and by the spinal marrow; the lungs have their suckers upon it everywhere, through the membranes and the blood-vessels. It, therefore, breathes under the attractions of the pulmonic air-pump. Like every other part it respires its own thoughts or objects. What these are, it does not behove us to inquire, but we may affirm generally, that they are those fluids which are the brains of the body and outward universe. The lungs breathe that which answers to lungs in nature, namely, the air. The heart breathes that which is the heart's in the system, namely, the blood; and each organ, as a rule, breathes its own corresponding fluids.

The *heart*, as we have just anticipated, breathes also with the lungs, and so manifestly, that physiology already contains many chapters upon the influence of the respiration upon the circulation. The pulmonary motions acting upon the heart and great vessels, cause the venous blood to return to the heart, and somewhat retard the outgoing arterial blood, during the inspirations; and *vice versâ;* and imprint upon the pulse at the fountain-head the force which is destined to supplant the pulse where the vessels enter the organs. By this means the ultimate intention is intimated from the beginning; the blood in its childhood is let into the secret of its destiny; and the sanguineous system is prepared at once for submission to the brains and lungs. The lungs then inaugurate the grand circulation into the life and habitudes of the rational body, animating the blood itself with the moving spirit of the atmospheres.

The *belly*, too, is in the human conspiracy; it would be dead to the rest if it did not breathe. The abdominal breath is the most physical of all, commanded by powerful muscles, and destined to suck in that large food upon which the belly lives, and whose pleasures it respires. As we have observed already, we need only lay the hand low down, and we shall feel our hunger moving and busy in the workings of its native cave.

The belly, however, not merely breathes its general atmosphere, the food, from the world of food lying in the stomach and intestines, but its organs and viscera breathe in each their peculiar blood, and breathe out their excretions. For each organ has a precise form and constitution, and like every other machine acts according to its construction. The power of the great steam engine, the lungs, is communicated to all, but each takes it in its own way. For example, when the liver is drawn out or breathes, and is filled with liver-thoughts and energies by its roused nerves, the expansion follows its make and texture; it is a motion of the machinery of the liver; and the purified blood on the one hand, and the bile on the other, are woven accordingly. So when the kidney is drawn out to act, it is a motion of the machinery of the kidney. The different machines moving in different ways, perform their functions, draw in their blood and manufacture it, exactly according to their build, each with a difference from the rest. There is no tyrannous influence of the lungs; their traction upon the gear of the organs is only the power necessary to set them to work, to enable them to revolve in their places, and to put forth their given genius for the commonwealth of which they are independent members.

Each organ of the body has, therefore, its own sphere, within which it is individual. It is true that its force comes from without, but then it is a force answering to that which it desires from within by the very nature of its nerves. It is, therefore, a rule, that the blood is merely carried by the heart and vessels to the doors of the organs, but is not intruded; for on the threshold of the organs it encounters another force, and is drawn inwards or sent outwards only at the times when the organ draws it or expels it. In a word, at the organs the jurisdiction of the heart and arteries ceases, and that of the organ itself begins.

To complete the empire of the respiration, we notice that the muscles and limbs breathe like the rest. During repose this is more difficult to show, but even then, if we attend carefully to the draw of the expansion passing from the belly down the legs, we shall find that the skin tends out in an inverse pyramid from the loins to the toes and heels; like trowsers tight at the bottom, but expanding and contracting above, and chiefly at the top. While, however, we are at rest, the respiration of the limbs is scarcely noticeable, beyond the parts of the arms and thighs immediately contiguous to the body. But when we rise into motion, and the will comes forth, the effect is different; and in powerful volitions and actions a limb of air becomes steel, runs rigorously down to our toes and fingers. The skin is braced so tight, that the muscles threaten to start through it, and the will in the same manner menaces to bare itself by throwing off the muscles. The clothes and the body fly out like concentric planetary rings in a rapid vortex. The man becomes more and more of air; he ceases to lie, he ceases to sit, he ceases to stand, and, like an elastic sphere bounding upon a point, the ball of the foot is his only contact with the ground. This is the extreme effect of the aëration of his limbs. He has become a bird for that moment, and can then fly through difficulties, which are the atmosphere of these great actions of the lungs.

The lungs, then, are consenting organs in muscular and gymnastic efforts, and precise muscles of breath or spirit lie under the muscles of flesh, and lend them force, hardness, and sphere in their operations.

It is also to be remarked, that as inspiration commences *à posteriori*, or from the muscular system, and as all the muscles concur to it more or less, so the inspiratory effort may commence from any part of the frame, and the breathing will be differenced according to the part. In ordinary normal breathing, the thoracic and intercostal muscles appear to begin the act; but in pleurisy the centre of operations is changed, the breathing becomes "abdominal," and the action upon the chest is secondary. In like manner, any muscle of the frame may take the lead in initiating the action, for all the muscles are connected together, and tend instinctively to influence each other. Thus we may have splenic breathing, or umbilical breathing, or hepatic breathing, according to the part of the surface which begins the inspiratory traction. Now, the spirit of any action is according to its beginning in the body. But this is too important a subject to be discussed within our present limits.

It is now, therefore, evident that the movements of the respiration are not confined to the chest, but are systemic motions pervading the head, body and limbs, and lying at the basis of the functions of the parts; and thus that bodily actions or functions are never created, but only shaped or formed out of a stock of motion given in the nature of things. Furthermore, as habits are no sooner engendered than they are written upon the body, and especially upon the nervous system, it is plain that this habit of reciprocal breathing is deeply inscribed as a second nature upon the animal textures, and that they tend to fall into it upon the least impulse given; according to the well-known laws of recurrence in the bodily frame. Thus, on the showing of facts, life may be defined as the progressive education of the organs and viscera into habits of breathing which contradistinguish them from dead organs.

What we have said might have been taken by analogy from the air as well as from the lungs. For the air also has the three functions; a chemical, by which it combines with other substances; a statical, by which it presses with so many pounds to the square inch; and a mechanical, by which it serves as a motor force whenever its columns are displaced or its volume agitated. The lungs, as we have now shown, correspond to, and make use of the air in all these three departments. We may, therefore, resume in saying, that the chemical powers of air are chemico-vital powers of lungs, and the mechanical powers of air, mechanico-vital powers of lungs.

We do not forget, in these observations, that breathing commences only at birth, and that another order of things prevails previously. But this different state does not contradict the views put forth. Were this the place, we might pursue the thread of science into that other and attractive but mysterious sphere of whose still spring the round of this life is but the first expansion. But we must be con tent with remarking, that, during embryonic existence, the main business is the work of formation; the body is then upon the stocks; and as the blood first and the heart afterwards are the builders of the body, the brain or nervous principle operates through them, and its movements keep pace with theirs; but after birth, the usage of the body is the main thing; the life becomes more than the meat; the body is now to be worn out in action; its growth and repair are secondary, and all with reference to its employment; and this, as we have seen, depends upon the lungs; wherefore, in the second case, the brain shifts its patronage and alters its step, and works through and with the lungs. It is what might be expected—that the brain should respire with the heart in the heart-epoch, which anticipates the period of birth, but in the epoch of the lungs, or conscious life, should sympathize or synchronize with those organs which have the ruling mission, and transfer its sceptre to the younger dynasty of the chest.

The views we have been considering find an agreeable response in established laws of nature, constituting a branch of the doctrine of universal attraction, whose appliance they show in the human frame. Nor is this an insignificant support that they receive. When a law is sure for one department, we have a right, assuming unity of system, to look for that law in every sphere, though modified in each by its new circumstances. So, if attraction be the most general law of the dead universe, we know that in a new sense it is the general law of organization, and also of the human or living universe. But, in the actions of the lungs we have found it omnipresent in the body; and have seen the spring of an attraction applied to the organs, which causes them to operate very much according to the Newtonian formula. Here then we join forces with the discoverer of material attraction, who regarded it as the immediate finger of God, freshly noting the solidity of whose wisdom, we find in the body that attraction is no abstract formula, but palpable and living lungs. Nor can we doubt than when analogy is better known, and can be boldly worked, the light that issues from the unfolded doors of the human body, will stream forth into the vault of nature, and kindle celestial physics with a breathing wisdom that never could come from inanimate things, even though their theatre be ancient night with its gorgeous pageant of stars.

The active or alternate attraction in the body, like all attraction, amounts also to a law of association—in this case the association of the organs. If each organ takes, and does, what it wants, each organ is conveniently placed to do and to take it. The organs which need the best blood, are so seated at the banquet that they obtain it naturally and necessarily. They are succeeded in punctilious rank by the rest, each having its attractions seconded by its place round the table. There is a society in our members. But this is such a subject, that we must be content with a glance at its stupendous proprieties. The order which it involves, could we open it but a little, is of visionary magnificence, and might make us into propagandists of the organization of the body. For the Divine Architect rests not in middling fitness, but now, as at the first, perfection is his way, and embodied truth is his everlasting child.

Surely, then, we say, at the risk of repetition, it is no longer difficult to see the fundamental importance of the lungs in the human body. Life consists in the peculiar faculties, passions, instincts, senses and actions, which our bodies execute; or life is spiritual motion. This cannot be founded upon physical inertia or dead body, but upon physical activity or living body. And this activity must be constant and pervading, lest life should be stopped by some lump of rest or carcase remaining on its hands. Motion or vibration, therefore, in various degrees, continually sways the organism, and shakes it out of the rank of death-like things. Thus it is always on the tremble and tiptoe; its motion its main essence, and ready for obedience, as a servant all ear, eye and sense, watching for command. This could not be the case if the body, or any part of it were at rest. The rest would require to be broken, and the body to be roused, before it could obey, and a thousand volitions would fail before one was brought into effect. But by means of the lungs which keep everything *on the move*, the man is ever ready for living operations. Thus, the quickness of the body's service depends entirely upon its response to the animations of the lungs. Or, life is founded upon motion, and the motion is evoked and maintained out of rest by physical life, animation, or in other words, pulmonary breathing.

Thus far, we recognize a scale of physiological truths pertaining to the respiration, and which we may distinguish into vegetable, animal and human. The doctrine which recognizes the lungs as providers of air, is on the vegetable level; well for it, if it does not think that it is talking about men, when it concerns only cabbages. For plants are like men in this particular, of taking in and giving out air. The doctrine which regards breathing as of use for motion, belongs properly to animal physiology. Lastly, that doctrine which considers the psychological part of breathing, or the manner in which the motion embodies, represents and carries out those faculties of thought, feeling and action, and those destinies that are peculiarly human, is proper human physiology. So each thing is named and characterized from its own essence, and from nothing either beneath it or above it.

This concludes our present study of the effects that the lungs produce upon the exterior, and thereby upon the interiors of the frame; we observe that they endow every organ with outward life, courage and spirit, and call forth its talents in its daily work by the influence of attraction. And the attraction being most general, is common to all the members, which therefore, conspire or breathe together for realizing the goods of life, and thence come under a genuine law of association.

Now as truths always point out duties, there is something immediately practical which arises from our view of the importance of the movements of the lungs. If each organ contributes its share to the ensemble of life, each demands a special care in the maintenance of health, which is the wealth of life. Much has been written, and justly, upon tight lacing, as injurious both to the development and stability of the body. But if our ideas be correct, the duty of leaving the chest and the body free, becomes tenfold more imperative than before. If motion be the essence of the life of the organs, and if it extends to the whole frame and to the limbs, then all articles of apparel may fairly be supervised and limited in their pressure, in order to give our persons their lawful liberty. In this case the emancipation of the body itself is a subject of individual and domestic politics of the utmost importance, and the science of every organ should wring a progressive Magna Charta of dress from the kings of fashion. It is another proof that we speak the truth, because it tends so directly, yet so newly, to reinforce our old duties, which is an excellent test of truth. There are in fact as many kinds of public health as there are different organs. There is one which should be represented at the board of fashion, as having a veto, and establishing a precedent upon whatever is enslaving in dress. And it is not to be doubted that in what we wear, equally as in what we are, grace, pleasure, and beauty are compatible with freedom, and with freedom only.*

But dress is not the only thing that coerces the frame; or rather I should say, the body itself is a dress which under certain circumstances may oppress and hinder the breathing. A "belly with good capon lined," is a garment difficult to ensoul. Over eating is a tyrant against motion. It impedes the play, not only of the lungs, but of the other members. A mass of crude food is like an avalanche of stones descending upon a country, which buries the soil under dead materials. How plainly do we see the small life in the scant breath of the unwieldy *bon vivant*, whose lungs have porter's work to do in lifting his disproportioned paunch. So it is that liberty and temperance are among the natural commandments of the lungs.

We have now spoken of the first commerce of the lungs with the body; it remains to consider some relations which they maintain with the senses and the other powers, that is to say, with the faculties of the brain, and by which they again influence the bodily organs.

Now the material senses inspire the body with its first proper life, and concur with the pulmonary *inspiration*. For beginning with *touch*, we find that pleasant contact which soothes the skin, is

* All parts of the body may be smothered or suffocated if confined. This is often seen in disease, and particularly in delicate and nervous females, who begin to gasp if there is the least pressure of physical restraint, and a touch sets them off into hysterical movements, the feeling of suffocation reacting from the circumferences or limbs towards the centres. Life in such persons is an exquisite balance which appreciates quantities of compulsion and restraint that make no sensible impression upon hardier organisms.

accompanied with full breaths, sometimes running into deep sighs if the sense be peculiarly grateful; and in extreme cases of the kind, inspiration almost obliterates expiration, which survives only in gasps and murmurs. Painful contact on the other hand straitens the lungs, and causes the breath to be held as long as possible. In short we *breathe in* the touches that delight us, but confine to their first place of invasion, and shut away from the vitals, the discords or agonies of our skins; and this by fixation or resoluteness of the lungs. Respiration then draws up the sense of touch towards the general sensorium. It also sucks in the sense of *taste* to the same goal. For taste lives when inspiration is proceeding, but when we breathe out, or stop the breath, sapid substances do not make their proper impressions. We keep back breath when we swallow drugs, and the nauseous taste is not drawn into our consciousness. At meals, however, we breath with satisfaction, for the circumstances are inspiring; and tending, as they do, to enlarge the man, they set his machineries in motion with a life of extra breaths. *Smell* is inspiration in its highest case; the nose is a lung planted upon the brain, to feed it with perceptions and excite it to operations. Air and scent are inseparable companions. To breathe therefore involves to smell, the one function swallowing the other up into the brain, and down to the bottom of the lungs. The motives to breathe furnished by these three senses make inspiration deeper and larger than it would otherwise be (p. 94), for pleasure takes great lungfuls; thus they animate the lungs with superior life, and the organization is opened by the senses through the lungs to a degree beyond what insensate lungs could effect. As for the senses of *hearing* and *sight*, the lungs do not so directly aid them, because light and sound are above their attractions. Their active offices terminate with the blood and the air; only their passive offices extend to the ether and the nervous system. For hearing and sight, so far as they are essentially acts of attention, are best transacted when the breath is held; and indeed impressive sights and sounds tend to suspend it. We observe then, with regard to the senses and their connection with the lungs, that touch and taste browse in the fields of inspiration; that smell, a winged sense, flits with ceaseless play between inspiration and expiration; and that sight and hearing, concurring often with suspension of the breath, live above the lungs in the airless calms of the brain. Touches and tastes we breathe in; smells we scent, or breathe in and out; and sights and sounds we do not breathe, but see and hear, athwart the air, either in spite or in the absence of its proper motions.

Thus much for the passive immission of the material or pulmonary senses. The senses however have an active condition in which their sensations are perceptions. In this state they partake of the common law of the two higher senses, and are awake and efficient at the times of suspended respiration. For active sense is a breathless power, and does not draw in body, but puts forth soul. Thus touch as a mental product is *tact:* it turns the tables upon its objects, makes itself critically harder than they, and resists and rejects, picks and chooses their impressions by deliberate inquest. The breath awaits while the steady-fingering thought explores, and then inspires, not whatever comes, but precise information. Let the reader observe himself when he is feeling for such information, and he will find his curiosity rejoicing in periods of suspended lungs. In active taste the same rule obtains. We no longer draw in the pleasant flavors by mouthfuls, but disparting the tongue for special acts, we make little sucks and respirations of the palate upon specimen morsels; we fill the decent sense with judgment, taking small account of pleasure; and holding the general breath, we calculate the result, undistracted by the lungs, in its smallest figures. *Tasting*, then, as contradistinguished from taste, is carried on in the intervals of common breathing. So also is *smelling*, which works its problems upon minute quantities of odors, shutting away the volumes; actively we exert our smell upon mere snatches of scent made to run hither and thither in the inquisition of the nose. And as we said before, we *hear* best in breathless attention, and *see* most observantly when the eye-thought gazes, unshaken and unprompted by the lungs.

It is also to be noticed that the *voice*, which consists of perceptions freed from the mind, and launched into the air, is made of the material of the expirations. The mind is breathed out into the social world by the expirations and their pauses, and not by the inspirations.*

The sum of these remarks is, that the exercise of the senses is rhythmical, chiming with some part of the respiration of the lungs; either with inspiration, expiration, or some level or pause of the one or the other at which the breath may be suspended. And as the senses belong to the brain, evidence is afforded that its animations, which comprise the senses, coincide with the respirations of the lungs.†

Passing to another sphere, we may glance at the connection of the lungs with the passions. On this theme it may be sufficient to say, that the breathing varies with every emotion; a circumstance which may be verified in experience, by noting the respiration at different times. If we could remove from the language of passion all reference to these organs, we should cancel I know not how much of its expressiveness. If we could take the variety of the breath itself away, the man, the bigger he was, would be the more an unmeaning lump. Where would mirth be, if it lost all its laughter? What would become of hope, if it had no dilated breast? What would be the

* Oratory especially requires the management of the breath, or the economical guidance of the expirations by the conceptions. If you spend your air too fast, a part of your in-coming air will go to pay off the extravagance, and you will probably be in nature's debt throughout the speech, presenting more or less of the phenomenon of a person who has "lost his breath." To "lose the breath" is to fall into an unnatural rapidity of inspiration and expiration, which will not be governed by the will, in which case the mind has *pro tanto* lost the power of dispensing the air.

† We have a further proof of the consentaneousness of the lung movements with the brain movements (pp. 62—65), in the fact that the voice, proceeding from the head of the lungs, is the voice of the mind, and images its thought or corresponds to its animations. But if in this exalted function of air such correspondence exists, does not the lung air correspond in its times with the brain spirit, equally with the larynx air, which is the voice? If the top of the pulmonic wind answers to the surface of the brain spirit, or, in other words, the voice to the mind of the moment, does not the correspondence run upwards and downwards, and does not deep call unto deep through silence more than through speech, and the spirit above to the spirit below through the lifetime as well as through the second? Do not our little harmonies swim in great harmonies, which are not ours only, but creation's and the Creator's? But upon this subject we do not dwell, because we purpose to treat of the voice on another occasion.

plight of love, bereft of its delicious sighs?* How could pride exist without its hardened chest and swollen throat? Or rage without his choking breaths? Or anger without his tempests? How should our poor weariness endure, if it had never a yawn to console it? And how would joy and gladness fail if their healthy bosoms did not swell with trembling airs of the clear blue firmament, eager to re-ascend in songs? But these are only a few of the presents that the lungs draw from the mighty winds to bestow upon their brethren, the passions. The law is this. Each infant or dawning passion disports itself first in the brain; attitudinizes there to the top of its bent in the chambers of imagery; observes and admires its goodly appearance in the mirrors of fancy, and is king uncontrolled in its own little cortical spheres. Then as the lungs are plastic as air, it descends into the theatre of resistance through their convenient mid-way, and shapes and crystalizes the wind for the moment into hardness and strength, softness or gentleness, sighs or fulness, or any of the other forms which the dramatic occasion requires, or the muscles and limbs demand as a ground for peculiar action. For each emotion it hews the body into a different block, wherewith the emotion pushes its way in the world. In a word, the lungs are the bodily arena of the passions; they give shape to our impulses, increase and deepen them, and begin to carry them into works. In inward gestures and deeply silent murmurings they first imprison the words and deeds that are at last to resound through history, and push the nations to their goal.

But to trace the special inhabitation of the passions or brain spirits in the breaths or lung spirits, will require a volume.† It may however be noticed that the inspiring passions concur with the pleasant

* Here we may remark that the spirit of the passions and actions, nay, of the states of man generally, may receive its formula from the breath of the lungs; for the breathing is a representative phenomenon, and is to action what words are to thought, and what tones or music are to feeling. If we hear the breathing of those whom we do not see, we infer to a certain degree what they are doing, and their general tranquillity, or the reverse. And this is a walk of observation that may be cultivated to almost any extent. In very susceptible persons, the inferences drawn from the breathing of others are wonderful.

† We have made some progress with such a work, but the field is of an unexpected magnitude.

senses, and are housed in the inspirations; that the depressing passions tend to lower or kill the breath; as extreme fear, for instance, which makes us aghast or ghostless, and causes the lungs to forget their reciprocations: and that the middle passions have a middle effect. And it may further be noted that the peculiar respirations which are the bodily spirits or tendencies of the several passions, have the office of provoking the latter, or reacting upon them. For example, in rage, does it not begin to fume and swell in the lungs? is not "the steam got up" in those locomotives; and does not the brain, with tempests in his hand, not only lash the body into the pace which answers to its own madness, but feed the madness out of the wind-swift speed? These passionate breaths, although not classified by science, are known to the observing, and interpret the underplay of the feelings, even when speech and smiles dissemble. In this field then the lungs have several offices. By concurring with the passions they raise the frame into each, or communicate it to the blood and secretions, enabling the mind and body to keep company through all changes, or to be impassioned together. By the same concurrence they amplify the field, and stimulate the fire of the passions, fanning it with the oxygen of their spacious movements. They also enlarge the material body to the scope of animal life, which is passion, causing stomach and liver to flame and expand with it; as we saw in the case of the senses, that they extend the enlarging breath of sense through the lungs to the same material organs. The lungs then lend the passions of the mind physical force, and the organs of the body passionate movements; and by this means they make the one and the other, or the brain and the viscera, into perfect bodily animals.

But the imagination also, which is the intellect of passion, builds especial houses in the breath, or, as it is said, forms air castles. These are its own expirations, in which it revels, for what it draws in is nothing to it, but what it breaths out is all. It does not however expire either to do or to die, but to run after its breaths as they sail through the air; not desiring to leave the world, but to propagate its image children in the universal imagery. The smoke of its lung-pipe keeps it busy with the plasma of a thousand twirls. It makes its objects out of its breath, and hence we locate it among the expirations. During such imagination, accordingly, the head is held up, and the breathing tube to the very mouth levelled like a barrel: words fly forth with arrowy straightness; the inspiration is inaudible though sufficient, but the man pants audibly towards the unseen, and each pant externizes more of the breath on which the faculty pulls and feeds. When the breath-palace is built, the laws of gravity bring it to the ground; whence air castles, as the frequent beginning of earth castles, are not to be despised; imagination being the proximate architect of the arts and sciences. We may formulize the respiration of this faculty by saying, that during its exercise the lungs take their airs to themselves just as the imagination represents its objects to itself externally. This lung conceit is one means by which the body holds its own sphere, and protects it amid the great fluctuations.

Respiration has also a peculiar relation to the intellectual processes, which lie, it will be found, in fixity of breath, proper *state* of lungs, or suspended respiration. Among other reasons for this is the fact, already pointed out, that inspiration is a means of drawing up the bodily sensations to the brain; for the body is as a sponge let down into the world, whose attraction upon the waves of material sense is exerted by pulmonary inspiration. But in proportion as these lower influences are admitted, often in the same proportion intellect is drugged, and sleeps in the cortical beds. We speak of a familiar fact. But because the mind has power over the lungs, it can handle the senses by their means, and prevent the floods of worldliness from penetrating to the upper sensoria. So also can it stop the mounting passions. This it does by suspending the breath, and cutting off the supplies of sense and animality. Or, to speak more anatomically, the brain at such times refuses to be invaded by the blood, which contains the turmoil of the lower life: the cortical spherules keep it at arm's length: for it is to be remembered that the brain expires concurrently with the lungs, and when the latter shut off the blood, the brain does the like. Hence it is that thought is still, and contemplation breathless: each involving, first, fixed breath, and second, a small expiring; and so on, until the thought is traversed, or the effort ends and begins anew. Deep thought, then, where not given directly by heaven, but conceded to human effort, is gained by the descent of a ladder of expirations, and the body dies down into intelligence by this scale: the best of such perceptions come from the confines of expiration and the grave, which lies at the bottom of the lungs. Intellect, therefore, in this light, is the capacity of standing and expiring, and living still; death to the body governing the body; an infinitesimal immortality into which thought expires and expires, to brighten and brighten its lives. To the senses, suspended animation is suspended consciousness: to the intellect suspended animation may be life, thought and supreme wakefulness. For it lives when the body is gasping: its chosen sons, as they drag the world onwards, are verily at their last gasp, in the acceptance of their own mortal immortality. The intellect, then, through the lungs, puts the body down under its palm; whispers to the sea of delirious sense, "Peace, be still!" and plays its melodies in the charmed air upon the whitest keys of silence.

Intellect touches so near upon trance, that the highest cases of either involve common phenomena, and exist in the same persons. But trance is complete suspension of the breath, sometimes for long periods. This suspension of the breath of the lungs involves the standing of the spirit of the brain, and the stand of this is the gaze of the intellectual eye, whose final and hard victory is *to see.* In some subjects, if the lungs are fixed, including the brain, the body can wait for breath and spirit for an indefinite time. It still lives, because there is a standing spirit in the body and a standing breath in the lungs, which partake of the fixity, or are charmed and entranced. These are cases in which life stands, and the proper spirit can go and return, because the animal state is safe and fixed. Nay, the time of the trance or separation counts for nothing. We do not know a limit to the duration of the body under these conditions; it is as a day miraculously prolonged, when our sun stands still upon Mount Gibeon and our moon in the Valley of Ajalon. Nor do we know a limit to the excursions of the intellect on these holidays, when it visits its celestial birthplace, secure of finding its lungs and factories ready to start into reciprocation at a moment's notice on its re-arrival.*

* The reader will notice that scientific experiments may be made by whoever pleases, upon the concurrences of the lungs with the mind and the body; and that

It must be recollected that in these conditions the lungs are not emptied of air; for expiration does not go to that extent, but plays upon a certain depth of the pulmonary reservoir, leaving the remainder undisturbed. The involuntary right of the lungs to the air is strongly asserted if we attempt to expire beyond a certain point. Otherwise, the body would lose its life size, the wedge of atmosphere would have ceased to open it, and the spirit crushed out of the body could not lift it a second time into the operations of breathing.

Moreover, in these conditions of suspended animation the chemical laws do not persist, but like the rest are *suspended.* Held breath concurs with held spirit, held blood,* held life and held time. The tissues, particles and fluids, and the wind in the lungs, are entranced; the body is absent from chemical corrosions as the mind from animal provocations. The air is not required for exchanging products with the blood, but for maintaining the level of the state, and serving as an elastic *animus* under the fixed attitude of the brain. And even when the air is expired in partial trance, it is not because it is vitiated, but for deepening the state, and as it were steering and standing in nearer to the shores and lighthouses of death.

We now see what this concurrence of the lungs bestows upon the organs, for they all stand when the lungs stand, taking up their places as fleshly eyes in the attitude and body of the intelligence.†

for this purpose no apparatus is necessary beyond observation on the one hand, and the possession of the human frame on the other. This, then, may be a universal science. In the same way we showed, in the Chapter on the Brain, (pp. 72, 73), that there are vivisections which may be studied without torturing animals, namely, the divisions and sects of a certain creature which cuts up its own species and its own brains for us every day.

* The heart indeed in itself, though not in its blood in the organs, is an exception to the universal concurrence of the frame with the breathing; for although influenced thereby, it is not reduced, like the organs, to the pulmonic rhythm. But as we shall show, it belongs to another regiment of natures, and to a different discipline from the lungs.

† The concurrence of the head with the body is provided in many ways; but the moving harmony of the lungs and the brains appears to be at the basis of all. Let us take an instance of this concurrence from the muscular system, and let the subject of the experiment be walking. Now let him fix the eyes in a gaze upon any object. Soon the walk becomes slower, and the body is brought to a pause, as it *seems*, voluntarily. If the gaze be continued under favorable circum-

It is the interpolation of the higher life and endurance into the organic movement; breathless but deathless moments set in the midst of the wear and tear of the tissues; chemical moth and rust ceasing their gnaw, and incorruptibility paramount in the corruptible. So the body represents the proper mind; the intellect sinks a shaft into the flesh, making it dramatic of the moments that we live beyond sense and passion. Man would not be embodied if that which is best in him were not bodily set forth. The lungs then introduce this transcendent representation, and the moral virtues that inhabit this order of intelligence commune with the organs through their means. They put down the body, give it the lesson of death or self-denial, and frame in it still windows of experience opening to the timeless state. They emancipate the mind for the occasion from the stimulus of the passions. In short, they embody the moral intellect, or give the frame a hyper-animal life not lying in physical

stances, the will thus brought into the topmost muscles (those of the eyes), and which has riveted the feet, will begin to rivet the muscular attention from above downwards. And although the eyes close, if the will be kept up, rigidity will invade the jaws, then the arms, and then the legs—producing a state like catalepsy. These facts are familiar to those who are acquainted with Mr. Braid's admirable discoveries in Hypnotism. It is upon this principle that lockjaw, attacking the high ranges of muscles about the jaw, runs down the inclined plane of the muscular system, locking it as it goes, and bringing muscle after muscle in tributary streams to constitute tetanus. And it is by acting on the still higher point of the muscles of the eyes, that this disease may be commanded from muscles loftier than its own origin. By the force of this inclined plane system, the expressions of the face tend naturally to produce the gestures of the arm and the postures of the body, which may be looked upon as torrents of will, that have come down at first in little streams from the mountain springs of fleshly action in the countenance.

Thus the muscular frame is all made for concurrence, and forces which act upon one part of it, tend to be diffused through it, as through a *continuum*, but with a difference of function according to the regions. The expression of the face, which is the dial plate of the general mind, is the mainspring and clockwork of the active body. Upon this principle, we see that smiles precede laughs, that clenched jaws go before clenched fists, and, in short, that expression not only anticipates but also stirs action.

These considerations furnish fresh evidence that the body is *solidaire;* that whatever the head does, the trunk does in its way, and the limbs in theirs: in short, that man is so formed as to act only in wholes, each full size. In this way we are constructed upon the principles of poetic unity, the mind and the body being but one volume written out in the rhymes of the brain and the lungs.

movement; and they intellectualize the body, or contribute their share towards constituting its peculiar humanity.

If sense, passion, and thought are in a certain dependence upon breathing, so also is action to at least an equal extent. All fineness of work, all *that* in art which comes out of the infinite delicacy of manhood as contrasted with animality, requires a peculiar breathlessness and expiring. To listen attentively to the finest and least obtrusive sounds, as with the stethoscope to the murmurs in the breast, or with mouth and ear to distant music, needs a hush that breathing disturbs; the common ear has to die and be born again, to exercise these delicate attentions. To take an aim at a rapid-flying or minute object, requires in like manner a breathless time and a steady act; the very pulse must receive from the stopped lungs a pressure of calm. To adjust the exquisite machinery of watches or other instruments, requires in the manipulator a motionless hover of his own central springs. Even to see and observe with an eye like the mind itself, necessitates a radiant pause. For the negative proof, the first actions and attempts of children are unsuccessful, being too quick, and full moreover of confusing breaths; the life has not fixed aërial space to play the game, but the scene itself flaps and flutters with alien wishes and thoughts. In short, the whole reverence of remark and deed depends upon the above conditions, and we lay it down as a general truth, that *every man requires to educate his breath for his business.* Bodily strength, mental strength, both lean upon our respirations.

The co-operation and state of the lungs in mental effort is represented on a large scale in their strain during parturition, in which they let out the air in groans from the relaxed and almost closed larynx. This, which is the type of all labor, as child-bearing is the image of all productiveness, is carried on by holding the breath, and determining it not towards the air but towards the obstacle; the portion of air whose spirit is broken by the effort escaping immediately afterwards in the form of a broken breath or groan. The air, which exercises everywhere a universal pressure, exerts in the body, when compressed by the muscles, a universal *push,* and is a medium in all our fruitful pangs, whether those by which children are born, or those by which thoughts, which are the mind's children.

For the brain is the womb of the soul, and the held breath during the effort of thinking tends to exclude the desired thought when the determination of all the parts strives towards the right part of the brain. After the effort comes the groan, which shows that the breath has no more will, but has done its work. It would seem that in labor, the rhythm of the uterus takes the lead (pp. 103, 117), in commencing the breathing, and the lungs are obliged to follow the strong contractions by shutting their apertures, and laboring precisely like the womb. The nervous system and the mind labor also at the time in the same ratio. There is neither the free child, the free mother, the free breath, or the free spirit, until the birth takes place; but the bondage of all is common and oppressive to insure the emancipation.

We remarked before (pp. 87, 111) that the respiration is divisible into four terms, namely, inspiration, the pause or satisfaction succeeding inspiration, then expiration, and then the deliberation or pause which follows expiration. And we have now shown that inspiration concurs with the *agrémens* of sense and feeling. This is the first motive of the lungs, or the pulmonary atom of the pleasures of the world, compounded however of two elements, the nose-breath and the mouth-breath, the former to please or inspire the mind of the brain, and the latter to please the mind of the body. This term, if persisted in, leads to *swoon*, from defect of expiration; whence swoon is the prolonged or compound atom of the pause after inspiration. The pleased lungs are so gluttonous of this world's life, that the world, bent upon equilibrium, swallows and drowns them in this swoon, which is the ocean of sensual satisfactions. Expiration, however, concurs with the spiritual life, and is the condition of intellect or of dying daily. And the pause which follows expiration, the refusal to breathe in from the surface, and the stand taken in the depths, is the atom which in its least form concurs with abstraction of thought, but when compounded, runs on into *trance.* The likeness of this in the animal world is hybernation. Thus every thought is a little trance, and every pleasure an initial swoon, as we shall presently see that each fair breath is a little life, whe-er of sleeping or waking. And thus if the breath is given in inspiration, the spirit is impressed upon expiration; for the spirit of humanity is not in the breath which is taken in, but in that which is given out; the former being planetary, but the latter psychical. There is, then, something beyond foul air which man breathes forth; for the air is charged with the vital movements; there is the character of the life wrought into the atmosphere, as drawn upon the organs. And here again we turn to chemistry, aud demand of it, besides the analysis of average breathings, the contents of the accidental breaths emitted in peculiar moments? We ask of it whether the breath of mercy is foul to the lungs of those over whom the mercy leans? Whether the laws of vitiated air hold here, or no? or whether there is an angel-galvanism by whose tension at such times the body and the air fly clean above matter and its pedantries?—Whether there is any antiseptic significance in the fact, that Jesus breathed upon his disciples, and said, receive ye the Holy Ghost? or in this that we swallow the breaths of those we love, and listen to the breaths of those we venerate? Whether the last breath of beloved friends, caught in all unsophisticated times, is exhaustively represented by the formula, $O+H+C$? Whether the blood and the body decompose in the same ratio during all states of the mind? or whether there are not moments, and degrees every moment, in the ratio of destruction; moments of immortality in which no waste occurs, and all intermediate grades between these and physical ruin and decay? And further whether there are not facts in human society, as of intimacy, closeness of persons, community of breaths, which show that the expirations of one man are in a cheerful and life-giving sense the inspirations of another? But chemistry can no more analyze human air, than animal air and vegetable air, but it throws them down before oxygen, hydrogen, and carbon, the insatiable Cerberus of the laboratory. On the contrary, we induce from larger facts, not otherwise accounted for, that the motions of the intellect and will, or the better faculties, are the salts of the human air, which varyingly wrest it from the gripe of the chemical laws. We induce that *humanized* (p 107) lungs have a duty to perform to the social sphere outside, and that the expirations from such pay back the world with usury for the simple air which the inspirations take away. This, however, cannot be confirmed from the steams of crowded assemblies, but from the closets of privacy, and from the exceptional facts and moments in which neighbor comes close to neighbor, and man hangs as a lover upon the breath of man.

Indeed it seems remarkable that the influence of the vegetable world upon climate, and of electricity upon the atmosphere, should be admitted, and that no influence of the human world of a similar but higher kind, should be suspected. Are the thought-movements and the will-movements sooner absorbed than the sound-movements? do they pattern and sculpture the air with less efficiency? or in what do their modifications end? Is the music of man's brain and lungs of no Orphic power in the tenseness of God's created harmony? But the time is not yet for these and similar questions; they are however as doves which float already in the poetic air, and the dry land of science is about to appear, upon which they can alight.

Quitting this consideration we have to say, that not only the moments but the lifetime are parted into breathing spaces; for the first breath and the last are the bounds of this existence, and the ends shape the means, or constitute the career itself into a series of breathings. These larger lines of breath consist of habitual modes of respiration answering to the tone of life, and constituting pulmonary morals, manners and customs. They are determined by the mixture of the four terms already specified in various proportions, and by the velocities and spiritual qualities which are carried into these. In this way the lungs move and associate individuals, as we before noticed that they move and associate the organs (p. 106); for only those who conspire or breathe alike are together in thought and intention; and the society of persons tends to last only so long as they have common respirations. The attempt to prolong companionship beyond these limits disarranges the springs of the organs; the presence of heterogeneous persons straitens our breath, or as we say, dispirits us. But we shall have to recur to this subject, of discord or want of tune in breathing, when we speak of Public Health, which means public association on the principles of the organs.

In mechanical coöperation the unanimity of breathing among the workmen is essential to oneness of effort. Hence the rude cheery work-shouts that sailors extemporize in weighing anchor, masons in hauling up blocks of stone, and so forth; and hence the adjunction of music to battalions, which require to have one spirit and step.

A mass that is to be as one man must breathe alike in its parts. The same thing is true of society, or unanimity in its higher departments. The heard breath of your neighbor is moreover regulative and contagious upon your own, and increases and realizes the union of which it is the effect, especially when the breath of all is represented in an audible rhythm. In this way the Eddas and poetries bind mankind into sheaves, being as common respirations or great world-tunes, the sum of beginnings of musical acts from the sailors upon the river of time.

And here we may observe that throughout life the lungs exercise the dramatic office of producing in the frame those motions which answer to the periods of existence. When the man is to sleep, the lungs give the effigy of sleep in the system, and the slumbering soul is imbedded in a slumbering body. When he is to awaken, the breath of morning sparkles from the lungs throughout him, and master and servant rise in a breath for their unanimous day's work. As a child, his innocent brains find a sisterly helpmate in his playful and peace-breathing lungs;* his blood and vitals, like his pretty face, are full of sweet and innocuous motions; his lungs transplant the childhood to his tissues; and soul and body, head and feet, he is all one child. The youthful spirit again and the youthful body are each the other's, and the bond between them is still lung, attraction, or the lover's link. When he is a man, his lungs too put away childish things; heart, liver, brain and bowels are engaged in manly movements; the breath of manhood strengthens him; his vitals are adult and personal; and the man lives well in an outer man who is the body of his powers and the servant of thoughts. Age steals upon him in the wants of a second childhood; he begins to breathe fainter; his old days are a young lesson of living above the air; and his last breath sets the body free as no longer able to move in his service. And so from the beginning to the end of life, the body conspires with the mind, through the friendly intervention of the lungs.

* Let the reader try to breathe like a child, and let the auditors of the breath decide whether he succeeds, or no. There is, indeed, in adult breath such a peopling of multitudinous thoughts, such a tramp of hardness and troubles, as does not cede to the attempt to act the infantine even for a moment.

But in speaking of the four parts of breath we have separated qualities which are not incapable of union. Thus we have regarded expiration as of spiritual significance, and inspiration as of sensual, whereas these two may be balanced, and the just lungs in consecutive moments may show them to be equal weights. Pleasure indeed makes inspiration, and energy and resolve animate expiration, but pleasure and energy are sometimes united in the *joy of work*, and then the inspiration and expiration are at one, and the man breathes *con amore*. Thus in what we call happy moments, when we do our little miracles, put in our least imitable touches, and sing our best songs, we breathe as if we breathed not; there is no greed on the one side of the lungs, or effort on the other, but levelness of taking in and giving out: the gold of inspiration is minted with the die of action, and it passes through expiration without a challenge, and so expiration itself becomes plenarily inspired. In this state both sides of the Janus of breath, peace and war, pleasure and energy, are combined in happiness. The highest moments and emotions are of this balanced order; innocence, peace, and the perfect qualities, produce equality on the two scales of the functions of the lungs. Man inhabits the world aright, in this equilibrium between his passions and his actions, whose hours are as the immortality of his childhood and the genius of his life. And innocence, peace, and all the sweet even-breathers glide down through a variety of states into that which is their compound atom, *sleep*, the fountain into which they descend, and from which again they arise like love born fresh from the morning ocean waters. In this sleep there are many depths of the level breathing; the child's, which scarcely stirs the surface of his tiny lake of breath;* the man's, which goes deeper, but always according to justice and equation. Thus in the fair proportion of the four terms, we locate the model states of waking and sleep: even-mind-

* The laws of the diffusion of gases are adequate to produce the function ordinarily assigned to expiration; but the motion of expiration, and its constant variations, are additional to this function, and to be looked upon as livingly mechanical, or, in other words, psychical phenomena. Moreover, the chemical impregnation of the air with the breath, is a different thing, both in regard to space or extension, and time or permanence, from the vital impregnation of the air with motions: for the material breath falls in dregs which soon pass away,

edness, which gives all things their places, and is the ever-vigilant balance of the soul; and even-bodiedness, which lays us along under our happy coverlets, and makes our slumber as still as our good conscience. This state is the rare complement of lop-sided pleasures and duties; of the swoons of delirious sense, and the trances of the ascetic soul: and he breathes best who most completely enjoys it.

We have now seen how fully the breathing is inhabited by the living powers, and how our breasts heave with our natures and our minds; in other words, how the faculties of the soul go up and down through the arches of respiration. We have seen what a thread of human life courses through the actions of the lungs, and by them is transferred to the organs. One function then of the lungs consists in equating the body with the soul, and momentaneously keeping up the equation. But these organs produce also the momentaneous connection between the psychical and corporeal frames. For if it is they that give motion to every organ; that attract material sense from the world towards the head; that represent all emotions, passions, and imaginations, and give them to the body; and by their power of station are the footstool of thought and will, first submission to which they likewise embody: if it is they that sleep with sleep and that wake with waking: again, if it is they that prepare the body by a model agency for every action; and furthermore, that draw down the real spirit of the brain into the body, or, in other words, pump us out of our spiritual reservoirs;—then it follows that it is the lungs which physically connect the organism with its animating soul. And what connects the lungs themselves with the same soul is, that their movements *correspond* with those of the brain; whence the feeling which we all have, that in breathing we are living.

We have endeavored throughout this doctrine to follow learned Vestiges, and to show that knowledge, like organization, may pass

while the mental breath endures, we know not how long: for as the poet says of the Forum—

"Still the eloquent air burns, breathes with Cicero."

Poor Byron! Even for the skeptics, as soon as they begin to sing, nature is either nothing, or haunted!

through "developments:" that science is sometime a cold-blooded animal, and sees respiration from a fishy point of view, as is the case with the existing physiology, whose doctrine lies motionless in the seas of knowledge, and without proper breathing takes in and gives out the little and casual air which is dissolved in the waters. But it is time for the scientific fish to undergo another stage, and putting off the piscine, to busy itself with warm-blooded motions. And, finally, the fish must become a man, and derive the pulmonary motions from another kind of warmth, which in old wisdom is the Soul. This will be something practical in the doctrine of "development."

For assuredly the lungs give our bodies a series of endowments which are not animal; comparative anatomy sheds no light upon these, unless you reckon the anatomy of the soul in the series of the comparisons. And this leads us to speak again of the lungs as space-makers; a function of which we are so jealous, that if we be confined in breath, or restraint put upon the chest—arms, legs, and every muscle fight with convulsive energy against the oppressor. For human liberty is doubly grounded, in the body, and in the soul. And in the body, the liberty, by virtue of the airy and opinionated lungs, is given attractively to every organ, as we showed before. Thus, each faculty has its proper size, or liberty, which is the air it breathes; and if it has it not, it dies. Breathing makes the living body bigger than the corpse. Sense makes the body roomy enough for a lustier exercise of powers. The passions dilate it to the scope and size of public strife. Thought again diminishes it, because thought does miracles *in minimis*, and alters worlds, if need be, from the throne of poverty, or from examples radiant through dungeon-walls. But free thought, by the blessing of God, is a liberty beyond liberty. In truth, each faculty and each fixed opinion, spaces the body to suit its own play; whence sects and parties wear their very bodies for liveries, and are dry or juicy, liberal or stinted, sensual or spirited, according to the openness that their tenets put into their lungs, and their lungs into their livers and frames.

Much has been written about the cause of the first breath, as though it had not the same cause as all the breaths, being derivable from no other source than the motion of the organic mind in the head. To be born alive, is to be born with a germ of mind related worldwards; to have such a spark, is to have a rhythmical motion of the brain directed bodywards, which motion cannot subsist or be promoted without a seconding rhythmical action of the lungs. There is no need of any other principle than the harmony of the lungs with the brain to account for the first act of breathing, which in fact, is the beginning of our life. We take the first breath because we choose, and we take the ten thousandth for the same reason; and when we do not choose, as in sleep, it is as if we did, because Providence backs our wills with similar wills of His own, then called souls, which fill up our intervals, and make our lives coherent.* Before all inquiries into the causes of beginnings in the body, there stand two inexorable axioms: 1. The soul; and, 2. The consentaneousness between it and the body. After this, the explanation of any given first effect, as, *e.g.*, breathing, lies in our knowledge of the functions of the effect, which account for our using it. The argument of the human body is like the body, *living;* or every physiological problem may be put thus: Why does the soul do so and so?

Before concluding, we revert (p. 95) for a moment to the statical function of the lungs, to remark how these organs distribute life into strata. The vulgar call the lungs *lights*, and so they are; for the belly gives us gravity and links us to the ground, but the lungs give us levity, and lift us towards the air. Erectness of attitude (p. 103) begins in the chest; we give ourselves the airs by which we strut, first in the breaths and last in the muscles. The second power of erectness is flight, such as we see in birds and insects, which conspire with the air so well, because their bones and tissues are open to it; besides which they can rarefy the air, both by their heat, and by the *cupping* action of their powerful muscles upon the closed cavities of the frame. Their feathers too are outward air cells, answering to the universality of their lungs within. Man

* Here it may be noted that the lungs correspond to both the cerebrum and cerebellum. For their breathing may be either involuntary or voluntary. In this respect they combine in a single organ the functions of the accidental and permanent life, or of the will and the nature (p. 70). They therefore cement the bond between the two brains by a marriage of their motions in the body.

seems heavy compared with the swallows and the eagles. Yet with faiths for his second lungs and sciences for his wings, he is the lightest of the tribes; and if he seems chained to the ground now, it is because these, *his* lives, have not been admitted into his bones. His lungs, which may hang in the air (p. 95), are a prophecy that his body, and bodies of his body, may be similarly suspended. For what are his lungs but a balloon corded down into his flesh, of which, when fully inflated by the spirit, his body is the car? The earth belongs to the human lungs as birthright and natural gift; they are "tied" to have it. The ocean belongs to the human lungs by the held breath of the diver, who is at once the fisherman and the fish. The air belongs to the human lungs by want, prophecy and science; by the leading of Him who has ascended already, and trod the lightness of the crystal climes. The spirit belongs to the human lungs, by their sails filled with every sense, passion and thought; by the trances of man, which are above the air, and by the breadth of the supernal life, which does not disdain the concurrence of the lungs. And peace and greater powers than these belong to them in all and through all, as the gift of God, who breathes his blessings upon his chosen.

It is good to look to the ordinary language of mankind, not only for the attestation of natural truths, but for their suggestion, because common sense transfers itself spontaneously into language, and common sense in every age, is the ground of the truths which can possibly be revealed to it. If there be no common sense to welcome a truth, that truth, however vivid, is lost in the darkness of the mind. It is of no use to speak it. If we set our ideas of the lungs before the glass of language, they receive, to say the least, a cordial welcome. For undoubtedly most of the words expressive of life, are borrowed by analogy, either from the atmosphere, or its organ the lungs. Thus, animation is the Latinized form of breathing; an animal, a living creature, is a breathing creature; an animated body is a breathing body; the soul also is an *anima*, a breath; the mind or disposition is an *animus*, also a wind or breath; we receive inspirations, which are the breathings in of high influences; we have aspirations therefrom, which are our voluntary breathings back after the good which has been shown to us; we are spirits, that is again to say breaths or attractions towards one Eternal Object, who, through our finite organism draws us to Himself. Finally, we read of the breath of life that was breathed into the nostrils of our first parents, and man became a living soul. Now we have attempted to show that these phrases are nature's own analogies; that the functions of the lungs are so identified with that motion which is the representative of life, that life cannot be imaged save in words borrowed from the lungs and their august ministrations. In this respect, I trust to have the whole of the depositories of past truth and genius, the old wives, upon my side, which will atone for the absence of others, because there are no better beginnings of physiology than these old wives' tales. Most of them are point blank true, and will come out more and more as they are wanted, constituting in time the last novelties of a sublimely scientific age.

But when shall the theme be ended of the atmosphere of the microcosm? Or what is the entire drift of these Æolian harmonies let out upon the sciences from the lungs? We can only further say, without pretending to a formula, that the planet-air is the double of the lung-air. For in the human sphere the viscera stand as trees with the lungs for leaves, and quiet absorption subsists in this vegetable degree. But furthermore, the superincumbent lung-air presses down upon the vitals, whence the physics and mechanics of the atmosphere run parallel with those of the lungs. But again, the lung-air has its system of winds, beginning whencesoever the will pleases, and blowing from the poles either of matter or spirit, love or hate, passion or consideration, and making the climate of the ruling state to travel through the whole body. Then also the lung-air is as full of the soul as the world is full of the sun. And lastly our little atmosphere in motion, like the great atmosphere, rubs our atoms against each other everywhere with electric rustle and collision, and by its congenial friction magnetizes the assembled parts, and they open the cases of the life-gems. But for the rest, the voice would cease to speak before these truths are exhausted that chant themselves up from the deep well of the lungs. For ever and anon as we listen we hear a more inward chorus.

At least we have seen that the body lives in alternation; that the voluntary moments of the machine are the breathings; that thus time is meted out in our frames, and we are introduced as finished clocks of time into a world of time—of measured changes. So it is that we are tuned to the universe, which exhibits the same play in its furnitures through all their fortunes. And so it is that we are fitted to our place in humanity, which expands and contracts from epoch to epoch, from idea to idea, from institution to institution. All things are as cycles of not an unending but an ever-during providential change. Their existence in time consists in their motion and change. Merely subsisting without moving onwards, would involve their rejection into the rear, among the shadows of the past. They live in the present by breathing as the universe is breathing. Moreover all things have their own space, their means of liberty, the play-ground of their being. And the wonder of the body lies in this, that it brings man into the whole order of the world, without surprise, because with full preparation. If he is to be subject to day and night, there is day and night already written upon his members; half his moments are a rest, even when work and thought are in their fullest power; his aims and desires have their gay, fresh morning, their high-flown noon, dubious twilight, meditative evening, and night of cessation or repose. If he is to ride in the train of the fourfold seasons, the reins which direct the procession pass through his own soul and system; he finds that there is a season for all things human; that the body has its spring, of refining delight and happiness; its good summer; its fruitful autumn; its winter, whether of needful rest, or unhappy torpidity; and this, on the minute scale of hours as well as in the circle of the threescore years and ten. If he is to live in the revolutions of his own societies, his mind and body are still at home, for they themselves are nothing but revolutions. And if he is to die—to expire at last—does he not die in his atoms, and expire many times every minute during his longest life; and carelessly lets go his breath after each inspiration, secure that the outward and inward powers are ordered to revive him from the fast embrace of this mimic death. So he is a genuine part of the series of nature; a true heir of time; a life depending on variety; a moving accident of progress; a being of alternate cycles; one to whom nothing is alien that happens or can happen in the wide creation. This is owing to his body first, and subsequently to his mind. The lungs are the divine provision which introduces and accommodates him to the world of change.

Let us then end by translating the lungs into thought and humanity. Every principle has a first name by which we lisp it in material representations. First it is an individual thing; next a being with relations; and at last it figures among the grand ends of existence. The world and man are the same principles, translated as they rise from the ground, into other atmospheres, or into more and more universal languages. Form in the lowest degree, means life in the second, and love in the third, and intellect in the highest; terms and things which are very diverse, and yet but one principle, full of resources, and showing its face through different windows of the universe. The soul, an inhabitant of all heights and climates, addresses the tongue of each to the creatures of the same; and one word is a brain, another the lungs, and so on through the hieroglyphical polyglot of the body. Every syllable there has its mission, to make mind, to support mind, and to alter it. Good is the interpreter of the whole, and truth is the interpretation.

What *now* are the lungs? They are a yawning hollow in the top of the man, the sides of which cavern are alive. The world enters them bodily. Tendency to vacuum, which nature abhors, is their spirit. They draw us out into more than we are, and we shrink back as nearly as we can into our old dimensions. Reluctant vegetation ceases as they open, and life is born, crying. They are an engine added to the body, under whose draw it becomes progressive, and every space therein is enlarged, to take in more, and to live more. They pull open all the solids, that all the fluids may enter them; and strain every nerve and blood-vessel into fresh activities and virtues. They give room and airiness to the inward parts, and set them to work in a daily larger sphere. Now these, in altered phrase, are also the functions of the understanding mind, which as it is opened, shows and causes new wants, and new wills and ways. Our wants are so many threatened *vacua*, according to the form of which, we open to the pressure of the truth, whether of nature or of heaven. The necessity to be more and better than we are, the divine dissatisfaction in which we live and move, is the germ of understanding; it is the want which can never be done, but is to breathe and ponder on through incessant ages. The understanding, like the lungs, is no wind, or shadow, but the substantive power and voice of all our wants, calling us away to larger lives and finer occupations. It is the consciousness of our position, and of how to change it; the solemn claim of the future on the present; the beckoning of the universe to the atom, to come up among the stars.

This is the *root* of the lungs, this is the mind which they carry, but mathematically speaking, what are the powers? Millions of vesicles make one lung; greater millions of lungs figured afresh make one humanity. Here they are *want* again; *first,* physical wants, necessities, the iron understandings which are the mothers of inventions; the looms and hammers which must ever ply, or we lapse into nakedness and starvation. *Second,* and based upon the first, all the necessities of public progress which strain and distress the time; domestic, political, social, and spiritual understandings; which show us, in bold imagination, ideal commonwealths, built all of white justice, and bid us strive, although we die, to reach them; Utopian societies, lovely and reciprocal, peaceful in the length of a redder sunshine, in plains beyond our travels' strength; our own and the world's infancy, painted with agonizing truth upon the stormy skies of manhood, or the dark cope of age; and with no desponding voice command us to be born again. These are signs and warnings, portents, or promises, in every understanding; there is no speech or language where their voice is not heard; attracting, commanding, or threatening, to one and the same intent; and airy or cloudy though they be, they stand in the breath of the eternal.

CHAPTER III.

ASSIMILATION AND ITS ORGANS.

ORGANIZED beings grow from, and subsist upon, matter which is extraneous to themselves, and the fitting of this matter for the purposes of their lives, constitutes the process of digestion in a wide sense. In the human body, the immediate end of digestion is the blood, which is the fountain or beginning of the whole solid fabric. The food, on the other hand, is the matter to be digested—to be converted into blood. Looking, then, at the food as the one end, and at the blood as the other, and noting the marked difference between the two, it may be anticipated that a long chain of means is necessary before the one can become the other. Where the diversity is great between any two things, their assimilation must be proportionately gradual, and their reconcilement gently successive; and the passage which either of them describes on its way to the other, will probably have many joints, and lead about through strange turnings to an unexpected union. Such is the case with the food itself, and especially with the portions of it that are introduced as chyle into the sanguineous current. Digestion, therefore, and its organs, present us with a new element in our studies of organization; with the element of *a series* pervading the body; in this case, a series terminating in the blood, and recommencing from the blood; and, as a consequence of the powers of series or gradualness, with the notion of assimilation, or the likening of one thing to another through a succession of changes occurring in a certain order, and exhibiting a play of harmonious differences and increasing likenesses running without a break through the whole. Let us, then, bear in mind, for the present, that successive series, and successive assimilation, are the meaning of the digestive act, and that these apply to the digestive organs, as well as to the digestive objects or the alimentary substances which we consume.

Let us also remark in the preliminaries, that if the substances of the external world were not inherently adapted for our sustenance, the possibility of any series of changes harmonizing them with our physical system, would be cut off; so that in the very fact of digestion we find a co-ordained fitness between man and nature. This we detect in all things the more and the better we look for it. It is a well of truth to draw upon in the sciences, and especially in human physiology.

We have said that organizations depend for growth and maintenance upon materials extraneous to themselves, and this is equally true of the vegetable as of the animal kingdom. There is, however, a difference between the methods of nutrition in the two cases. The plant or tree is fixed by its roots into the soil, and imbibes at their extremities the nutritious juices; it is imprisoned in a particular spot, and its supplies depend upon the fertility of a very limited area; although by its leaves it extends into the movable atmosphere, and at the summit of its powers begins to take advantage of the new principle of locomotion. In the animal, on the other hand, motion is the pedestal of life; the pillars which support it are engines of movement, and the ground under it is fluid; its roots also are turned inwards, collected and associated, and constitute repositories or stomachs, into which alimentary substances are carried by proper animal actions. The range and freedom of animal existence immensely excel the strait security of the lower nature; the precarious income of life is far better than the small certainties of vegetation. Thus, while the stately tree subsists on a few square yards of earth, the animal which it shades, and the bird that lodges in its branches, choose their food from wide districts, and are only confined by the barriers of nature, as the shores of the ocean, or the limits of the climate. Moreover, the vegetable, with few exceptions, draws its sap from the underground—from the dark scurf of the mineral kingdom; whereas, the animal takes its nutrient juices from among the children of air, light, and motion; from the succulent tops and fruits of nature; from the results of an elaborate previous digestion in the bosom of the earth, the plant, or even the animal frame itself.

Much more are these superiorities true of man physically, or rather of the physical man considered as a member of society; for apart from society man is not a man, but the most destitute of animals; an animal without instincts; with a germ of reason never to be expanded; with wants never to be known to himself—never to serve as ends of action: his intellect, a sad surmise of a false destiny and lost estate; his jaws, hands and limbs embruted and maimed in the attempt to awaken his dead mother, nature, and in contesting his mights with his fellow-creatures, the beasts. Man, however, as in this late age we are to observe him—man as the unit of society, is infinitely more locomotive than any animal in its natural condition, and as all men are travelers either by proxy or in person, each as a centre can draw his supplies from all. The human home has one universal season, and one universal climate. The produce of every zone and month is for the board where toil is compensated and industry refreshed. For man alone, the universal animal, can wield the powers of fire, the universal element, whereby seasons, latitudes, and altitudes are levelled into one genial temperature; nay, whereby every spot may in time bear its harvest of men, and contribute its proper merchandise to even the poorest brother of the social table. Man alone can command the architecture that will hold the domestic hearth, and on the inevitable model of his own frame, build a house in which he can use, and yet shut out, the universe and its atmospheres. And man alone, that is to say, the social man alone, can want, and duly conceive and invent, that which is digestion going forth into nature as a creative art, namely, cookery, which by recondite processes of division and combination, by cunning varieties of shape, by the insinuation of subtle flavors; by tincturings with precious spice as with vegetable flames; by fluids extracted and added again, absorbed, dissolving and surrounding; by the discovery and cementing of new amities between different substances, provinces and kingdoms of nature; by the old truth of wine, and the reasonable order of service; in short, by the superior unity which it produces in the eatable world; also by a new birth of feelings, properly termed convivial, which run between food and friendship, and make eating festive; all through the conjunction of our Promethean with our culinary fire; raises up new powers and species of food to the human frame; and indeed performs by machinery a part of the work of assimilation; enriching the sense of taste with a world of profound objects: and making it the refined participator, percipient and stimulus of the most exquisite operations of digestion.

Man then, as the universal eater, enters from his own faculties into the natural viands, and gives them a social form, and thereby a thousand new aromas answering to as many possible tastes in his wonderful constitution; and therefore his food is as different from that of animals in quality, as it is plainly different in quantity and resource. How wise should not reason become in comparison with instinct, in order to our making a right use of so vast an apparatus of nutrition! Is it surprising if the prodigy of human digestion too often sinks into a perverse development, or if diseases that happen never to simple animals, are engendered in the course of the indefinite appetites of man?

A controversy that may one day be of importance, and whose data seem coëval with history, requires a passing mention while we are speaking of human food. It has been held by many individuals, and even by sects, that vegetable substances are our natural and proper aliment, and that our taste for the flesh of animals is an acquired and a morbid appetite, the gratification of which unmans us in our better part, aggravates whatever is low and fierce in our characters, and discourages our highest and gentlest affections, and our calmest reasons. As to what may be natural to man, the argument is suspect. An old writer has pithily remarked, that "many things which would be preternatural in a natural state, are natural enough in the preternatural state in which we live at present." Human nature indeed is always changing by its own act and deed—by its own choice of change; and no change in which it concurs is to it artificial, but it remains human *nature* still. The career of mankind is a line and chain of new human natures, and nothing is so natural to us now as artifice itself. For the rest, experienced anatomists and physiologists, reasoning from the teeth, and from the comparative properties of the intestinal tube, its length, and so forth, are confident that the human being is omnivorous, and they have the historical and geographical fact, if not the right upon their side. There seems to be a series of aliments required by the different races from the equator to the pole; the vegetable predominating at the equator, and running by a lessening scale down the sides of the globe; the animal commencing, if I may so speak, at the polar end, and likewise running by a lessening scale, modulated according to climates, towards its minimum in the torrid zones. Thus we have the highly baked white meats of India passing through long gradations to the raw red flesh and blubber* of the Arctic regions; and again by a curious inversion, the highly flavored vegetable dishes of India decline in the same manner towards the insipid vegetable cookery of the north. In the tropics, taste is the great allurement, and cookery there, as in China and Hindostan, is in its most varied perfection; in the frigid zones, hunger is the predominating sense, and readiness and quantity satisfy it best. As measured by the sense of taste, the hot countries are the tongue of the world, delicate, liguriating, fastidious, in which hunger is near to thirst; the other regions are the successive portions of the alimentary canal, ravenous and void at last; little appreciating quality, but loving impletion. The tropical man lives upon the sun, which gives him its warmth, and he

* Although fatty substances seem to be taken in cold climates upon the principle of supplying fuel to the animal fire, yet oils or vegetable fats are consumed in large quantities in warmer countries, and especially in the summer season. The diet in Italy, for example, is largely mixed with oil. The frigid zones provide animal fats in abundance, the torrid yield vegetable fats, rich oily nuts, &c.; and undoubtedly the staple of the soil is a kind of measure of the diet of the inhabitants. In cold latitudes fat makes heat, and keeps out cold, for a well-lined skin is an important ingredient of warmth, and is the basis of warm clothing. But in hot parts oily aliment is taken upon a different principle. It is demanded in the latter case as a corrective to the watery, pungent, acid and cold things which are so refreshing to the frame. On this principle it is used with salads, cucumbers, &c. &c. It acts mechanically or sensationally, communicating its own smoothness to the otherwise irritated stomach and intestines; or giving a soothing organic feeling to the parts, which is communicated to the nervous centres. All foods have this contagion in addition to their other effects; they propagate their own *feel* to the body. The pleasures of taste depend partly on this *feel*, as well as upon proper gustation; thus the crunch of nuts and the smoothness of cream are agreeable apart from the taste. We cannot but think that this demulcent use of oil is important in a medical point of view, in spasmodic actions of the intestines. It is said to have been remarkably successful in Spain in the treatment of cholera, which involves intestinal insurrection of the greatest intensity.

requires to be strongly attracted to material food; while at the extremities of the planet food is the gross fuel of life, which is constantly consumed in large quantity, and must be constantly supplied.

What is true of the world at large holds equally of its countries, which partake of the roundness of time in the whirl of the fluent seasons. Summer is our torrid zone, and winter, our frigid; and we feed as Esquimaux in our nocturnal solstice, and are as abstemious Hindoos in our melting dog-days. The planetary dinner-table has its various latitudes and longitudes, and plant and animal and mineral and wine are grown around it and set upon it, according to the map of taste in the spherical appetite of our race. There is the great ecliptic of thirst, with its lesser circles, the path and parallels of the burning sun; and there are the poles of hunger, also with circles, terminating in the equator of languid appetite and easy satiety. For hunger is the child of cold and night, and comes upwards from the all-swallowing ground; but thirst descends from above, and is born of the solar rays. The fluid and the solid are of inverse genealogies, and different centres, and their primordial wants are as much opposed as their sources. Thirst lives in the throat, at the summit of the alimentary tube; substantial hunger low down; and the two run upwards and downwards, intermingling their desires, are happily blended in the stomach, and each is lost in each at the extremes. And so the man is solid, or fluid, which you please; the blood so liquid in its vessels, becomes bone in the bones, and flesh in the muscles; the human form looked at from within, is an ever-changing fountain; seen from without it is one steady crystal, congealed and unmoving, though rolling swiftly still along the line of years.

Let it however be observed, that hunger and thirst are strong terms, and the things themselves are too feverish provocations for civilized man. They are incompatible with the sense of taste in its epicureanism, and their gratification is of a very bodily order. The savage man, like a boa constrictor, would swallow his animals whole, if his gullet would let him. This is to cheat the taste with unmanageable objects, as though we should give an estate to a child. On the other hand civilization, house-building, warm apartments and kitchen fires, well-stored larders, and especially exemption from rude toil, abolish these extreme caricatures; and keeping appetite down to a middling level by the rote of meals, and thus taking away the incentives to ravenous haste, they allow the mind to tutor and variegate the tongue, and to substitute the harmonies and melodies of deliberate gustation for such unseemly bolting. Under this direction, hunger becomes polite; a long-drawn, many-colored taste; the tongue, like a skillful instrument, holds its notes: and thirst, redeemed from drowning, rises from the throat to the tongue and lips, and full of discrimination becomes the gladdening love of all delicious flavors. At the same time there is this benefit also, that we can always descend to the lower condition, and find an agreeable variety in the plainest fare.

But to recur for a moment to the vegetarians (not to do them an injustice), although we accept the testimony of the anatomists and physiologists, and the dictation of facts, as of value for the present, yet it must be admitted that it is not conclusive for the future; for in a being mutable like man, capable of improvement and of deterioration, with power to alter his mind, and therefore his brains and his body, it is difficult to say to what extent his anatomy may have conformed to his habits, good or evil. No doubt our frames have changed with the times since the world began. Existing customs and organisms are not fixed points to limit the truth, or to govern the future. A new appetite for flesh, conceived in the mind, and daily gratified, and of consequence daily strengthened, could not fail in the course of generations to mould the consumer to the desired end; and in short to make animal food natural to the human constitution, by modifying the latter. That it may fairly be called natural now, is evident, for where is the race that abstains from flesh unless either religion, or strict necessity, forbids its use? The question therefore, like other administrative questions, is one of times, and wants, and wise expediency. If by other sorts of temperance the members of society find their thoughts calmed and deepened, their senses refined, their emotions more constant, powerful and peaceful, it is hard to say to what new sublimities temperance may not aspire; what fresh interpretations it may not assume, or how it may not assault the carnivorous man. Every herb and every fruit may presently be for our meat, as in the days of our first parents. In the meantime, however, man is strictly and potently omnivorous in the nineteenth century; and we may well question whether the vegetable kingdom is at present sufficiently comprehended to supply the varied qualities of food which are necessary for our support; for how few of the said herbs and fruits continue to be eatable. We cannot banquet like the first men, until we get back the golden earth. As it is, the encroachment of drugs and poisons has driven our esculents into a penfold, and we are fortunate if from the exceptions to the kingdoms, we can furnish our table. Assuredly the woods, fields, and gardens must be more humane, which no doubt they are willing to re-become; the stomach also must put forth the hands of a more inventive agriculture, before the vegetarian crusaders can be allowed to wave their leafy flag over the city of the cooks. But by dabbling too much, or prematurely, in their limping pultaceous diet, we should become not the children, but the abortions of Paradise.

It has been suggested by Dr. Prout, that the milk which is our first aliment—an aliment certainly human, if not animal—and which is the food prepared by nature herself, is the type of all food whatever; that it contains certain alimentitious principles in combination, which are but repeated, and their combination re-attempted, in the most elaborate *cuisine*. His analysis of the components of the milk, is, into oily, saccharine, and albuminous matter; and after investigating our food and its combinations, he finds that his view regarding milk is borne out by the instinctive tastes and artificial cookery of mankind. According to this, all our meals are but aspirations to our original milk. There may however be different analyses of milk besides the partly chemical one he has mentioned, and it remains to be seen whether other views will equally comport with this reference of food to the lacteal type. We can scarcely doubt that the idea has a truth; it wears so organic an aspect, and takes its stand with such powerful innocence at the head of a science of alimentation. Moreover nature, the mighty mother, offers herself breastwise to all her little natures; she swells in landscape and undulating hill with mammary tenderness; each creation is a dug held forth to a younger creature; and milk is thus again a symbol of the food and feeding which are everywhere.

But let us pass onwards from the food which is the matter, to the fluids which are the medium, and to the organs which are the instruments, of digestion. Now there is nothing more general in animal life than the digestive apparatus, because matter is the largest, if not the greatest, fact in the material universe. Every creature which is here, must be made of something, and maintained by something, or must be landlord of itself. Every part, and every faculty, of every being inhabiting the planet, must be duly clothed and ballasted with stuff derived from the earth, or it would have no operation in the body, or upon the body, much less upon the external world. Hence the stomach is an organ of the first importance to all mortals. You may take away brain and nervous system, and leave their place to be supplied by the fluxions and imponderables of nature; you may take away the lungs, and consign their office to the circumambient lifeless atmospheres; you may abstract the heart with its blood-vessels, and commit the dull, gluey circulation to the almost mechanical and chemical laws of affinity that obtain in vegetables: and notwithstanding all this, you may still have an animal being remaining. In short, there are animals which are nothing but stomachs, but there are no animals which are nothing but brains. In the human race also the stomach is of the same paramount importance; its existence, and due impletion, are the first or last conditions of the existence of the individual; they are the basis of humanity, and nothing is so sublime but it rests upon them, and must perish out of this world if they cease, and otherwise follow their vicissitudes. The assured feeding of the nations, is a question that involves in its settlement all other questions, and postpones sublimities until necessities are complied with. Jewelled goblets there are besides, but this earthen cup must be satisfied before the other vessels of the man can begin to be filled.

The food, consisting of matters from the animal, vegetable, and mineral kingdoms, is at first received by the lips, which present to it a more and more delicately sensitive surface from their beginning at the external skin towards the roots of the teeth; and especially where the sight is not employed, it is apprehended by a sense of touch so gradually fine, as if touch itself were passing into taste, as the food is passing to the tongue. It is next transmitted to the teeth arranged in manifest row and series. What is received by the front or incisor teeth, is delicately treated, or minced, and the fine things set free go at once to the tip of the tongue, which is waiting close behind to receive them. At the sides of the mouth the molar teeth stand ready to grind their portion, and their milling surfaces become more and more severe and powerful from before to behind; *they* also are a distinct series of structures, and involve a series of operations; for every new tool has a new action. Intermediate between the mincing and grinding teeth are the canine, so large in carnivorous animals, which both thrust and cut the food, and submit it to the molar action. It is good to believe, that the juices separated by the different teeth go primarily to the part of the tongue alongside those teeth, and are especially to the taste of that part; for the neighborliness of the body is not useless but functional.

All this time the food has not been merely reduced, and its juices set free, but animalized also; and this, from the very porches of the mouth. Even before the first nutrient fluids are expressed from it, a living fluid, the saliva, has come out of the body to receive them. There is a series of salivary glands running from the lips throughout the tube. The sight of the food, the action of mastication, the pleasantness of the morsels, and the suctorial power of the tongue, draw out the saliva from the respective glands in the mouth, as it is wanted to moisten the organs, and to penetrate and dissolve the food. Especially do the emotions call out the attentive saliva, and the mouth waters with appetency. Sight and fancy wherewith it is full, and which it obeys in the first place, have fed it with anticipatory fire, and schooled it for its duties. Moreover the saliva is no menial, but the immediate product of the blood proceeding to the head and the organs of the senses, and if not summoned into the mouth, goes its course towards the brain; and being thus descended from the blood, and akin to the blood, it is clearly an excellent medium between the blood and the new food, whose finest portions, under its guidance, are themselves to be educated into blood. The expressed food is the new guest which is to be inaugurated into the duties of the household; the blood is the royal table itself; the saliva is the commissioned master of the ordinances, who busies himself to instruct the food in the laws of the place, and in the conditions of its hospitality.

The saliva, like everything else in this system, exhibits a play of varieties. In the mouth this is manifest to all. Do we not feel that in the front of the mouth the saliva is thin and trickling; in the back part, as it approaches the throat, more and more viscid and tenacious? How this difference in thickening is produced, we need not now inquire, but may simply note the fact of an orderly series existing even in this small compass.

The food already upon the tongue, carved by the front teeth, pierced by the middle teeth, and ground by the back teeth; also saturated by the saliva; affords the sense of taste to the sensorial papillæ of the tongue, which tongue-like themselves, protrude and exert themselves to enjoy it; and this nimble member seeks and suffers the pleasures of the time with infinite agitation and emotion. What is the sense of taste? Is it merely that abstract thing, an influence, made of nothing but metaphysical motions? Do tasty substances knock at the door of the organ, and leave their names, without going in themselves? It may be difficult to demonstrate by anatomy the absorption of juices by the tongue, yet facts show that such absorption takes place. The sudden recruiting of bodily and nervous power by matters taken into the system; the effect produced by wine and other fluids when only held in the mouth; the fact that the tongue is the beginning of the great absorbent organs, and must, therefore, in consistency begin the absorption; also the variously-formed prominences or papillæ upon the tongue, which are poorly accounted for by physiologists, because they overlook the right of the tongue to taste in that real sense which *tasting* implies; moreover, the certainty that there are in creation no abstract influences or impressions unaccompanied by streams of tangible stuff: all these reasons establish, that the tongue enjoys an antepast of the food; drafts its best essences, recruit after recruit, into the system, in union with the finest saliva; and only sends down into the stomach the portions which are exhausted for the mouth.

But current science tells us, that there is no tasting in the tongue, and no feeding in the stomach, but that the man is nourished entirely from the lower belly—from the intestinal tube. A physiology that confesses to living so grossly, can have little enjoyment of refined truths—small sympathy with the good things of the world. It is the very servility of the senses, to make the man dine in his own kitchen; to bring the repast laboriously down from the drawing-room of the mouth and the saloon of the stomach, to the place of sanded floors and wooden trenchers. It is, indeed, a matter of taste. For the reasons stated, we cling to the upper story, and the other opinion.

As soon as the exhaustive feast of the tongue is ended, the food is prepared for its next destination, and an act of swallowing takes place; the tongue rolls the morsel back into the pharynx, a chamber intermediate between the mouth and the gullet; the pharynx, successively contracting, moves it down into the latter; and this, taking up the contraction, forwards the ball from point to point, until it reaches the upper or cardiac orifice of the stomach. The viscid saliva of the back of the mouth, and, moreover, all the fluids supplied, or drank, either sheathe the food, or lubricate the passage, and make the transmission easy, and almost spontaneous. Arrived at the entrance to the stomach, which is shut by a circular muscle, the food opens the latter, and passes into the great cavity of the organ. The passage from the mouth to the stomach is accomplished with rapidity, yet perchance there is a graduated absorption which takes place more swiftly still, and leaves not even an apparent break in the absorbent function following the absorbent structure.

At the back of the mouth, under ordinary circumstances, the food becomes lost to our consciousness; but as in the tongue we have found manifest taste with invisible feeding or absorption, so here the matter is reversed, and we have henceforth to contemplate manifest absorption with a latent sense of taste. Generalizing the functions of the elongated parts we are enumerating, we may consider that the function of taste or quality reigns throughout it, equally with the function of eating or quantity; that to the alimentary organs, from beginning to end, there can be no substance without taste; and, on the other hand, no taste without substance.

In the stomach, judging by what there is done, what a scene we are about to enter! What a palatial kitchen, and more than monasterial refectory! The sipping of aromatic nectar, the brief and elegant repast of that Apicius, the tongue, are supplanted at this lower board by eating and drinking in downright earnest. What a profusion of covers is made and laid! What a variety of solvents, sauces, and condiments, both springing up at call from the blood, and raining down from the mouth, into the natural patines of the meats! What a quenching of desires—what an end and goal of the world is here! No wonder; for the stomach sits for four or five assiduous hours at the same meal that the dainty tongue will dispatch in a twentieth portion of the time. For the stomach is bound to supply the extended body, while the tongue wafts only faëry gifts to the close and spiritual brain.

The stomach, anatomically speaking, is a vaulted chamber consisting of three walls or coats common to the whole tube. Its inner wall is made up of little compartments placed side by side, and which open into its cavity, and differ in construction in different parts. These are beset by a mesh of the smallest blood-vessels. The inside of the stomach forms a kind of honeycomb surface crowded with little mouths, and when the organ is roused, red and turgid with blood. At this time also numerous little points or papillæ waken up upon the membrane, and bring forth a dissolvent liquid termed the gastric juice. These honeycomb structures are the stomachs of the stomach; the natural components of the organ.

The muscular coat is the middle wall of the stomach. Its fibres run circularly, spirally, vortically, in all the writhings of stomachic taste. They are the moving arms of the stomach, which enable it to lay hold of the food, and to work and agitate it. The principle of the mouth, the jaws, and the fingers, here ensouls a sheet of membrane, which extemporizes shapes of every required variety. Thus this seemingly simple organ erects itself into a thousand different apartments, in each of which a peculiar digestion, and a peculiar assimilation, is proceeding. This encameration, insured by motion in the human stomach, is the fixed condition of the part in some animals, as the camel, the ox, &c., in which one of the stomachs is formed of deep pouches, lodging the food. The body-kitchen of man and carnivorous animals is levelled away when the stomach's meal is done; while the kitchen of creatures which chew the cud, is a permanent building, in order that eating may consume the bulk of their waking hours.

So much for the muscular coat of this series, whose activity constantly builds the plastic walls into new and ever-varying chambers, in which the functions of the place are carried on; for every passing wrinkle on the surface is a fresh thought and artifice of digestion and assimilation.

But as we are not on a medical quest, let us henceforth rather pursue the food that is assimilated, and the organs which take, and convey it, than the portion which is rejected, or the extensive system of public works that underlies the thoroughfares of the human city. Respecting this department it will be sufficient to note the principle, that series governs here as elsewhere; that what is renounced as useless by the first cavity becomes the especial food of the second; that the leavings of the second are the table of the third; and finally that what is of no use to the animal kingdom becomes the support of the vegetable world. Thus the purposes of the aliment are drawn forth one by one in regular order, and with a thrift that can go no further, the last use returns to the first, in the fertility of the earth, in the precious yield of the plant, and in the support of the animal, which reverts to the human body by this circle of renunciation and absorption.

Let us now note the distinct divisions or joints of the alimentary tube, which prove that a series of different digestions and assimilations are performed therein. Observe the mouth, which is guarded by the lips, and then by the teeth, against any unmeet intrusion of the food. This is the type of what takes place five times over in the lower parts of the series. Thus the entrance to the pharynx and gullet is analogously guarded. The entrance to the stomach is secured by strong muscular lips, which are only opened by the muscular wave flowing down the œsophagus, and playing upon the wedge of food. The exit from the stomach is a similar lip or ring, which is likewise opened by the successive action of the stomach and the aliment. And so forth. The cause therefore which determines the passage of the food from cavity to cavity, is that the first cavity has assimilated its portion, and digested its portion, and having no longer any affinity for the latter, the walls contract against it, and

13

send it down stairs, where it is received or wanted by the next chamber; which in like manner exhausts, digests, and detrudes it. In this wise the successive digestion by the agency of the successive muscular coats, is the means of sending the aliment from stage to stage of its useful journey.

In health however there is no intrusion from above downwards, which is not commanded by an equal attraction from beneath. The food lies upon the plate, and the man and the mouth voluntarily take it. Eating and drinking are *quasi* voluntary all the way down as in this first instance at the top. Whatever violence of supply is superadded, is the answer to a greater urgency of demand; for the intestines become bigger as they proceed, and more and more ravenous. The lower belly, hollow by nature, calls to the upper with loud Halloos. But the consequence of any force unsolicited from below, is seen in a lively image when food is thrust down our throats without our consent, in which case we cannot swallow it. And just as jealous as the man, is the stomach, and are the intestines, of having anything thrust down their throats. So food to which we are averse, rouses the hostility of the whole line, and is successively unswallowed by part after part; for swallowing in its finer meanings, blends with digestion. Many facts prove this case, and also that a series of instincts or voices are located in the tube, which command or refuse particular aliments.

But the series of parts has a series of powers whose existence is written upon the forms of the organs. The velvety palpation of the food by the lips, its reduction by the teeth, the warm squeeze of parts and juices by the tongue, are again the exemplars of all the forces and actions which succeed them. There is however nothing so violently physical in any other part of the tube as in the mouth, but motion, warmth, long delay, incessant working, and the semi-chemical forces of certain fluids, perform changes which are in reality far more sweeping, without the assistance of any rude agency. Nevertheless from the stomach the muscular powers and opposing surfaces become more forcible and urgent, and thrash out the last grain from the chaff with somewhat of indignant exasperation.

As the point of this series of muscular forces there is a corresponding series of solvent fluids. These are the wheels of the digestive chariot, carrying its passengers either inwards as food, or outwards as drainage, for food and drainage are the same problem applied to two different worlds, the animal and the vegetable, and constipation whether in a town or an individual, whether in gut or in sewer, is a limit to the circle of nutrition. Now of all the solvent and locomotive fluids, the saliva, so called from its salty properties, is the head and the type. It is the vehicle and the flux of the particles which it liberates from the food. These it dilutes, softens, and dissolves; draws forth their essences; alters them into suitable forms; sheathes the latter, and enables them to glide through the pores which carry them to their destination in the chyle or the blood. These are organic properties of the saliva, as declared by its acts. But each joint of the tube has its specific saliva, in addition to the previous fluid that descends from above. The saliva of the stomach is the gastric juice. This saliva the stomach can digest, and leave it mild, or grind it sharp, according to the materials upon which it is to act. It can vary the fluid, and suit the condiment, to its own want and palate. Moreover it can modify the heat which is disengaged, and alter the shape and intensity of its own culinary fire, fueling it well, and stirring it up, as the viands require. This the stomach chiefly effects by the opening of the saliva itself, when the vital fire comes cheerily forth, and eats up the chemical heat which is also disengaged. The saliva of the next compartment is the threefold contribution of the liver, the gall-bladder and the pancreas. This triple-headed saliva is more raging and terribly chemical than what went before it, and wastes and melts the food with gross heat and uncompromising persecution. Nor are there wanting other fluids suited in a series to all the offices of these elongated tables. The point to be kept in memory is, that there exists a commonwealth of salivas, and that the saliva of each part is derived into the next when its proper uses have been accomplished. These uses have been already enumerated in part. In the first place they consist in the business of absorption, of which they are the primary medium; in the last, in the work of rejection, of which they are at once the measure and the stimulant.

We now come to the series of matters realized or absorbed. The food of the mouth is the freest, comprising the particles which only require the moving opportunity offered by the teeth and the tongue, to quit their former attachments in the substance they belonged to, and to rise at once into the new society whither they are bound. They put off their bodies with nimble ease, and are presently at home among the particles and occupations of the human form. The grosser molecules never miss them, or know how they go in, or whither they have gone. The most living saliva, wing-footed as Mercury, spirits them in by the secret paths which it has trodden perhaps often, and of its own nature knows so well.

The food of the stomach, immeasurably more bulky, and subject to greater and longer probations, yet passes immediately into the hungry blood: the little veins which stand open-throated on every portion of the distinctly ventriculated surface, carry crowds of these slower individuals into the bosom of the abdominal circulation. The alimentary mass reduced into chyme in the stomach, yields its reluctant vintage directly to the blood. The possibility of an immediate reception of the food by the blood, appears for the most part to end with the stomach; the way for the next order of particles is long and difficult, and the preparation corresponding. Beyond the stomach a lower order of vessels than the veins receives the food, namely, the lacteals; there is an intermediate school between the food and the blood, namely, the chyle. These lacteals arising from the alimentary tube, but decreasing in number in its lower parts, where there is less of the milky chyle to absorb, converge from the intestines to the receptacle of the chyle, a small reservoir seated in front of the lower vertebræ; this reservoir is then continued upwards in a fine pipe called the thoracic duct, and runs all the way to the left side of the neck, where it runs into the fork of a great vein that pours its blood directly into the right side of the heart. A long passage for the chyle, contrasted with the short cut which the higher portions of the food enjoy, to their desired haven in the blood! We infer from this, the comparative imperfection of the chyle, vexed as it is with such abundant trials, and ultimately let in through the intervention of a peculiar saliva, termed the lymph. This lymph is brought by the lymphatics from all parts of the frame to the same receptacle and duct: it is the old spirit of the blood serving to inaugurate the new body, and thus is the last of the salivas, which digests and introduces the chyle itself as the other salivary fluids digest and alter the food.

As a corollary of this we may infer, that delicious, pleasant and agreeable foods contain a native series of offerings to our intestine wants. Fruits, aromatic and luscious, hold their delights the loosest of all, and give them away at the first solicitation. Their nectars claim instant kindred with the tongue and the oral saliva. Nature has cooked them, and they need no mixture, nor artificial fire: the grape and the pineapple are a sauce unto themselves, and are baked and roasted and boiled in the sunlight. They are at the top of their life at the table; their niceness is not foreign, nor does their beauty depend upon disguise. By feeding the eyes with bloom and loveliness, they call forth a chaster saliva into the mouth to welcome and introduce them; different from the carnal gush which savory meats engender. They are flasks of the spiritual blood of the earth, of the kith of our tree of life, and nearer to it than aught besides, unless it be the mother's milk. The term *fruit* implies that which is for use, or which has attained its own object, and seeks its place in another system. Fruits therefore hang before our mouths, and tempt us by nature's sweetest wiles; as it were the nipples of her bosom, which still run pure with rills of the milk of her ancient kindness. They belong essentially to mouth digestion, which is mere melting.

Meats belong to the lower man — to the blood and the chyle. Animal life has diverted them to itself, and the spirit of the beast has to be exorcised ere they can enter the human body. They demand a long and severe process of reformation and excretion; artful fire and elaborate treatment before their preoccupation is put aside. They are proper to the belly, and are the taskmasters of digestion. Vegetables are, in these respects, intermediate between meats and fruits; and milk, eggs, and the like products, which are the fruits that animals yield us, are also intermediates put forth from the animal side. As we said before, the modern man requires all these viands to supply his different natures, seated one after another adown the long hungers of his entrails.

Let us then assimilate this idea in passing onwards, that every step of the alimentary tube has its distinct food; and let us be sure that these great varieties are necessary to the maintenance of the human blood, that the spirit may come down into the world thereby, through its own mysterious process of avatars and incarnations.

But in this extensive system, for aught that we have hitherto said, the parts might be deficient in unity, and want active combination. Have we then left out no fact in our treatment of these straggling organs? Undoubtedly, for we have said nothing of the motion that we feel pervading the belly, and whose use should be as palpable as its existence: we have taken no account of the alternate movements of the lungs, "which press and actuate the inferior viscera as the atmospheres press and actuate the earth." Now the grand effects of the body depend upon this actuation of the lungs: it is the source of mechanical power to the whole of the organs. This is manifest in the alimentary tube. When portions of this tube are exposed to sight, or removed from the rest, they still exhibit a creeping or peristaltic motion. This is their last individual effort, when they have lost the support of the neighboring organs, and the fulcrum of the skin and muscles; and it is vague and indeterminate. They still indeed move for a brief space, though no lungs draw them, because everything in the body is constructed to do of itself, as far as may be, whatever other parts do for it; so little is man's work physical, and so helpful are all things to all things. But this individual life, though it exists, cannot of itself maintain itself. Before, however, the machine is broken into, the movement of these parts is rhythmical and exact, chiming with the motions of the breath. For the gullet runs down through the lungs, and necessarily obeys their attractions, and as necessarily rolls onward its motion to the subsequent structures. As for the stomach and abdominal organs, we know by touch that they are subject to reciprocal breathings coming down from above. The same is the case with the lacteals and the thoracic duct, which latter runs up, as the gullet down, through the active space of the chest—the air-pump of the frame. Thus the digestive series, from the mouth of the lips to the mouth of the subclavian vein, and including the myriads of mouths between the two, eats and drinks (for the two are no longer two in assimilation) in alternate moments, as the lungs draw in their food, the air; and in the next moment swallows along the whole line; then eats again; and so forth. It is in this way that the lungs are continually drawing up the food to the blood, which is seconded by every point of the tube, successively contracting and expelling it; somewhat as the arteries contract upon the blood, and produce the pulse. And by the same agency, through the general framework of the body, the lungs harmonize and combine these creeping parts and their motions, peristaltic and vermicular, into one coördinate system.

We have seen above that the work of assimilation terminates in the blood, in the introduction of new elements into the living circle, and we have to note that the blood itself, so soon as it is formed, begins to undergo the process of regeneration, or to be eaten by higher purposes and powers, and charged with impurities that must be put aside. Thus as there is a scale of stages leading upwards from the food to the blood, there is a parallel scale leading downwards from the blood, and consisting of its excretions, or the various salivas, which come out of the system near the spots where the new materials are entering it, and uniting with the latter, as the similar old with the similar young, lead them to a certain distance through the first schools that conduct them towards their destination in the organs.

What is the end in view in all this digestion and assimilation? In one sense, the formation of the blood; but let us advance beyond this answer, and inquire after the human end for which we instinctively, and our bodies organically, appropriate the substances of the external universe. The end is, that we may live in the world by means of a body derived from the world, and representing the world. Each place has its laws, and when we are at Rome we must do as the Romans. To be full freedmen of nature, we require to be allied to her by our constitution; to marry into all her royal families; and to take a body from every kingdom, in order that we may enter, inhabit, appreciate and understand it. The sense of taste, and that alimentary series which we have been considering, afford us our material embodiment, by which we are brought into fellowship with the mineral, vegetable and animal kingdoms; the sense of smell and the pulmonary series draw down our aërial food, by which we gain the freedom of the atmosphere; finally, the brain is the governor of assimilation, and compasses on our behalf the ethers and spaces of the mundane system, introducing us to the illimitable spheres. The human body, as Pan's last flock, crops every nature that it touches. The highest organization also coincides in this with the lowest, that it is an omnivorous stomach, which has the senses and faculties of human kind domesticated as tastes in its comprehensive cellwork. And so by a just divination, the word *taste* is synonymous with whatever is refined and of good fashion in the objects of our knowledge.

And in the process of assimilation, and the parallel process of nutrition, we are to consider that day by day, and year by year, the spiritual essence is assiduous to constitute the body in its likeness; this being the reason of the vast series of materials supplied by the earth and the atmospheres; of the various influences brought to bear; of the manifold changes and purifications that occur momentaneously in the body itself; of the stupendous chemistry and growth carried on there:—all with a view to bring forth, and support, a piece of nature, which shall availably correspond to the soul, to which, however, nothing in the world can more than conveniently approximate. This ever-nearing, ever-distant correspondency of the soul, is the human body, whose form or plan is spiritual, or from the spiritual, and only the matter of its seeming is material.

It is wonderful indeed how this correspondence is maintained during our quickest variations. We sleep and wrestle, work and play, weep and laugh, with the same body, which answers equally to all these states. Every passion, as a human extense, uses the person of this dramatic frame, in which all that we do, or can do, is contemplated from the first. This is notorious from the countenance, which assimilates with the momentary spirit. The world also and circumstance are as flexible and correspondent: according to our wants, they feed us with good or evil, and bend about to the whole exigency and gladiatorship of the spirit. The soul distils the food not into chemicals and gases, but incarnations of its moods. Anger sops in the same dish with love, and both are confirmed at the one supper. This rapid ensoulment of the alimental chain is especially marked in the saliva, which as we noticed before is a running commissary between the mind and the food. The soul glitters down upon its salts, and smites them with the reigning love beam. They become as different at different times and in different persons, as the billing of the dove from the bite of the rattlesnake, or the sweetest milk from the deadliest poison. There is saliva full of care and sourness, which eats, not the food, but the stomach itself. There is saliva charged with contempt, which is spit upon meanness, and carries the badge from soul to soul where it lights. There is the saliva of disgust, which is vomited from the loathing blood, and avenges our disgust upon the ground. There is the spittle of self-complacency, elicited by the happy tongue, and too good not to be swallowed. There is the saliva of luxury, which runs greedily after the meats, and loses its life of immoderateness. There is the saliva of rage, which foams violently forth upon the beard, and that of haste and hurry, which froths and sputters. There is the saliva of grief, hard to get down, and full of choking. There is the mouth of fear, from which the saliva is frightened, and the dry tongue cleaves to the palate. There are lovers' kisses, in which soul is fluid unto soul. In the lower animals the saliva becomes actively poisonous, sometimes by nature, sometimes by vehemence and passion. We might extend the instances to every well-marked affection, and point out the sympathy of the saliva with each. It is, however, sufficient to observe, that this medium between the food and the blood is the abode of an ever-varying life, according to which it summons from the food the pabulum of that very life, making the new elements of the body into vehicles and implements of the inclinations. Were it not for this, every meal would disarrange the harmony between the body and the soul, by furnishing materials that could not become incorporated.*

Recurring for a moment to the act of spitting, it teaches us that the casting out of evil spirits is a branch of bodily truths. The excretions are emotional, *i. e.*, they are excretions of the emotions themselves, and as they pass off, they help to rid the body of disturbances. Speech is an excretion from an overburdened heart;

* We postpone speaking of the psychology of the abdominal organs until we treat of the heart, when we shall offer some general remarks on the life and feelings of the viscera. In anticipation we assure the reader, that these parts belong to the soul as much as the brain itself.

we tell our sorrows and halve them. Tears abate the grief which they carry and signify, for they roll it out. The excretion of the bile affords a similar vehicle and relief for other strongly troublous states. The saliva, full of melted feelings, is very busy in these clearances. Filthy odors assailing the sense and striking foul upon the mind, are rejected and forgotten by the satisfaction of spitting. It is not merely the material but the animal sense, not only odor but idea, that is thus cast out: were it the former alone, the nose ought to be blown and not the mouth cleared: whereas stenches cause the holding of the nose: the fingers nip it to take no more, while the downright mouth acquits us of that which is already. The motion concerned in rejection goes some way towards the riddance of the mind; hence inner annoyances of various kinds cause shocks, muscular twitches, efforts, throes, and in short bodily struggles, by which their effect is not only symbolized, but in part thrown away. Nature begins the cure of diseases in their own first results upon the instincts. It is where no excretion or movement occur that trouble works with terrible energy upon the vitals: the dry eye of great grief is nearly insane; its motionless attitude is the frost of catalepsy. But the subject of "casting out" requires a volume. As a practice it is resorted to unconsciously every day in medicine, which stimulates the excretions. We are most familiar with its *rationale* in our Saviour's casting out of the devils, and their entrance into the herd of swine, as it were the carnal salivary medium which conducted Satan to Satan, or spat the devils into the deep. This miracle is no isolated case, but has little miracles of a similar kind and from the same God, happening constantly in our bodies, the name of whose troubles also is legion. I do not know whether this be not the prime law of excrementitious rejection. It is morally true also, for we have the power of clearing the mind through its nutrient tube, the memory, when evil contents come in, by refusing to think of them, and rejecting them. The excretional herd is always ready, and the infernals beseech to pass into the pigs as soon as the whole man is against them, or the divine power is fully present.

We have further testimony to the ensoulment of the tube in the play of the emotions upon the appetite of hunger. Who does not know that the stomach may be hungry, and provision at hand, and yet the receipt of painful news, a tale of horror, the arising of disputes, or the suggestion of anxieties, may, in a moment, not only expunge the sensation of hunger, but produce a loathing worse than heavy satiety? The organic mind will not extend its miseries by feeding them. But when it is in peaceful outflow, respiring unhindered action, the appropriation of food which supports it is delightful. *Cæteris paribus*, peaceful workers have the best appetites, and can consume the most with benefit. Peace, indeed, is the eating-time of whatever is best in our informations; it is the spiritual world of the satisfactions of conviviality. On the other hand, the time of trial is the time of fasting, and it is only afterwards that we are again an hungered.

Let us then remember, that every permanent passion that we cherish, dines with us at every table; and that as for temporary conditions, tranquillity can eat, but strong emotions go to suspend the want of food, that they may use the time for their own work.

The subject of food subdivides itself into quantity and quality. A certain assured amount of provision—a decent minimum—is the ground of further wants. This annuls those distressing anxieties that consume the stomach, and make it the seat of care instead of exhilaration for a large portion of our fellow men. Under these circumstances the body feeds upon itself, only miserable sensations are alive, and the mind has neither leisure nor wish to pursue its own avocations. But when the first demand is satisfied, quality and variety are the next necessities to be considered. And here there is room for a new gastronomy, to instruct us under all circumstances and seasons, what nutrient matters, and what artificial compounds and alterations of these, will enable the body to carry forward happily the various works required at our hands. We know that the mind can modify the frame to almost any extent by the manner of feeding it: by the substances introduced we already produce the baser conditions, of fatness, intoxication, stupidity, ferocity; and it must be the business of a charitable science to reverse the direction, and to feed the industries and virtues with their daily bread, from among the riches in this kind which the earth is instructed to yield. There is not an emotion however retiring, not a thought however fine, not an action however skillful, but may be carried into further and more exquisite perfection by some helpmate from the social board, when humanity there enlarges its heart, and brightens its calculating cheerfulness.

With respect to diet, however, it is as various as days and moods, and in the largest sense no food as digestible or indigestible *per se*, but according to persons, times, and circumstances. Private judgment, where it exists, has full rights here; where it does not exist, the patient comes under dietetic rules. Fixed habits of diet otherwise are a prison to changeable man, and curtail his versatility in every sphere. His body ought to oscillate from the middle ordinary point to fasting on the one hand, and to conviviality on the other, in order that all his faculties may be helped by steering the appetites in the direction where their power lies. Judicious fasting especially is to the structure of the mind what held breath and expiration is to its movements; the abstraction of food corresponds to the abstraction of thought; without it there are no times of ascetic intellectual separation, and without these, an important part of man is unrepresented. Moreover, without times of fasting, the tongue is immersed and lost in its objects, whereas fasting places the whole of diet at a distance, sees it clearly and appreciates it sharply, and introduces the criticism of intellect into the gastrosophic sphere. Cooks, therefore, require these self-denials for their own business, as well as other Christians for theirs. In short, we find that fasting answers to spiritual eating, conviviality to human eating, and ordinary meals, pleasant and calm, to the regulated private senses and sensibilities; which three spheres contribute to make up the visceral man.

Nothing is more important than society for the useful pleasures of the table. "Chatted food," the proverb says, "is half digested." As a rule, the solitary man does not reap full advantage from his cheer. Setting aside that single blessedness is opposed to economy and abundance, it also leaves out convivial mirth, which otherwise pervades the body, and gives sunshine and activity to its operations. In the breaking of bread our better eyes are opened, and the truths of communion are made known to us. We all know how very much good company enhances the enjoyments of the palate; this is an open sign of the delightful influence that it exerts upon the entire work of assimilation. Ancient habitudes might be imitated in these respects. The hospitalities of other times enabled the guests to digest hard things for which their successors have no stomachs: courage and clanship and bold ambition haunted the boars' heads and smoking beeves, and horns of mead and of wine. The revelers were firmer in friendship, brighter in honor, softer in love, and stronger in battle, for the spirits which descended upon the hall. We, too, must dine as our forefathers, only the board should be newly laid to invite down the later graces of labor, justice, and peace. Our fight for to-day is not with men, but with the enemies and difficulties of all men. Our honors are the extensions of the tone of peace, disarming the world by the dispersion of offences. It is a new supper of gathered rich and poor that will refresh us in this second epoch. Nor on grounds of mere utility should the song of the bard and the replication of the jester be absent from a work in which a merry complacency is the springtide of the season, and the oil and smoothness of the whole machine.

Let us remember, therefore, in passing from the subject, that this intestine public question implies the dispensation to all of the due quantity of the proper quality of food, subject to the law of variety, and under the best circumstances, internal and external, or mental and convivial; and further that precarious modes of life, to many temperaments, are incompatible with a healthy assimilation.

We have postponed to this place the subject of wine as a part of diet, because the case of stimulants rests on human life, and not otherwise on physiological laws. Alcohol in its various forms acts specifically upon the brain and animal nature, themselves the stimulants of the other systems of the body. Teetotalism on this account takes rank with vegetarianism, as both of them tend to reduce the animal powers.

What is called "total abstinence" has claims which deserve to be admitted. The abstainer on principle is generally workful to an extraordinary degree; has his senses about him, such as they are; is equally cool and collected at all parts of the day; feels little irritability from current events, but bears and forbears well. This is while health and strength last. And if he be capable of fanaticism, or kindly speaking, faith in abstinence, he may be a strong man through life on cold water. His strength will be in proportion to his dose of faith. A batch of abstinence soldiers working in emulation against a batch accustomed to stimulants, will generally be the conquerors. New systems, and especially self-denials, have the advantage of enlisting faith, the wonder-worker. The victories of Mohammedanism were due in part to the combination of a religious faith with a faith in abstinence; a union of two powerful springs affecting the soul and the body. Torrents of passion which had been wont to vent themselves in pleasures, were suddenly stopped off, and they burst through another channel, in faith and energetic fighting. Faiths, however, wear out in many cases, and the truth of things is the ultimate level, unaffected by mortal enthusiasm.

Successful abstinence shows that stimulation is a law of existence, for an abstinence neither hereditary nor stimulating is not kept up. The most sober people have their "pocket pistols," and take their own stimulants as neat as they can get them. For there are two sides of the cellar, two decanters of spirit, the body and the soul. Take away the bodily decanter, and life itself must furnish an excitement that will be equivalent. There are other stimulants besides drink that cheat us of our senses, other drunkenness than that of the public house. Teetotalism might be drunk with its own cause, with the additional indecorum of exhibiting its disgrace in Exeter Hall.

Abstinence excludes temperance or the faculty of balance, which communicates with reason, the temperance of the upper degree. For the sake of the evil it bans both the good and the evil. It is the suicide of choice. Similarly, vowed celibacy excludes chastity, and is a knot tied in the will against both cleanness and uncleanness. This is not healing, but castration. But there are those who require these extirpations, at least with our present means of cure. "If thine eye offend thee, pluck it out and cast it from thee; if thy right hand offend thee, cut it off. It is better for thee to enter into life," &c. &c. Thence we note that total abstinence is a thing commanded, and a means entering into an ultimate plan of fulness of life. But there is an *if* in the case—"If thine eye offend thee!"

Teetotalism reasons without this *if*, and brandishes its surgical vow over temperate and intemperate alike. It goes to science and morals for corroborations. It says that intoxication is poisoning, and that poisons are like themselves in their least doses as in their greatest. There is a mistake here founded upon an etymology. Poison is one thing, and stimulus is another. Poisons destroy the structure, or subvert the functions of the body; stimuli kindle it into life and exhaust it into repose, or even death if their action be excessive. The sleep of the night is nature's recovery from the excitement of the day. The sleep of death is the spirit's recovery from the lifetime. Our machines are meant to wear out, and stimuli are the wearers. The organs of the body and mind live by stimuli, which in temperance animate, and in excess destroy them. Light is the stimulus of the eye, but its intensity will extinguish sight; yet is it no poison even when its glare is destructive. We do not "totally abstain" from light, though a part of our brethren have weak eyes, and are ordered into dark rooms. Sound, which in voice and music makes the ear alive, deadens hearing when too loud, and destroys the sense. In short, the sensible world is one great excitement to carry man beyond his first organic water. Joy, too, the wine of the soul, will kill by its abundance and unexpectedness, and yet it is next of kin to the life that its overmuchness withers. High truth intoxicates those not fit to drink it; causing oftentimes madness from its misapprehension and abuse; causing still more frequently need of rest, to recover from its dazzling revelations. We repeat that man lives by stimuli, any of which, administered in too great quantity, too often, or too fast, may cause destruction or suspension of life. Yet none of them is, therefore, a poison. Just as little can we so denominate alcohol, from the fact of its producing intoxication or death. For every stimulus carried to excess has the like effects, and, in all the cases, the excess is reprehensible, but the stimulus natural. Our Saxon word Drunkenness bears no poisonous sense; it is merely the far-gone past participle of Drink made substantive.

In truth poison differs from stimulus as medicine from food, for poisons in little doses are medicines, and food in its greatest concentration is stimulus. The plainest food will kill in too great quantity. And then again, medicinal substances, as coffee, tea, &c.,

come into dietetic use. Yet we cannot infer that food and medicine are the same thing, though they touch each other and are not incompatible at the extremes.

However this may be, the nomenclature of a subject from its abuses is inadmissible: we might as well name rich and provocative viands from gluttony as good wine from drunkenness. We might call fruits after diarrhœa, plum-pudding after vomiting, or peas after flatulence. But who would have evil courage enough to go through with such a dictionary?

But here the chemists occur, and tell us that alcohol is never absorbed; they have smelt the brains of sots, and the reek of unaltered liquor was manifest in the tissues. Thence they argue that stimulants are poisons which the body naturally refuses. We distrust this argument from drunkards to humanity and from the dead to the living. Rather let good men's boards be canvassed, where temperance reigns, and let it be seen whether there is no assimilation of wine. Surely the flow of soul can drink the measured cup, and fill its mood better than when the intellectual fire is dry. It is at banquets like Plato's that wine is vindicated. Their guests show the scope of human assimilation. What this is *in extenso* may be briefly sketched in an English fashion. The laborer wants the support of labor in good solid food, which under the magic of assimilation becomes muscle and sinew. The artisan wants nicer feeding, to support and not overbear the *finesse* of his fingers. Intellectual artisans need greater temperance still. We are now speaking of the hours of work. As for stimulants, the laborer may moderately mingle them with his food; the hayfield and cornfield, especially at cutting and reaping times, are wisely wet with cans of wholesome harvest ale. But as we rise in the quality of labor, work and stimulants are more incompatible; for the edge of the eye must be sharp and hard to bear the weight of the soul behind it. So much for work, which thus becomes ascetic, or shall we say, stimulating in itself, and intolerant of secondary excitements. The case alters in play hours, when a new set of faculties and feelings, and a new set of assimilations begin. The spirit of play mates with the spirit of wine; the pleasant emotions and the brilliant saws and dreams of society, like wine-lilies naturally rock upon the cup, and dip their spirity roots into the beakers. The imaginative skies are vinous then; Valhalla has its mead, and great Odin never eats, but all sustenance is liquor to Allfather, who drinks only wine. Elysium too would be a poor Elysium without nectar and ambrosia. The case is purely one of assimilation. If the life can drink the wine, and make life of it, then the wine is food; if the life is overtopped by the wine, which lies in pools in the reeking stomach and above the swampy brains, then there is excess, sensuality, or spiritual drowning. It is like the case of the horse and his rider. Many are thrown, and break their bodies; many should never get on horseback; some can ride where their animal is not too spirited; and some can back any charger even were it Bucephalus. But it is to be noted that allowance is only to persons in health, and not to those who have bodily ailments, diseased imaginations, or predominant lusts, or other maladies which want a physician.

It is now therefore obvious that we cannot take the personal experience of any man as to abstinence, beyond the limits of his own temperament, business and pleasures. The different functions to which our employments convert our organs, require for the organs different stimulation to put them "into condition" for their work. A Newton may require to fast and agonize before he can end his fearful mathematics, but that is no reason why our grooms and gardeners, with no end of the kind, should go through the like discipline. The adjustment must be left to experience, but it cannot come from an author's study.

Thus the power of assimilating wine is various in different persons, and in the same person at different times, and a flexible sense is necessary, to adjust the indulgence to the occasion. Observation proves this, for we can drink more at some times than at others, and sometimes with impunity and great restoration; at other times we feel ill consequences which show that we have not judged wisely at the board.

Above all things it is plain that cutting up dead sots will not settle the question. On the contrary, we must dissect all the drinkers, temperate and intemperate, and the same man at many different times, to elicit the living law of temperance. And first we must dissect ourselves. We know already that the habitual abuser of stimulants is full to the brim, and will run over with each fresh potation without assimilating it. He has no digestion left for common food, and therefore in point of assimilation his case proves nothing. It is the casual use of wine according to times and seasons, as Solomon says, and according to the feelings of the hour, which is life-giving, healthy and enjoyable. But whatever is good becomes, out of its place, an evil. Air which sustains us, native and exhilarating in the chest, is a torment and insurrection in the belly. The spirit and consciousness of the brain, too much poured forth upon a spot of skin, are sensitiveness, or agony, according to the degree. The blood thrown down from its vessels is a foreign body and black death in the system. Continence is the case in which forces are safe. And so with wine, which now and always has unflawed goblets in some men's brains, from which it is safely drank by their powers, and these of the very highest. The ideal then is, to emulate those chaste but intense qualities that enfranchise the favored guests of nature to dip their cups in the rubies.

And if wine is good to drink, it need not be drank on pretexts. Men have drunk it from the beginning for that which is the best and the worst of reasons—because they like it. "Wine maketh glad the heart of man"—there lies the fortress of its usage. To the wise it is the adjunct of society; the launch of the mind from the care and hinderance of the day; the wheel of emotion; the preparator of inventive idea; the blandness of every sense obedient to the best impulses of the hours when labor is done. Its use is to deepen ease and pleasure on high-tides and at harvest-homes, when endurance is not required; for delight has important functions, and originates life as it were afresh from a childhood of sportive feeling, which must recur at seasons for the most of men, or motive itself would stop. A *second* use is, to enable us to surmount seasons of physical and moral depression, and to keep up the life-mark to a constant level, influenced as little as possible by the circumstances of the hour. Also to show to age by occasions, that its youth lies still within it, and may be found like a spring in a dry land with the thyrsus for a divining rod. A *third* use is, to soften us; to make us kinder than our reason, and more admissive than our candor, and to enable us to begin larger sympathies and associations from a state in which the feelings are warm and plastic. A *fourth* use is, to save the resources of mental excitement by a succedaneous excitement of another kind, or to balance the animation of the soul by the animation of the body, so that life may be pleasant as well as profitable, and the pleasure be reckoned among the profits. A *fifth* use is, to stimulate thoughts, and to reveal men's powers to themselves and their fellows, for *in vino veritas*, and intimacy is born of the blood of the grape. But is it not unworthy of us to pour joy's aid from a decanter, or to count upon "circumstances" for a delight which the soul alone should furnish? Oh, no! for by God's blessing the world is a circumstance; our friends are circumstances; our wax-lights and gaieties likewise; and all these are stimuli and touch the being within us; and where then is the limit to the application of art and nature to the soul? At least, however, our doctrine is dangerous; but then fire is dangerous, and love is dangerous, and life with its responsibilities is very dangerous. All strong things are perils to one whose honor's path is over hair-breadth bridges and along giddy precipices. A *sixth* use is, to make the body more easily industrious in work times. This is the test of temperance and the proof of the other uses. That wine is good for us which has no fumes, but which leaves us to sing over our daily labors with ruddier cheeks, purer feelings and brighter eyes than water can bestow. The *seventh* use is, in this highest form of assimilation, to symbolize the highest form of communion, according to the Testament which our Saviour left, and to stand on the altar as the representative of spiritual truth. All foods, as we have shown before, feed the soul, and this, on the principles of a universal symbolism: this then is the highest use of bread and wine—to be taken and assimilated in the ever-new spirit of the kingdom of heaven.

The corollary that we draw, is, that total abstinence contains no universal argument; that it is an admirable strait-waistcoat for many of us; that abstinence is a needful discipline for every one at the most of times, and then coincident with temperance; but that moreover wine is an indispensable gift of heaven, and the use of it to the sane an inalienable matter of private judgment, into which abstinence-leagues, though backed by medicine and chemistry, will find it impossible to intrude.

Time has not yet perhaps elapsed sufficient to show the ultimate results which follow to those who steadily maintain the abstinence pledge. We can therefore only conjecture that the effect will be, to dry up somewhat of those feelings which give the body an extraordinary life when they enter it well. And as these feelings are the horses of our greater power, and when skillfully ridden accomplish times and spaces that are impossible without them, we suppose that the teetotallers will be somewhat pedestrian and prosy, excepting in the first vigorous days of their self-denials. But as human nature contains all wants represented in particular persons, there are probably born abstainers from, as well as born requirers of, wine. And if some of the most ordinary because passionless men are in the former class, they seem to have with them the men of peaceful power and supermundane continence, with whom stimulus is less the law of life than a spiritual tranquillity passing common understanding.

The current disadvantages of vowed abstinence, supposing it not to be natural also, appear to us to lie in a certain dilution of the powers; a certain want of sleep in the faculties; an unending character in the days, or a want of difference between the evening and the morning; also in a certain rigidity of reason, and a loss of those spiritual chances which are a part of the empire of clairvoyance and Providence in the mind; and in an undue incessant drift towards public, utilitarian, and money-making enterprise, to compensate for the loss of that stimulus of the heart which to the generality is of festive growth. But we would speak modestly in a case involving the experience of others.

As for temperance, it may range in respect of times from a daily moderation of wine to a single annual glass dropped into the diet. And as to quantity also, instinct and experience are its compass. But whatever the natural regime, the practice of temperance is a divine blessing second only to the use of reason. For temperance is the whetstone of our faculty of observation, and the axe of reform, which is to hew away forever at one vice after another, is nowhere so well sharpened as on its square and eager sides. Temperance again is the home of the virtues and powers, which shelters them in their nonage, and which they enlarge and furnish in their maturity, until it includes the images, pictures and tastes of them all. Or again it is the native land of man as different from the beasts. Or to speak without figure, temperance, including eating as well as drinking, is the foundation of our refinement, as involving constant acts of physical judgment or bodily wisdom. Intemperance is the devil opposite to this angel, temperance. It dulls every sense, burns out passion prematurely, and turns the mild light of intelligence, as it flickers towards extinction, into a horrid reproach against a swinehood which is reeling down to disciplines of which total abstinence is but a shadow. For intemperance fosters and aggravates nearly every disease that flesh is heir to, and sharpens the power and sting of every sin, nay, calls out fresh legions of infernals: the ghastly troops of malady and wickedness deploy before it, and muster, as on a field day of death, where excess of wine is a prevalent vice among the people.*

But now to recur to our main subject, the vastness of the assimilative function may be comprehended from the great agencies set in motion by the palate and the stomach. The skin or cloth-making spirit enjoys a considerable realm, but the spirit of the stomach is the owner of three-fourths of the world's navies, the ultimate landlord of three-fourths of the cultivable earth. Every sea is benetted with the lacteals of the social intestines; every road is laid down from men to men, as a vessel for food. Agriculture and commerce in their staples are an instinctive obedience to the claims of the belly. Armies are the guardians of its interests; and the dynasties of a thousand years are transmitted in security, or rock and dissolve, according to the dinners of the people. Such is the material force conferred upon hunger, thirst and assimilation. And in working out this destiny, whereby the globe is remodeled upon the basement story of the human body, and converges to our mouths by trains of produce from every climate, man grows socially also, and

* To those who wish to know what can be said chemically, medically and experimentally in favor of total abstinence, we recommend the perusal of Dr. Carpenter's essay, *On the Use and Abuse of Alcoholic Liquors in Health and Disease;* 8vo., London, 1850.

becomes assimilated to his fellows; the merchant behest carries him abroad, and informs him from other races; the aroma of his brother's industry clings to every stuff that he brings; forbidding waste as the loss of humanity and virtue. Thus the organs that assimilate our food, cover themselves with sensibilities, and assimilate our fellows. And thus we note that the spade-way or plough-track of the husbandman, the paths of ships over the sea, of caravans over the desert, and the road with its refinements, are only magnified images of that which goes on with intense smallness in the assimilative organs.

It is not in the lower kingdoms only that assimilation and digestion are proceeding. The plant, it is true, assimilates the mineral, and assimilates the atmosphere, the fruitful soil being the amalgam of their twofold natures in the one case, and the active aroma in the other. The animal again digests all beneath, and fertilizes all. Above the lowest nature each thing is eater and meat, end and beginning in succession. The external world in its extent is progressive assimilation and refinement. Through every change, by a secret providence, the surface of the planet is steadily fitting itself to sustain a grander edifice of society. For this, the primitive forests and their inhabitants have been industriously making, and shedding, their frames, in unreckoned generations. For this, the little flowers have been working since the first were self-sown from the miraculous garden. For this, the earth has rested uncropped in the balmiest latitudes, and still the sun pours the tropical spirit on her unexhausted islands. For this, an aboriginal savage tenantry lease but as hunters the future corn-lands of a long-deferred civilization. The human body, also—fallow and in great part tenantless as the planet—shall it not refine its organization century after century, and become the microcosm of a new mind, to be connected with it entirely, and to inhabit and cultivate it entirely?

And what is the growth of this mind itself, but renewed digestion and assimilation? In this again the creation is our food, but which enters through the mouths of the senses. Touch, taste, smell, hearing, sight, carry inwards their several matters of information to the nutrient reservoirs of the memory, where by the active imagination they are raised to some uniformity of life, and being cast into imagery, are extracted by the understanding, and the upshot is referred to the judgment and the will. The broadest common sense strikes home the first; is received at once without any process, and identified with the life of the mind; the details, difficulties and ambiguities of sense, which seem to suggest no present action, are long and passively retained in the entrance, and only come back to mind through other and oblique considerations; being not the chief, though the bulkiest sustenance of the human understanding.

Passing from bare consciousness to practical education—from the mind to the man—What is education but an assimilative career? The full social form is the blood into which we are to enter; the nature of the child, or the roughness of the adult, is the material to be admitted or refined—Delight and curiosity, with tiny gestures and sparkling eyes, come tripping forth to their lessons' time in the classes of existence. The cradle and the mother are one organic school; the nursery is another; the school-proper is another; the workman's probationary bench, and the student's table, are another; the life-calling is another still; and there is no finish to education, because there is no end to the refinement of mutual good works, or to the closer friendship of the human family. That grand individual, mankind, true to the spirit of Him who evokes it—can it less than reflect what is "infinite in conjunction, and eternal in perpetuity?"

Does this throw any light of probability on the dim hereafter? Somewhat of a luminous hope seems to overshadow and tremble around us, while we follow the analogies that proclaim the oneness of God's laws in nature and in man. Are these primordial laws so divine that they govern with their own flexibility even in the future life? Are we attracted thither to feed a mightier organization? Is the good received and welcomed, and the ill renounced, with a selection more discriminating, a rejection more total, and a wisdom more unconceived and irresistible, than even in that human form which comes from the spirit, and returns to the spirit? Or rather, is the principle one and the same in both cases? Let us lean on nature's arm, and follow the analogy until we have better lights: the rather because analogy itself is assimilation.

But to condense and finish.—The possibility of assimilation lies in the fact, that the universe runs man-ward from its source. The means to assimilation are implanted faculties in the soul, the body and the mind, of imitation, emulation and order. The use of assimilation consists in renewing all things upon their highest models, and by their best examples. Accordingly in the physical man, it is the bringing together of the ends of the earth, to be built in the temple of the body. In the mind, it is the translation of nature into thought, and of matter into spirit. In the household, it is the grouping of young and old upon larger affections into consolidated kindreds. In the state, it is that love and warrant of the commonwealth which reconstructs our private lives into the elements of an advancing society. And in the individual and the race, as the part and the whole of existence, it is that supernal fire which burns to make us more and more from the dust of the earth in the image and likeness of the Divinity. The steps of these assimilations are miracles; the emulation of every nature to other and better than it is; the death of the old, and the emancipation and resurrection of the new; the conversion of all brass into gold, of all grubs into butterflies; of earth into passions; of men into spirits;—we see not the stages here; because they are easier, smaller, purer and faster than our light; but neither do we refuse to credit them, for the result is humanity.

Therefore it is thus: at one end the universe is the quarry, at the other is the heaven of heavens already shapen and ever shaping; and God is the sculptor; and between lie all time and form moulded as they emerge into new heavens and new earths, the likeness of the wisdom uncreated. And all together rises now, because in the new covenant no temple made with hands, but immensity is the mercy-seat.

CHAPTER IV.

THE HUMAN HEART.

THE brains animate the body with intention and purpose, and the lungs give it corresponding motion, as the active spirit of the organs, and the basis of the operations of the will; the heart, as the blood's executive power, gives corporeal substance to the frame, inasmuch as the body itself arises from the blood. The existence of the human machine depends upon the heart, but its usage upon the lungs and brains. The heart is the source whence the finished blood descends to the organs throughout the system; it is the immediate administrator of the supply of nutrition to the body. And as the life is in the blood, the heart is the agent for bestowing that life upon the organization, and for giving every man a temperament, or peculiarity of animal being, secondary to and seconding that nervous life which he receives from the spirit of the brain. In a word, the heart or blood determines the fleshly tenement. Let what powers there may act upon us from within, or from without, we are made of no other stuff, and carry no other body, than comes from the fountain of our blood. We put it to what use we please or can, but the body itself is given, limited, constituted by the life-blood poured forth by the heart.

We have then to consider the heart as the centre of the blood-system; as a vessel suited to the whole composition of the blood; as the forceful agent in various motions whereby the circulation is perpetuated, or whereby the end of bodily life coincides with the beginning, and the animal circle is completed: also as the isthmus which, according to its build, receives the wave and shock of the passions advancing bodyward on the one side, and transmits them in modified vibrations to the expectant tide of blood on the other side. In short, we have to regard the heart in its own peculiar relations to physics, physiology and psychology.

The anatomical heart is a conical, hollow, muscular organ, lying obliquely in the chest between the two lungs, the base of the cone pointing upwards in the direction of the right shoulder, the apex pointing to the space between the fifth and sixth ribs. It rests upon the tendinous portion of the diaphragm, which is the partition between the chest and abdomen; and it is encased in a peculiar bag or capsule, the pericardium, which consists of two layers; the outer fibrous, by which the pericardium is attached to the great vessels at the root of the heart; the inner, a serous layer, continuous with the serous membrane that covers the outer surface of the heart. The cavity between the heart and pericardium, thus lined by a serous covering, generally contains more or less fluid, whereby the heart is lubricated on the outside, and its local motions are rendered easy.

The heart comprises four cavities, two auricles and two ventricles, one auricle and one ventricle being on each side, and the right pair of cavities being devoted to the circulation of the venous blood, the left pair, to that of the arterial blood. The auricles are at the top, constituting the base of the heart; the ventricles form the apex; the latter are much stronger than the auricles, consisting of very thick muscular walls, the reason of which we shall see presently.

The heart is a peculiar muscle, and when any of its four cavities contract, they have the power of expelling their contents, the force of the expulsion being the prime mover of the circulation of the blood. We may begin the circle where we please, and we shall find that it returns into itself. Starting for instance from the left ventricle, we see that the blood is driven by the contraction of that cavity into the aorta, the highway which leads into all the arteries of the body; through these the blood is discharged into the veins, which unite to form at last only two great trunks, the venæ cavæ, which are again the thresholds of the heart, and debouch into the right auricle. The blood which has now passed from the left ventricle to the right auricle, has still a journey to make before it completes its course. Accordingly from the right auricle it is forced into the right ventricle, and by the right ventricle, into the pulmonary artery, which conveys it to the lungs, where it ramifies through the multiple branches of that artery, and whence it is brought back by the pulmonary veins, forming ultimately four large trunks, which discharge themselves into the heart's left auricle, by the contraction of which the blood is next forced into the left ventricle, to the place from which we began.

We have sketched out the circulation before treating more minutely of the structure of the heart, in order that we may have the blood for a guide, and proceed from the uses of the organ to the anatomical structure. We shall however again recur to the circulation in greater detail.

The left ventricle of the heart, a powerful contractile cavity, has the task of projecting the florid or arterial blood which has just traversed the lungs, through the aorta over the whole body; upwards, to the head, downwards to the feet; this office it discharges by a quick act of contraction, or as it is termed, systole, which propels the blood into the aorta. The aorta, like the other large arteries, is elastic and muscular, and tends constantly to assume its smallest calibre, in consequence of which it moves the blood forwards whereever a free space is found. But when the heart again expands or performs diastole, the blood would regurgitate from the aortic tube into the ventricle, were there not a provision against this in the presence of three semilunar valves, little crescents of the lining membrane of the aorta, which swell out into pouches when the blood regurgitates, and close the passage. Their function in this respect is constantly called for, and the sudden stop of the blood by the valve gives rise to a *click* which may be heard among the other sounds of the heart.

The volume of blood we are considering is now fairly in the artery, and like a slippery ball it eludes the successive pressure of the vessels, and flies onwards in its course. The opening of the artery by the injected wave, and its contraction upon the same, ensues like a rapid undulation along the whole line, and constitutes the pulse. Eighty times *per* minute the quiver of this stroke permeates the life tree of the body and its infinite ramifications. The station, the tram and the passengers are all locomotives on this railway. First, the heart closes, and its out-thrown blood opens the artery; then the artery itself successively closes and opens down the entire extent; and the result is, the propagation of the wave from the centres to the circumferences of the system. The heart in successive moments forces life upon the arteries; the arteries, by an even pressure, tend to contraction or death, and thereby diffuse the life, or make it universal. Organic beings are expanded *ab extra*, but contract of themselves. Were it not for pressure from without, the contractility of all things would result in general death. But the influx of life opens the narrow into the wide, and ultimately effects a compromise, whereby contraction or individuality rules conjointly with expansion or receptiveness, and the two together perpetuate the commonwealth. This is the indispensable strife between progress and conservation, here represented by the heart and arteries.

The arteries, forming a tree whose stem is the aorta, terminate by their capillary twigs in the veins, which form another tree corresponding to the former, but the reverse of it in motion or function; for the blood that runs from the largest to the lesser and least arteries, returns to the heart through first the least, then the lesser, and then the largest veins. The arterial pulse is quite lost in the minute branches and twigs of the arteries, and the blood passing into the veins presents one continuous flow not manifestly influenced by the beating of the heart. Nevertheless it receives the force of the heart, which is the grand cause of the venous circulation, there being many secondary causes promotive of the same effect, and particularly the respiratory movements. The blood in the veins is prevented from running back, both by the *vis à tergo*, and during muscular efforts by a set of valves, something like those in the aorta, and which are placed at short intervals in many of the veins, and determine the wavering current onwards. Arrived in the venæ cavæ, the heads of the veins, the blood receives a new stroke from the muscular strength added to the cavæ, and presses with all its forces into the right auricle of the heart, which thus receives the last of the blood, and in the words of Harvey, "is the first part of the heart to live, and the last to die." Thus intruded, it distends the auricle; which, when it has borne the distention to the utmost, begins to resist, to react, to contract; it does contract and expels the intruded blood. Whither does it expel it? Not of course into the venæ cavæ, excepting in the slight proportion between the whole force of the oncoming venous blood and the contrary force of the contracting auricle; for the steady pressure of the blood in the cavæ has already been sufficient to open and command the auricle. The latter then drives the blood into the ventricle, and so is relieved and contracts; and now the ventricle expands, and the blood which it contains shuts upwards a triple valve, the tricuspid, which is attached by tendinous cords to the muscular columns of the heart; and the ventricle, reacting against its own forcible expansion, bearing it no longer, throws out the blood, prevented from regurgitating into the auricle by the tricuspid valve, into the pulmonary artery, which is an artery carrying venous blood. The blood thus injected into the pulmonary artery, is in its turn prevented from reflux by three semilunar valves placed at its commencement, and it circulates through another or second circle of arteries and veins, which constitute the pulmonic circulation in contradistinction to the general circulation upon which we have already expatiated. By the twigs or ends of the pulmonary arteries it is returned into the twigs or beginnings of the pulmonary veins; and we may somewhat appreciate the reason of an artery in this system carrying venous blood, and a vein carrying arterial blood, when we learn, that this system is in an important respect the inverse of the general system; inasmuch as the blood becomes de-arterialized, dark and venous in the capillary vessels of the general circulation, but becomes re-arterialized and crimson in the capillaries of the lungs or pulmonary system.

It is not now necessary to consider the circulation of the lungs, although it may be observed in passing, that the pulmonary arteries and veins, running through those energetic air-pumps, can hardly beat at any other times than during inspiration and expiration; and the pulse or stroke of the left ventricle can only act to the root of the lungs, and fill the beginning of the pulmonary artery, as a reservoir from which the lungs, at their own intervals, drink in the accumulated blood. One function of the pericardium or heart-bag lies in emancipating the heart from the power of the respiratory motions; for the heart lies in the midst of the lungs, grasped by their two arms, the pulmonary arteries and veins, and were it naked or unprotected, it would be drawn into the pulmonary votex, in which case its involuntary life would cease, it would receive the immediate play of the mind just like the lungs, and the spontaneous order of the body would be lost; in a word, the pulse would coincide with the respiration. Even as it is, the respiration exerts a marked influence upon the pericardium, though the heart is not further affected thereby than as it receives the general movements of the lungs and brains, while at the same time it exerts under them its own movements, or maintains its individuality.

We have now pursued the blood into the trunks of the pulmonary veins, where it is still impelled by the *vis à tergo*, and moreover drawn by the inviting lungs, until it is poured by the four trunks of those veins into the left auricle, which it opens and distends. The auricle now reacts or contracts, and squeezes the blood whither there is the smallest resistance, that is to say, into the left ventricle, which, when filled, in its turn contracts, and its blood shutting back the mitral valves placed between it and its auricle, is driven forwards into the aorta; again to perform the same revolutions, and to perpetuate life by incessant cycles of formation, destruction and reformation.

From this second view of the circulation we may follow the account given by one physiologist, namely, that the cause of the alternate motion of the heart is the action and reaction of the blood and the vessels; that the immediate cause of the motion of the ventricles is the action of the blood and the auricles; and the immediate cause of the motion of the auricles is the action of the blood and the venæ cavæ; further that the immediate cause of the motion of the vena cava is each particular branch of it; the immediate cause of the motion of the branch is each particular twig of it, and of each capillary tube thereto appended; the immediate cause of the motion of each venous twig is the action of the little arterial vessel which empties itself into it; and the immediate cause of this action is the action of the branch, of the trunk, and finally of the heart: wherefore the cause of the heart's motion is continuous, and like the blood itself runs in a circle from the left ventricle, throughout the blood-system, to the right ventricle; showing that there is not a point in the system but contributes to the motion of the heart. We see from this whirl or world of immediate causes how necessarily a motion once begun in these living wheels, rolls onwards, circling round, and how slight the resistance is, where all the parts contribute in their places to reciprocation and equilibrium. Under these provisions the smallest touch awakens the organism into its beautiful motions, emulous so far as nature can be, of everlasting existence and immortal life.

But if we delve a little under the human organism, we shall find that the circulation of the blood by the heart, is based upon a natural or spontaneous tendency to circulation in the blood itself, and that as in the case of the nervous system (p. 31), there is an automatic life at the foundation in every part, the fluids as well as the solids, to which higher stories or more measured powers are afterwards superadded. Thus the sap circulates in plants, and the blood in many of the lower animals without any heart to propel it. The fluid runs by attraction to the spot where it is wanted, and forms an uninterrupted fibre of supply, which is continually wound off into the loom of the organs; and it is indifferent whether we look upon it as fluid or solid, for the one end draws the middle and the other end, as if the current of life were a series of coherent threads; while on the other hand the portions wanted for deposit, drop out of the chain when and whither the want pulls them; for want itself, in its phases in this sphere, is their magnetism and their string.

The heart, nevertheless, though based upon all that is hearty, magnetic or occultly impulsive in the animate and inanimate worlds, is itself the essence of the human circulation, just as the supreme or rational brain is the essence of the human nervous system with its animation. We may indeed say that there are two essences to every progressive being, the beginning and the end of it, or the germ and the ideal. The germ gives the body and the ideal the spirit, which latter is to alter the body, made already with a capacity to be altered, into its own likeness. The ideal, or what is the same thing, the uppermost addition, as in the vascular system, the heart, in the nervous system, the mind and the cerebrum, become the essence of their respective orders; for in a progressive nature, it is not that which is, but that which becomes, that comports with the moving series, or comes into the view of ends. Moreover, the essence or peculiar capacity is that which distinguishes each organization from all others. The essence of man is that mind which he possesses in discommunity with any animal; the essence of his lungs is their idealization by that mind, and the peculiar physical structure which capacitates them for this becoming; the essentialness of his heart is thus likewise. Hence it is unscientific to regard the spine as the essential part of the nervous system in any animals but those in which the spine is the highest part; or the heartless circulation as essential in any but acardiac systems. Cerebrate organizations are created for a brain, and long for a brain, and acephalous monsters, though they may exist, cannot continue; so also organizations fitted for a heart, though they may maintain life for a time with a defective heart, yet cannot become adult, or travel on the road of ends, for their essence has failed them. This is not a matter of words, but a difference of ends and methods, involving the question of whether the sciences shall be founded upon distinctions or upon confusion; of whether they shall walk upon their feet, or stand upon their heads; and of whether the high, the moving and the intelligible shall give the cue to induction, or the stationary, the occult and the low.

And here we may remark that in organizations it is the additions *ab extra* that become of all importance in the long run. Thus nature, at first vertebral and serpentine, becomes capital and human by the addition of parts which snakes and tortoises can dispense with. And these extraneous organisms, not at first necessary to mere existence, but necessary to the ends of existence, become the essentials of useful knowledge, because they show the aims of facts, or the ends of all developments. On the other hand, the basis which nature supplies to be built upon, is of no importance except for the building, just as the vertebral column is insignificant excepting for and to the head. Precisely in the same manner, the magnetic, sap-like, and even animal forces of the circulation, are of no importance in our bodies excepting as the ground upon which the last essence, the human heart, is to be built and chambered. Man at any rate is a distinct subject, and that which is manly in his heart is all that belongs to its human physiology. The rest is animality, vegetable physiology or general physics, and had better be sought after in its purity among the beasts, the plants, or the loadstones.

This theme requires to be dwelt upon from the current misapprehensions respecting it. For at present it is held, that that which is essential in the human body is that which the body possesses in common with hydatids and zoöphytes—that a digestive sac or cell is the essence of mankind. And the reason given is, that all but stomach and nutrition can be dispensed with, and that these can go on for a time on their own account without the brain and nerves. It is perfectly true that these foundations are as independent as they can be, and that they are automatic. The very flesh is alive *per se*, and carnally ensouled, before the better life is added to it by the brain. But in developments the development becomes everything, and covers and gradually buries the matrix out of which it sprang. And if the development does not come, the barren matrix cannot last or work. No one ever heard of an acephalous monster going to church or taking a hand at whist; on the other hand, it is walked out of the ranks sexlessly and shabbily into the unconsecrated piece of the next churchyard. And all headless things which come up among those which have heads, must take the same fate; headless sciences among the rest. Moral headlessness is in the same category; those who make the belly and the flesh essential because they are the last things that can be done without, and throw away the higher parts because they are merely additions, and some function continues without them, go to the worms by their own claims. For just that which is first and easiest lost and lived without, is the pearl of the human mind.

To return from this digression. At one end of the circulatory system is the heart; at the other a class of vessels termed, from their hair-like fineness, the capillaries. The heart, as we before observed, sends from its left ventricle a grand arterial arch, the aorta, which in its turn produces stems, branches and twigs, and these last the capillary tubes, an intermediate field, which is the end of the arteries and the beginning of the veins. In these almost invisible capillaries, nature, "greatest in the least things," carries on some of her most wonderful operations. The blood which in the larger arteries is a medley volume of red globules floating in serum, and only comes into indiscriminate contact with the sides of the vessels, as it runs down from the larger into the lesser tubes, becomes more and more distinctly divided, and in the capillaries the red globules are in the highest state of discrimination, fit for supplying from their bosoms living nutriment and warmth, and whatever depends upon the most individual endowments. In the great vessels the blood is in public, exercising general offices, and thronging the body to keep it wide and open; in the capillaries it is at home, its freedom, power, and fluidity raised to the greatest pitch, and subject to the still influences of a higher spirit than its own. Accordingly the capillaries are more plainly under the direction of the brain, than the larger vessels; their exquisite nervous coats undergo greater proportional changes in calibre, and enjoy more individuality in differents parts than the Herculean aorta and its tubes; the emotions of the nerves being almost omnipotent in their influence over these kindred and subtle disciples. For the most part the blood-globules march in single file through the capillaries; but under the influence of cold, stimulants, or inflammation, the latter will either close, so as apparently to admit no red particles, or will enlarge considerably enough to give passage to several rows abreast. A familiar instance of the power of the mind or nerves upon this system, is seen in the phenomenon of blushing, in which the modest emotions, giving all place to the pressure about them, instantaneously relax an outspread field of capillaries, which lose their straight-laced dignity, and red blood is injected into vessels that before carried only colorless fluid, or in which the amount of red blood is now so greatly increased as to glow through the skin of the face and neck, and tinge them with apparent crimson. If we generalize this common fact, we observe what a prodigious sway the brain must exercise over these blood-vessels. The capillary system is coëxtensive with the body, for looking from the heart as a centre, the framework of every organ and part consists of capillaries. These, the brain and nerves have the power of expanding and constringing in any place, or any number of places, in a moment; of producing secretions or stopping secretions at will; of varying the life that comes from without to any particle of the frame. In short, the circulation in its active and flexible part, is held in the leashes of the brain, and the accommodation of the blood system to the existing state of the body and mind depends mainly on the nerves acting on the capillaries. Nay, the texture of the heart itself is capillary, and by consequence that private force and freedom which belong to these quick vessels, lies in the very core of the heart as the general force. In view of this, we might begin with these ambidextrous and complaisant parts, and derive the power of the heart itself from the array of their combined individualities. And if there exists even in the cold life of plants a natural attraction of particled fluids to their destination, and an answerable force which creates the vegetable current, much more in the human frame with its powerful lungs and brains, is there an attraction of the particles of the blood towards the capillaries—an attraction which draws them at an accelerated ratio the nearer they approach those original conduits older than the heart, where their uses and sacrifices are to be accomplished. It has indeed been supposed by some that the blood is freest in the great vessels, but that its globules scrape against the sides of the capillaries, and lose their power by friction; and this has been supported by microscopic views; but then on the other hand it is impossible to place any living sheet of membrane, *e. g.*, the web of a frog's foot, under the microscope, without changing the conditions, setting disturbing emotions at work upon the circulation, and destroying the equilibrium between the globules and their homes. And even if this be done during the insensibility of the animal, the case is not normal still; the anæsthesia itself is a new element; and moreover there are organic emotions as well as conscious ones, which go on under inflictions just as though the pained animal were at the back of them (p. 31). Even dead leather crinkles and writhes over the fire as though it were full of burning agony. Rational induction alone can then decide the question. And this ignores the idea that our blood experiences new physical difficulties when it nears the goal of its existence, or is hindered more and more by material clogs as it comes within hail of the spirit of the brain. It is true that the blood dies in the capillaries, and its ruddy frame turns blue, but it is never so tenderly alive as on its death-bed, never so near to its real ends, and in its ascent from the earthy heart it wings itself with speed, until at the last, in its final place and secret hour, it is all spontaneity and calmness. It is wrong then to speculate upon any rigidity of the tubes in the balance of healthy life: if the globule tends to impinge, the tube will tend to yield before it, and the contact will be nothing more than the support and maintenance by both sides of any inclination which may exist on either.

Let us now proceed to the use of the circulation. The first use is, the formation or determination of the body. The second is, the nutrition, maintenance or reformation of the body. The third is, its prolonged vitality, or perpetual stimulation. These three purposes mutually suppose each other. The formation of the body is effected by the germinal fluids, determined into membranes, canals, and ultimately into vessels with coats; and as the blood-vessels are the last expression of these, we speak of them as the framers of the body; for the solid comes out of the fluid at first, and the substantial body grows from the blood even in the adult state; the capillaries moulding the organs; which, however, when they are built, work on their own account, and govern their blood, calling into existence the second use of the circulation, or the nutrition and reformation of the body. The progress of discovery upon this point, tends to establish the individuality not only of every organ, but of the elements or smallest impersonations of the organs. The deposition of the solids from the blood does not take place immediately or by intrusion, but the organs put forth fine cells or stomachs from their several absorbing surfaces, which take what they want from the capillary circulation, and make it over to the organ through a fresh similar organization. In short, the body grows from the vital blood by its own vitality, as a plant grows from the sappy meltings of the ground; its development is due inside, and in every part, to its own attractions, and not to the rush or thrust of the blood. The increment is not like placing brick upon brick from without, each with no peculiar relation to the building, but every brick is a cellule or little building of itself, similar to the edifice of which it forms part.

This is a higher idea of nutrition than that which it supplanted, and deserves to be called a vegetable idea, in distinction from the other, which was a mineral idea. We must, however, remember, that in the human body we are on the stair of endless ends, and that no fixed idea will last out of its place and time. We must also note, that we are reasoning in a sphere where the fluids are next to all; where the toughness of vegetation is a disease, and where fresh formations are extemporized with mushroom rapidity, and dissolutions take place with equal speed. Cells in this case melt like mist into their original currents, and the only set of analogues which do not fail us, are the changes of the mind itself, whose velocity belongs by transference to the body in its higher and healthier moods. Thus the organs are nourished as well and rapidly as though the blood was forced into them by the pulses, but in those moments of moments they have deliberated, judged, chosen, and lastly acted, by shooting forth a cellular gauze, wise to let in the exact quantity and quality that they require.

The rule then is, that the heart and its powers act only to the door of the organs, and no further, after which the organs take from the proffered blood their own demand. This is true, as we showed in speaking of the lungs (p. 102), of the largest tissues of the body; it is true, as observation shows, of the smallest parts of the tissues. In a word, attraction is the law; and it exists between all the fluids and their respective destinations and uses; it is a true animal magnetism; and in this high form there is a manifest propulsion or heart to second it on the one side, and a manifest invitation or lungs to create it on the other: to which we may add, that the fluid itself is so natured as to run of its own accord away from the parts which do not want it, and away to those which do. Thus, in this system of divine convenience, where every tendency is trebly gratified, the blood propelled to any organ is no longer the heart's, or to be denominated from the heart, but it belongs to the organ, part, or particle, wherever it may be; it is always sailing upwards and inwards to deeper purposes, and taking new names and new liberties; just as sensation rises from the eye into the brain, is adopted into intellect and faculty, and walks at last, unknown to the lower scene, in the breadth and color of the sky. It is thus, that the red blood from the heart mounts into the region of the capillaries and organs, that new world where the nervous system hangs its ethereal expanse over the vascular.

The third use of the circulation consists in giving life or nervous fluid to the tissues. This presupposes that such a fluid proceeds from the brain through the nerves, and is shed perpetually into the body, and especially into the blood. The red globules of the latter are its most living parts, and may be likened to little caskets in which this nervous fluid is carried about, and dispensed as life to the tissues, to which it gives a natural, though engrafted capacity to be acted upon by the brains and lungs, and to react on their own account. The freeing of this nervous fluid in the capillaries is as a millionfold torch that kindles the decompounded elements of the globules, and produces a graduated scale of heats, of which the first or living fire is the nervous fluid itself, instinct with its organic emotion, and the excrement, residue or *caput mortuum* is the oxygen and carbon which the blood and tissues yield to the curious chemist; the difference between which two is greater than that between the smoking ashes of a burnt house and heaven's lightning.

Were it not for this inspiration by the brain, the blood could not be humanly alive, or in other words, the soul could not associate with it. With respect to the change that takes place at the ends of the arteries, or in the capillary circulation, where the blood loses its arterial color, and purples or becomes venous, the manner of it is sufficiently obscure, and has only been investigated at present from chemical grounds. But looking at the blood itself as organic, or vitally compounded and mechanical, it is easy to see, that in the place where it has to yield up any of its constituent parts that the successive organisms require, it must lose its principle of combination in order to allow it to pass into other forms. In short, it must undergo a vital decomposition for the purposes of the body; and if its arterial glitter depends upon the spirit shining through it as through an organic face, then when this spirit escapes and comes out, and the subordination of parts is lost, the comparatively lifeless hue of venous blood will be assumed. Furthermore, we may infer that the nerve spirit, which is the charioteer of the globule, and its principle of organization (p. 55), when its heat or desire of blood-organization is ended, and when it comes down among the other elements, will kindle with that sensible heat which is experienced in the lowest sphere; in other words, that in the body, the intellectual or organific heat is in a lower degree the parent of true animal heat. This we find to be the case on a large scale by the glow of zeal, passion and affection in the cheeks and the body; and these large instances, in which the whole body displays itself as one globule of impassioned blood, give the only analogues we can use consistently with our purpose of reasoning not upon the dead but upon the living blood (p. 181). The difference then between the arterial and the venous blood is, in the first place, that the arterial is organized, spirited or impassioned, and just like ourselves in such a case, is ruddy with the fervor of a soul, while the venous blood is disorganized, chaotic and aimless, and like ourselves again under these circumstances, is murky in its hue and leaden in its gait. In the second place, the arterial is full blood, but the venous has yielded up many of its elements, and has no spirit or nerve fire, but the residue of this is taken up by the lymphatics, while the apathetic body of the blood is derived into the tardy veins. Such is an approximative organic account of the change.

We have now treated of the use of the heart to the general circulation, and if this were all, the heart would be a simple forcing pump, as *Chambers's Journal* declared it to be; "a good heart," in their words, would be "a good forcing pump," and "good heartedness" would either mean "good force-pumpishness," or nothing at all. But when we glance at what is taking place around us; at the subjects which are extending their limits; at the old things that are again brought forward by our growing love of fairness, because they have been previously dismissed without a hearing, which says nothing against them, and much against their judges; at the new things which are thronging into notice in shapes that cannot be neglected—when we glance at this we may be expected, in the face of any shabby idea of nature or her Author, to propound the question, Is that all? and to cut short the men who say to us, It is only this, and It is only that, by a decision that no finite mind has a right to palm one such *only* anywhere upon nature. For our part, when we look at the human frame, we are always impelled to put this question, and in the same breath to answer, that there is more and ever more to be known about it, not in the way of niggling additions and grains of scientific sand, but in great principles, in new tracts of knowledge, underlying, overspreading, and surrounding that tiny edifice of books and cards where we are so comfortably at home. Yes, at this very hour methinks there is a good and guiding genius standing just beyond the range of every conception that has entered the world, and beckoning us forwards in words familiar from childhood:

> "There are more things in heaven and earth, Horatio,
> Than are dreamt of in your philosophy;"

and then as we pitter out, "'Tis strange, 'tis wondrous strange," the same apostle to the Saxons speaks yet again in mother English:

> "And therefore as a stranger give it welcome."

There are points in the structure of the heart of which we have said nothing hitherto, but we must now describe what are called the coronary vessels, which are supposed to nourish the heart with blood. They are called coronary from *corona*, a crown, because they run in crowns or coronal circles around the heart. They arise from the aorta close beside the semilunar valves, and running around the base of the heart, and sending branches down the lines of partition between its fourfold chambers, they form a kind of vascular cage-work in which it is contained. The coronary veins, said to begin from the minutest twigs of the coronary arteries, by their considerable branches for the most part accompany those of the arteries, and discharge themselves by one, two, or three orifices into the right auricle.

The interior of the four cavities of the heart is not a smooth even surface, but is rendered extremely irregular by muscular columns, projections and partitions; it is scooped, channeled, and caverned; besides which, on the walls of the cavities there are minute openings, the foramina of Thebesius, which are supposed to be the mouths of little veins.

Everything in nature, and especially the cardinal movements of living systems, are designed for more than one use; for, unlike rest, motion is a steed which can have innumerable riders. We are, therefore, certain that the movements of the heart are of other uses besides the propulsion of the general blood, unless it could be shown that the provisions of the heart are exhausted in this propulsion; which cannot be done. On looking further at the heart itself, we find that its working is also employed to insure the commixture of the elements of the blood, especially in the right ventricle, which receives all the venous or disintegrated blood coming from the most divers organs; all the chyle arriving from the food; moreover, all the lymph; and lastly, the nerve-spirit streaming in from the jugular veins, which contain a far greater proportion of this than the rest of the venous blood. This triple scale of elements, the blood, the chyle and the nervous fluid, are worked, kneaded and commingled by the right auricle and ventricle: is it too much to grant that the motion of those cavities is intended to do that which it plainly does; or that there is an end answered by the commixtion. There are not two answers to the question. The right ventricle, then, we will say, after a quaint authority, is the great vat of the blood-system, in which the fluids are mingled to form the ingredients of the red blood; after which the mixture is sent through the lungs, to be skimmed of whatever comes forth into the air, and to be subsequently passed into the left auricle and ventricle.

We are aware that views derived from the forms and powers of the organs, are out of fashion, yet in the living body they are logical and physiological, which the chemical notions are not. There are two ways of looking at organic subjects, which should be carefully distinguished by the mind, and carefully united in the sciences. First, there is the investigation of the form or structure, at rest, or in a dead state; and this gives an osseous basis to our knowledge: but as permanently resting subjects are on the road to death, and dead things are on the road to decomposition, this method leads, in process of time, and in continuity of doctrine, through the land of bad smells to sheer mineral chemistry. The second method is the investigation of the movements, functions and deeds of organic subjects; the examination of what they do, and their judgment by their fruits; and the facts which this supplies are as flesh upon the dry bones of the former knowledge. Let us apply these two methods to the blood, both to illustrate what they are, and, at the same time, to pursue our inductions respecting the offices of the heart. In the first method, the examination of the dead, dying or abnormally affected blood (p. 179) by the eye or microscope, shows that it consists of red globules in a transparent serum; and as the blood finally dies, it undergoes a series of changes, forming a clot or coagulum in which most of the red particles are entangled. Thereafter, purely chemical effects succeed, and we have a disengagement of gases, and a further alteration of the clot. Such is the first method and its chronicle. We observe that it is most perfect when the examination begins, and at that time holds on the subject under the eye; but gradually and not slowly the subject changes, and a set of phenomena present themselves, which, if taken for living appearances are mere delusion, for they belong strictly to disorganization.

The second method with the blood observes it in motion, in order to gain the hint and image of thought, and afterwards and especially learns what it accomplishes in the body, in order to gain the scope and details of the thought again. It is as when you first see a man, and take his impress on your memory, and afterwards from his observed actions you put his character into him, and find what his person is and means. For it is the deeds of men and things which by time's benefit range themselves into their intelligible vitals. Who can care whether the blood contains minnikin particle within particle *ad infinitum*, without these can be tallied off against something that the whole blood achieves for the human body? Otherwise they are the decomposition of our observing powers, and rot the organic sciences. At best such microscopic observations are the visions of the underworld, the empire of the dwarfs. But what then is the motion of the blood? It is, so far as we see, the streaming of an incessant population of globules through the vessels; the body is a city of active life-bloods, moving like the nations and peoples of a whole planetary system at once, through every atom of space and time which the system allows it. And what are the offices of the blood? We answer that it deposits the elements for every organ; that it perpetually deals out the parts into their respective places; that it is the body itself in a fluid state. Thus, everything in the body contributes to our notion of the blood, and the man is a representation of its powers and tendencies as well as of its substance. We may define it as a compound of all the simple elements of the organization, and as the globules are its most living portion, this is pre-eminently true of them. The eye sees nothing of this its character, as neither does the eye see in the brain the faculties by which men invent the arts; but that such a wealth lies in the blood the mind's eye knows, for how could the body come out of the blood if it were not first involved within it? We have now, then, arrived at a certain knowledge of this manlike globule, as being a group of the principles of the solid organs and tissues, resolvable by disintegration as it circulates, through the attractive peculiar doors of the organs* (p. 181), into each part to which it passes. For, on its unwearied round it gives heart to heart, lungs to lungs, liver to liver, kidney to kidney, and like to like everywhere; and what is left in each case, forms the venous blood, the lymph, the deader heat, and the several excretions. Moreover, one set of glands compounds the blood while the other destroys it, and in its perpetual life, death and resurrection, it images the destiny of him whose bodily existence it constitutes.

With such views of the blood, added to what we derive from the eye and the microscope, we require a vast machinery adequate to produce this composition, and we are driven to look to every organ, and first to those where the blood is contained, for its contribution to the result. Shall the heart be excluded from the privilege of blood-making? Though a large, it is a highly complex structure, full of special cavities and conduits, edged and jagged machinery of tendons, and fine muscular limbs and fingers. Its two sides contain different kinds and qualities of blood: is there no communication between them, no intermediate compounds to edify the little temple of the globule? Why is there an association of the sides of the heart, and a community in their substance, if there is no society in their functions, and no reciprocation of their goods? Are the two sides of this channel also, natural enemies? In the right ventricle there is the chyle, the venous blood of the body, the venous blood and spirit of the brain: is it not rational to infer that the left side of the heart furnishes model globules and middle essences to unite these heterogeneous parts?

We must touch this matter slightly, and perhaps therefore ob-

* We must take care not to let the vegetable idea of cells grow to any woodenness, or interfere with the animal idea of freedom and instant fluidity. The cells in life are mere instant refrigerations of the steam, liable to be vapor again in a moment between the strokes of the life-engine.

securely; however, the main guidance is, that the blood consists of an orderly involution of the elements of the bodily organs: this will give light upon what would otherwise be dark. We now recur for a few moments to the coronary vessels, or to the circulation in the walls of the heart.

It is a curious fact that nearly all the old anatomists, and some also of the moderns, have suspected a puzzle in these coronary vessels. They come from the aorta and run backwards to the heart. In a certain proportion of cases estimated at 5 in 20, one or more of their orifices lies behind the semilunar valves, and such orifices it is clear cannot receive the stream propelled from the heart, because it lays down the valve flat upon them, and effectually closes them. As therefore nature's law must be constant, it was argued, that what holds of one orifice must hold of all; and that the blood runs back into the coronaries from the aorta when the heart's contraction ceases. This was Boerhaave's opinion. Morgagni, a more practical anatomist, was more cautious, and requested others to decide the too difficult problem. Another view was now propounded by the celebrated Swedenborg. He argued that the raising up of the semilunar valves during the contraction of the heart, when the blood is expelled into the aorta, precludes its passage then into the coronaries; and that the stretching of the coronaries, and their pressure by the full aorta, contributes to the same preclusion. Moreover, that to suppose the heart supplied with blood by regurgitation from the aorta, would be to ascribe to the latter a new action different from what it exerts upon the other blood-vessels; nay to claim for it, after the discharge of its functions, a stronger, inverted and retrograde action upon a body the most muscular of any. These considerations led him to infer, that the coronary arteries do not arise from, but terminate in the aorta; that they are veins relatively to the heart, although running into the beginning of the arteries of the body. The doctrine in brief is this:—that the heart as the head of the vessels and the fountain of the blood, itself requires the firstling blood for the exercise of its noble offices, and cannot hold its life by tenure from one of its own arteries, which would be to invert all ideas of the order of nature. The heart is already full of blood, and if fluids, or fluid persons, like solid persons, move with greater velocity in proportion to their life, the best blood in this race will continually outrun the rest, and always first in the heart, will skirt along its porous walls. Now what structures do we find upon those walls, but caverns, jagged cavities, and at the bottom of these a number of little holes, the foramina of Thebesius. Into these caverns, then, miniature ventricles in the great ventricle, hearts of the heart, the quickest blood is received, and the pores open with all their hearts to take it in. And when the heart contracts, it drives out the general blood of the body into the grand aorta, but its own particular blood, detained in the cavernous lacunæ, it squeezes, slippery with spirit, through its walls into its muscular substance, and thence onwards and outwards to the surface, into the coronary arteries and the coronary veins, from which there is a reflux, when necessary, into the auricles and ventricles.

It was also held that various currents of blood exist in the heart, and in short a multiple communion, one object of which is, the production of a successive order or series of stages in the blood itself, fitting it for its manifold operations. This was brought to bear upon another subject, namely, the influence of the mind upon the heart. We must here spend a few lines in considering this well-known circumstance, on which we shall have much to say presently.

From the oldest times the sympathy between the mind and the heart has been acknowledged. The records of disease likewise show, that the heart is affected and altered by the state of the mind, and *vice versâ;* and that powerful feelings will cause palpitation, fainting, and even sudden death from their influence upon the heart. Now the heart is the centre of the sanguineous system, the organ from which the motions of the blood begin, and the bed in which its pressure terminates. And the reader will recollect (p. 178) that the brain, the organ of the mind, carries the influences of the mind along the nerves to every part of the capillary circulation, and produces in the capillary body, for it is the body, throughout the day, a motion restless and ever-changing like the fluctuations of the atmosphere. This is the nature of our human mind. If we recur to the instance of blushing and universalize it, we easily understand what is meant (p. 178). But in this continuous fluid system, every inconstancy in the circulating current produces its effect upon the centre—upon the heart; and in this way the whole play of the mind becomes transferred to this wonderful organ. Again, every muscle that moves, whether in breathing, standing, walking, manual labor, eating, or in whatever other way, throws the venous blood forcibly and newly upon the heart, and affects its condition. Of this we have evidence in the operation of bleeding, in which, when the arm is tied to prevent the blood from returning through the veins to the heart, and the fist is rapidly doubled, or even the fingers worked about, the current squirts in forcible jets from the opened vein. What then must be the squirt of the venous blood heartward, in a wrestle of the whole body, or during a rapid run? Moreover, where the frame is enfeebled by long disease, and the patient is bed-ridden, the motion of rising into the erect position sends dangerous jets to the heart, in many cases oppressing it beyond recovery, and leading to swoon and death. And where there is disorder of the heart, the muscular exertion of undressing and stepping into bed, will cause prolonged anguish, until the heart's circulation is equilibrated. These are proofs that the state of the circulation affects that of the heart, and that the movements of the body affect the circulation; and we have already seen that the mind is continually playing upon the capillaries, and the capillaries referring their disturbances to the heart. The question arises, Is there any provision in the heart to enable it to maintain its own constancy in the midst of the fluctuations of the blood, and to make it the head, ratio or balance, as well as the heart of the too mobile circulation?

This problem has not been unnoticed by physiologists, and Mr. Abernethy in particular endeavored to connect the solution of one part of it with the foramina of Thebesius. It is here that Swedenborg's doctrine of these foramina and the coronaries, finds its strongest present attestations. According to this view, the varying quantities of blood returned upon the heart find an outlet through the walls of the heart itself, and equilibrium is thus maintained by the coronary vessels; so that the heart *plus* the coronaries equals all the forces of the circulation; while the heart *minus* the coronaries is a comparatively regular force uninfluenced by the general state of the system. Were it not for such provision, the heart would be at the mercy of extraneous influences; the most important organ of the trunk would have no stability, but would in the end yield, and be distended into a bladder or membrane, incapable of anything but the most passive recipiency.

It is contended on this view that there is a representation of the mind by the heart in the manner in which the latter equilibrates the blood by the channels of the coronaries; for the passages may vary in different hearts. Thus in case the blood has a tendency to run out of the heart through these avenues into the aorta, it represents a want of firmness and courage; and universally during fear the systemic arteries empty themselves, and the blood runs away into the veins. On the other hand, when the blood tends to keep in the heart, and to press back into the right auricle, it indicates a firmness and strength of the nervous, arterial and venous systems. And in this way these animal qualities are based upon the construction, tendencies or habitual channeling of the heart. Now we know from our sensations that different feelings cause different actions in our breasts; for example, that in hard and firm resolve, the heart seems to stand its ground, and not to let one soldier-globule ooze away; that in moments of timidity the pit-a-pat of flight and disarray seizes, and the ear-drum beats the inglorious tune of "devil take the hindmost:" also that in melting moods the heart goes with the eyes and lips, slipping and trickling away from its station as it were a tear. Further, we know that courage and fear are constitutional to certain persons. And can we doubt that their constant action upon the heart, implies in the first place a corresponding fabric in that organ, and in the second, a continual alteration of the fabric, as the mental state and circumstances vary. For it is to be said that the principles of the mind will govern even the heart, and make cowards brave in the second nature and strength of conscience, or abash the lion-hearted when that higher spirit of courage is gone.

While we are upon this theme we will take the opportunity to digress for a moment, to dwell again upon the consistency of the body as exemplified in this passion of fear; a consistency which we have already shown to some extent while treating of the lungs (pp. 108—119). Now in fear the heart is bloodless, for the blood, as we said before, has run away from it; the lungs are aghast or ghostless; the brain is mindless, and consciousness gives place to fainting; and the man embodies all this by himself either sinking down, or running away. The fact that we avoid what threatens us, often with an instinctive passion called fear, formulizes the action of all our parts in this state. The spirit or mind running away from the brain in mortal chilly streams down the back bone; sight deserting the swoony eye; hearing leaving the ear; spirit running off also by the hair, and the hair standing from the head; the head and all parts horribly wide out and erect for a moment; the tendency to universal displacement seen in head, eyes, tongue, arms, and legs flying out and away from each other; the wind shrieking wildly forth from the lungs; the blood rushing pellmell from the heart; the excretions running from the bowels and the bladder; heat falling through abysms of cold, and life, which is courage, perspiring from the skin in big drops of cowardice;—these are all the same passion in different parts and appearances. In all, the man and the organs run, or tend to run, from the place of terror, which is not only the particular locality, but the body itself; wherefore death, or running away from the body, is not an infrequent effect of fear. Our knowledge, that fright produces *running away*, carries us through the effects of fear upon all the organs, and we need no other principle, but only the details of this, to explain the state of the blood in the case, or indeed of any of the parts, whether solid, fluid, or mental. We may now generalize further, and affirm that the broadest consequences of every passion and living state march through sphere after sphere of the body, and deposit themselves in fresh but consistent shapes as they visit fresh provinces. Thus love, which clasps its objects to its own bosom, draws closer the parts of the loving brain, and makes harmony of thought; it knits the blood into new relations, and as the newly-kinned globules touch each other, the heart becomes its body's delight. And so each state of man is a human frame complete. The unlearned world may follow this knowledge, deep and depth-seeking, using broad sights as an organon, and never becoming microscopic, or resting in anything less than limb, trunk and head. Common life is the college to teach live physiology.

Returning now to our immediate subject, we observe that the theory that the heart alters and amplifies the blood, is supported by the analogies of the principal organs. For example, the brain is not the mere centre of the nerves; it is not a simple turn of fibres in which sensations are converted into motions, but new faculties are there piled upon the summit of sensation, and the brain is a commanding head to the nervous system. So the larynx is not a simple enlargement of the windpipe, but a super-addition of gifts, by which plain air becomes discourse, and is cast into words of meaning, the vocal symbols of intelligence. So the tongue is not a mere thickening of the unconscious gullet and stomach, but a capital organ in which sensation is added to the other functions of the intestinal canal. And so man is not merely an eclectic centre of the world, but he is a spiritual world also, and a set of miracles, if he chooses, playing their will with animalities. In fine, wherever there is a head, it does not differ from its subordinate parts in size and situation alone, but also has a freer life than they, and exercises supreme functions additional to theirs—functions both more in quantity and novel in quality. Therefore, to recapitulate, the heart itself both commixes the elements of the blood; builds them up in a regular series; and levels and balances the general circulation; and all this, in addition to the functions which it performs for the arterial system below it, of propelling the blood through the body; and for the nervous system above it, of receiving and representing the fluxions and passions of the mind. In this way we have a first draught of a beginning, a middle and an end, to some purposes of the human heart.

Here we conclude our first view of the heart, which we have found to be more than a cross road, or convergence of vessels; in fact, to be the metropolitan city of the blood, the interests and business of whose home-inhabitants are of primary importance in the system. We have then now three great divisions of this subject, viz., the current through the lungs, the current through the body, and the inhabitation and uses of the most privileged blood in the heart itself. In other words, we have the doctrine of the circulation from the moderns, and a justification, more modern still, of the flux and reflux of the blood, heart-felt and intuitively seen by the genius of the ancients.

As to the doctrine of the circulation which was demonstrated by our great countryman, Harvey, it newly teaches us the import of the circle, by life and substance added to the strictness of mathematics. It is also a cardinal instance in a line of universal truths. For there is a circle of all things, as there is a circulation in the human body. Not a fluid is contained in our frames, but according to its perfection aspires to circulate on the model of the circulation of the blood. There is a grand current from the fluid to the solid, and from the solid back into the fluid; a circulation of perpetual life—"formation, destruction, and reformation." (p. 174.) There is a circulation from the universe into the body, through the food, the skin, the lungs, the senses, the brain; a circulation back again from the food, the skin, the lungs, the actions, and the mind itself. And the world is an everlasting circulation. The mineral ascends into the vegetable, and both into the animal, and all into man; and man's body descends into the dust, and completes its circle there. In short, whereever we go, we meet this old emblem of eternity: the Midgard-serpent with his tail in his mouth hoops the whole world round; ends and beginnings meet, and nature is bending round from her first issues towards her source; like the weapon of the Australian, she comes back into the hand that flings her, and the human body is a permanence of her cycles, which are the pulses of our hearts.

The heart, in common with the other organs, is the subject of a twofold discourse, and has two sets of books and votaries appropriated to its consideration. Indeed we may say that the present world has two hearts, which have very little to do with each other. There is the one heart of which Shakspeare is an interpreter, and the other where Harvey reigns. Two Englishmen have been high priests in the service of these two organs, and it would seem to belong to the same race, to mix the flames of the altars into one common pyre and ascension. Shall Cupid then learn anatomy, and the ace of hearts, transfixed with his ancient dart, stand for something in Carpenter's Physiology, or in the dissecting room? The one heart, it is to be observed, is much older than the other; the heart of love beats through human tongues before sciences were born; its affirmers are that great cloud of men now above and beyond us, who lost to individual ken, make the cope of the past blue and immense; mythology, poetry, and language itself are the bright points in this firmament, which still ray down to us the same message about this primitive and perennial heart. Every man is still valued in this sense by his heart. Every feeling comes from it and goes to it. Resolve stands in it, or melts away from it; hope deferred makes it sick; fortune sits upon the wheel of its capacities; it makes the breast by which man touches man, or comes fairly forth from its cage on great occasions, when heart touches heart. The most touching thing in the world, it is the most tangible too; it feels before the fingers, and pulls the words from the speaker's tongue by an anticipated hearing. We should rather say, that all this is attributed to it since the beginning of time. Nor is the attribution lessened to-day, but the air of the nineteenth century vibrates with this heart and its properties wherever free or common speech endures. But we cannot overlook the fact, that another heart has come upon the carpet.

The scientific heart is that hollow muscle of which we have spoken in the foregoing pages; four rooms with nobody living in them; and the hollow muscle has not been slow to suggest, that the ancient heart is a figure of speech, and only exists metaphorically. Meantime, however, the latter cedes nothing of its prevalence, but the words which express it are guarded by the whole atmosphere of life, and keep their places under a weight of forty-five passionate pounds to the square inch. Heavy dictates of sense will not allow them to be evaporated.

Hence arises imperfection and struggle. For an animal with two hearts is lower in the scale than an animal with one, or in which the two are twined cordially into a single organ. And then for struggle, science, which has so much to learn by heart, makes a continual enemy by setting up another heart beside that which has to learn it. But if the two could come together, science would rise into warm-blooded life, and memory, its register, would enlist a new set of feelings in its service, and would become long and tenacious like the heart itself in the higher sense. To say nothing at present of other advantages.

It is, however, a fair question, notwithstanding the tyranny of common speech, whether we be justified, and to what extent, in assigning feelings to the heart, and making a heart of the feelings? Whether the quickness of words be according to a method that the structural heart has known and sanctioned? In short, whether common sense is a great instinctive anatomist, or not? Certainly, if it could be shown that the passions belong to the heart, with all their vocabulary; with the heat they receive from heaven, or summon from the abyss; with the power they shoot through limb and brain; with their play and balance at the core of society; with their issues, circuitous and direct; with their countless insinuations, successions and intermixtures; with their lava that lies at some depth under the coldest action, and sustains the vaulted breast of man upon an oven of flames; if they and their progeny, we say, could be charged upon the heart, the scalpel would have new artifices to employ, to get a sight of these wonderful natives, and the professors of death might well be startled to see the children of fire walking among the sciences. The question, therefore, becomes the more pregnant from the new labors which an affirmative answer would enjoin, and from the alarm which proof itself would cause to all but common people.

What then is the nature of the evidence that the feelings live in the heart? The evidence itself, we reply, if considered apart from language, is a mere matter of feeling. Herein lies the strength and weakness of the case, so far. The evidence, true to the organ, is circular: the heart is a self-supplying knot of affirmations. *Stat pro ratione voluntas*, is the heavy hammer of this logician. *It is because it is*, is childish and hearty—a ring of wilful fire; there is no reason in it any more than in the heart, which is a YEA, YEA, of everlasting man, approving himself by living and by feeling.

Feeling, however, thus affirmative and infinitely irritable, and inviolable in its circle, gives no knowledge of the subject, and where knowledge is in question, though feeling will not yield, yet of itself it cannot conquer. If the case rested here, we could not infer, except remotely, a constant connection between the heart and the emotions. We might, or might not, remember the pulses of our own passions; we might notice our acquaintances beating and striking their breasts when the furies were at them; and so a few violent instances might fix themselves, in which feeling and the heart were together. But feelings have, moreover, tongues, and are the best of talkers; they are notorious for hitting the nail upon the head; they make all men into their poets, and are the authors and founders of languages. The words which convey and assign the feelings are masterpieces of justice and felicity, and hold the sheer perceptions of our brightest moments. They shine with suggestion from age to age. In language, therefore, feeling becomes a staid and intelligible substance, and when the feeling is past, we note what it was by the hearty words that it uttered. Moreover language is a common product, and chronicles the feelings of the world; for the soul talks to be heard, and therefore speaks by a vocal compact. It is therefore no solitary sound, but the voice of mankind to which we now listen, and which identifies the heart with the feelings.

There is no point on which language is more trustworthy; for the heart itself is physically as well as feelingly at the bottom of the wells of language. So near is it to the lungs, that the words in which it signalizes itself are like the bubblings of its own blood. If the heart were wrong in every other synthesis, we should still expect it to be right here. And so when a chorus of nations and tongues chimes forth that their hearts have feelings, we believe from a triple persuasion that those hearts know best about it, and have made them say it; and we take them at their word.

So far, therefore, the case proceeds upon the joint testimony of feeling and speech, and we may now say, upon the witness of the heart and lungs themselves speaking from the life. They were not summoned to give their testimony, but it was extant in their existence. Gesture and speech, which are heart and lungs marched into the world, without hesitation identify heart and feeling.

It is on the same grounds that we aver the whole of what we know best respecting the living body; as that the body contains the soul; that the mind is in the head, and then in the brain; and that the senses are in their organs. And in truth, to doubt of these inhabitations or connections would depopulate the physical frame of its lives, and strikiug out common sense from the scientific faculties, would float the body away from its cables, without a crew, a pilot, or a destination.

Pathology also, or the science of disease, is equally clear upon the point we are maintaining; for violent feelings not only agitate, but may kill the heart in a moment; in short, broken hearts are medical facts, and the tearing of the organ is often coincident with agonized feelings.

But why insist upon a fact which nobody denies? Our answer is, that truths are not well treated when they are only not denied: we desire that these greatest truths of the heart should not simply be assented to, and then passed over, but used as keys to its organization. We desire that every feeling which warms the bosom, should find a place in the scientific heart, and give it the same life which it gives to its human prototype. We desire to conciliate Shakspeare and Harvey, that the genius of the one may cohabit with the genius of the other; that man's real life may not be missing from his blood; and on the other hand, that the doctrine of the circulation may make a prouder orbit, and gain its rightful swoop through life and history. For ever the world is a chaos of truths, but fluctuating and inapprehensible; but when they are fixed, the dry land appears, and habitable ground or proper creation begins, a centre is struck whence order flows; and now we essay to fix this floating allegation of the heart to the feelings, that it may become moored to a solid bottom, and gathering up all its parts and particles, present a sward to the sun of knowledge, whose light and heat dwell with man alone.

What then is the voice of common experience as to the feelings which are assigned to the heart? Evidently the heart stands for the affections, and the man devoid of natural affection is said to be "without a heart." Our first business is, to dissect the verdict of language; and the result may be stated as follows. The friend is a man with a heart; friendship is one of the affections commonly denominated by the organ. The good mother has a heart which beats towards her offspring. The lover has a heart, and is a heart, towards his love. The citizen's heart is for his birth-place and his country: he has a public affection or love; a sense that he, and a certain space with its contents, are warm and related to each other. These are the chief natural feelings, to be short of which is to be morally disgraced or diseased, and cut off from the bonds of healthy mankind. We are also commanded to love God "with our whole heart," but this is no part of the mere nature of which we are now speaking. There are innumerable other attributions of feelings to the heart, but they are either subordinate, or may be classified under the foregoing.

Each of these feelings is a warmth or fire peculiar to itself; each gives a different glow in the breast; each shines with its own life in its going forth. Yet they are one inseparably twisted ray, which seen from its end is the quadrine star of human nature. They are the withes and band-makers of our societies; and they not only draw their own kind about them, but are grappled each to each in the fibrous motives of a mutual self-preservation; kin and kind, parents and partners—they are one man, clasping, and clasped by, his fellows in the fourfold magnetism of nature.

Let us see then whether these feelings have any correspondence with the fleshly organ; in other words, whether the flesh be intelligibly alive, and whether their signatures be legibly written upon the muscular tables.

We stated in the foregoing pages that the heart, by the blood-vessels, is everywhere present in the body, and that the frame, in one point of view, is a double tree of arteries and veins. Assuming that feeling and heart are synonymous, each arterial space is of course a part of the extended firmament of feeling. The organs of the senses, for example, are a fivefold feeling of the external world, receiving its impressions of five kinds through these channels. The sensories, however, are neither more nor less than so many blood-works, constructed and maintained by the circulating streams. Sensations are received, not in dead organs, galvanized by the brain, but in a bed of structural desire, where the mind meets and marries them, and carries them to the head. The heart then produces, by continuity, and at a distance, these animal tendencies to five classes of external objects.

We do not deny the empire and influence of the brain. Each part of the body, however, is alive, and each, so far as possible, is independent. The heart, and all the feelings, live from the brain at last, but the brain has sunk its capital in building them, and they are no longer convertible; they are not brain, but heart and feelings. They are alive, as their architect is alive; and we cannot too often repeat, that there is nothing but life in the body, and no life without feeling, or *quasi*-feeling. We say *quasi*-feeling to express, that even where the feeling is inscrutable and unconscious, we dare not, in reason's interest, give it up as feeling. Where all seems not only dead to us, but contradictory to our life, we still know that we are at work with the same versatile tools; we know by our brains that we are feeling in our very bones, although we never feel our bones but when the nail-prints of pain or inflammation are shown to us. So far as we carry this feeling into our studies, so far we are exploring the living body; so far as we do not, we are groping in dead flesh, and making a science of corpses.

We do not now investigate the problems of embryology. The order in which the body is formed, is one thing; the order in which it subsists and acts, is another. The king is not less a king because he was once elected out of his own subjects; and the history of his elevation is a distinct subject from that of his functions. Let us cut off questions, and take limited fields to cultivate. In these pages we treat of the adult heart, and what it is, leaving a thousand problems to be treated at other times, and by other persons.

To proceed from one expanse of feelings to another, the blood-space which we call the lungs is the bodily affection for those two ends which we term air and thought, whence the synonymous expression of both by the word spirit. The coincidence of thinking and breathing, the representation of what goes on in the mind by what is proceeding in the respiration, is conclusive as to the fact here set forth. The lungs therefore are the love of air on the one hand; the bodily love of intellect on the other; the processes by which they inspire the atmosphere being the same as those by which they rouse and inspire the mind; thus deep breath will produce deep attention; at all events, a careful minding of the air as it is drawn in, if no more intellectual object be present to the brain. It is here to be remarked, that whether we say that organs have feelings, likings, or loves, it amounts to the same thing: for feelings are agreeable or disagreeable; and an agreeable feeling once experienced, leads to a desire to obtain it again at another time, which desire, the child of love, henceforth becomes the active point in the feeling. Therefore the eye is impregnated with the love of the visible world, and the skin, of the tangible world, from the first sensations onwards; and to gratify these eye-loves and touch-loves, the body is set in motion in the service of their organs. It will here be remarked again that we are speaking of full-grown senses and sensations, and that we by no means enter upon any question connected with the growth and genesis of these feelings.

We have now therefore located sense, bodily-felt intelligence, and the feelings, respectively in the organs of the senses, in the lungs, and in the heart itself, and we have seen that the organic system of thought is parallel in the three cases, and that the same reason, namely, feeling and speech, which has caught light in the eye, and hearing in the ear, has also surprised spirit in the lungs, and love in the heart.* This short analysis suffices to mark out certain distinct continents in the psychological map, and to show that the living body, though all compacted of feeling, yet distributes it countrywise; the heart being the central land, where accordingly the feelings proper inhabit, whilst the term feeling in the other parts is changed for that of sense, respiration, &c. Terms however shift according to the occasion when our course lies through new contexts and expediences.

We now then assume it as indisputable that the feelings dwell in some sense in the heart, for experience dictates this conclusion; we feel their correspondence with a certain glow, beating, and sense in the breast, and this unfailing correspondence it is, that forces us to say that the cause is present and agent where the effect is felt. Language, to which we have referred, is the voice of this well-known fact. Let us now adopt this law of correspondence as an instrument, and proceed to apply it further.

The physical heart, we will presume, lies before us, and as we have now exhausted the present information to be derived from our feelings, we can only regard the heart in its physiological functions. The question is, whether these exhibit any correspondence with the emotions; whether auricle and ventricle, and their bloods, are expressive of passion, somewhat as the face is expressive; whether, in short, the structures of the heart in action are not a countenance

* Let it be borne in mind that we use the term love in no limited or sexual sense, but as embracing those active central impulsions that connect mankind everywhere to their human objects.

in which the play of the feelings may be detected? If so, we shall make matters of feeling into fresh objects of sight and sense. Let us try. The attempt is one of synthesis, or the putting together of the pieces of language and passion on the one hand, and of the pieces of heart or blood on the other, in one doctrinal machine. In such a composition, it is manifest that fittingness will be the test of truth; and that if the love-heart interlocks with the fleshly heart by the dove-tails of a just analogy, what we at present propose will be accomplished. We shall see the veritable clasps by which the two are grappled in word and in fact.

Certainly the heart shows all the signs of loving the blood, which is the fearful and recognized symbol and casket of human life, for it grasps at the blood eighty times every minute. With quadrumanous hands it clutches the passing life-stream. If the life is indeed the blood, and is in the blood, as the Bible says, then the heart grasping the blood, is the very love of life, and in our case of human life. Its eagerness is apparent from its work, as the busy hands of men show that they also are *con amore* in their occupations. And if we love the blood of our race and kindred, and embrace it through the skin and outworks of their bodies, much more does the heart love its own blood, which it squeezes hot and naked with its ruddy fingers. Our own affections, which we interpret from their actions, are far-off types of this affection in the heart; which moreover shows an answerableness to love's common signs, such as no other organ testifies. There is a fiery and as it were abstract purity in this passion of our bosom's lord. If *we* love ourselves, the *heart* loves the life which is the self of self; and this love it shows by grasp after grasp, by a zeal which never sleeps, by fresh manipulations of the blood with every varying feeling; in short, by all the signs which show that we ourselves are in active and impassioned pursuit of our objects; but these signs, raised to a constancy that belongs to no will, but only to nature, or the fatal will of will; and exempted from that fatigue which makes night and day into the blessing of mankind; for the heart is its own day, and works in the fire-factories whether the outward man be turned from it, as in sleeping, or revolve round to consciousness of its influences, when his morning feelings seem to rise. The heart then corresponds to the love of human life by its everlasting grasps and embracings of human blood. By its deeds we know it.

It may however be said of this reason, that when we see other persons acting, we infer that they have feelings of a particular kind, because in similar actions of our own we experience these feelings ourselves: but that the works of our hearts are not a parallel case, and therefore we cannot argue that hearts have feelings. To this we answer, *first*, that hearts, as we proved before, have feelings, and we will not be dispossessed of a truth which we have got. Moreover we are not arguing that question, but are investigating how hearts show their feelings; and we have now found that they do this as we ourselves do. But *secondly*, we do not say that the case between our feelings and theirs is parallel, but correspondent and like; and of the attribution of feelings and life in this way, common sense gives many examples. Why do we attribute passions to the tiger or the dog, or gentle feelings to the lamb, but because we know that passion and feeling may descend many stages beneath our consciousness, and alter so that we can never experience them, and yet be passion and feeling still? All we contend for is, that the heart is similarly circumstanced; that it is in no sense dead, but as the old anatomists said, "the animal in the animal;" in which case we treat it as we treat the animal kingdom, and infer life, feeling, instinct, ends, as the account of its operations. Our own experience and faculties are therefore as fair an organon for our study of the heart and the other viscera, as for the investigation of natural history, to all whose subjects, feeling and thought give their own lives, in the certainty that they will fit, and more than fit, the case. And in proportion as the feeling is broad and common, and the thought scientific, the better does this method succeed in exploring the actions and habits of the lower creatures. We design then here at first to show, that we may safely, and must inevitably, transplant that life which we understand, into the heart, as we have already carried it into zoology.

Dismissing this general sign, of the heart's eagerness, passion or love for the blood, and noticing that it is common to all the four chambers, we have now to come to details, or to the specific auricles and ventricles.

Let us commence from the right auricle, which is "the first part of the heart to live, and the last to die;" the primordial feeling and the latest passion of our bodies. And first we have to ask, what it is that this hand would seize? For the hand has many objects, but wherever it closes, it has a desire of possession, and the grasp is proportioned to the object. And so of the heart, which is the visceral will, or the inmost hand of the body. Now vessels and their contents signify one and the same thing, under active and passive conditions, and as each part desires to perpetuate its life, each grasps at its *alter ego* in the fluid form. The fluid, therefore, is the index of the solid which holds it; the organic cup answers to the cordial; the ruby of the chalice is the wine in a mineral metamorphosis. We have, therefore, to interpret the desires of this auricle by noticing their objects. What are they? The old blood of the body, the elderly blood of the brain, itself wise or cerebrated by its visit to those upper regions, the return blood of the heart, and the white young blood, or the conjoint chyle and lymph, meet together in one common chamber. The end and beginning of life are there represented; at the point of completion of the circle the extremes of existence touch. It is the house of the heart where the elders behold their posterity about them. All that could die of the old blood during its last generation or circulation, has been put aside through the medium of many secretions, and the activity of numerous organs, and in the heart again it is a mere abstract or passion, immortal for at least another circle: hence the old blood is but the old in the young, about to continue the gyre through another curriculum of ages. Father-love and mother-love, or the love of race, naturally exists in such a group; the bond which seizes the inmates is that which makes families out of individuals. The tide of feelings sets in from this first grasp of the heart, whose contents in the right auricle are embraced by the family tie. The free blood receives the impression, and is a patriarchal clan. The right auricle, the first parent of the blood, sends down parental love, as the first river of life, through all its generations, and also recruits itself every moment from its latest offspring. On opening this chamber, then, we see the perspective of race in its various phases; processions of parents and children, the everlasting progeny of the heart:

man has met man by his first points of contact; the old are as gods and influences to the young, and the blood of our hearts is no longer vague and venous, but it is housed, and feels in the home the powers of time descending from behind and from above, and giving it the first force of past and future in the attachment of race. We gather therefore, from the correspondence, that this fire of natural affection plays upon this home and its inhabitants, and teaches them to be posterity. We call this the hereditary auricle, into which the blood flows by the pressure of fate, in the same way as generations descend from the sources of parentage.

The blood is now no longer indeterminate, but has received the contagion of one life; the first cord of love or union has been passed around it, and it is full of the household warmth; the right auricle, the ancestor of all, has laid hands upon its generations. But what in the meantime has happened? The solid has grasped at the fluid, the love at its object; but as between solid and fluid, where the solid is an open circular channel, it is plain that the total object can never be caught; the attempt at seizure forces it into progress: the maid pursued by the urgent lover is turned into a stream by the friendly gods, just as he seems about to overtake her. The right ventricle receives the one-lived blood, and fills with it. What is the character of this new object of the heart, at which it is next to grasp?

Into the right auricle several streams of blood distinctly emptied themselves, old from the body and the brain, middle-aged from the heart itself—though this in small quantity compared with the rest in the hereditary cavity—and infantile blood or chyle in the current of the rest. It was family which the heart desired, and the family tie which its contraction took and gave. The feelings which we have in our breasts were there at work, and are always there at work, *in minimis*, upon our blood, making us naturally into parents from our first drops upwards. In the right ventricles we have another stage. All the life or blood which is not permanently familiar, eludes the grasp of the auricle, and belongs to another chamber. In the right ventricle there are no distinct streams from different sources, but its blood enters it by a single great orifice, in one uniform gush. The right ventricle has been aptly termed the mixing-vessel or chaos of the heart. Parentage and non-age have both disappeared, and the area is equality and fraternity. The common feelings of man to man, philanthropy or the friendly affections, are at work here; those feelings which know of no distinctions, but only of brotherhood, toleration, and even-handed intercourse. In the family loves the tie is unequal, descending from parents to children, but in no similar proportion reciprocated; this being necessary in the beginning of the circulation, in which progress as a spring of pressure is the wheel that sets the rest in motion. In the philanthropic passions, however, the tie is double, coming from both sides; hand grasps hand; the muscular contraction of the organ is closer; the ventricle of our friendship is of twofold strength. Here then we have the blood-population in the fiery palace of the heart, themselves all feeling, with no distinction of high or low, old or young, father or child, and what direction can the feelings take but that of universal community, of friendship in its various phases? The mere apposition of lives in such a place, and under these circumstances of unrestraint, is sufficient to produce the relation that throws down the walls of other distinctions between man and man. We augur then that the friendly emotions play especially upon the right ventricle; that it is there they are felt, and there they live, and enter the blood, giving it this tie momentaneously, as a needful element in the constitution of the bodily society.

In these investigations let us never lose sight of the keeping of the subject; of the fluid which we are pursuing, and which is *both* blood and life; and of the organ, which is *both* heart and feeling, that is to say, heart in both senses. Blood in the heart is different from blood in the head or the belly; in the former case it is alive with passion; in the latter, it hungers and thirsts; while in the head it is subordinated to spirit. Each organ has its *genius loci*, which possesses everything, even the most transient guest, in the organ. To come then within the sphere of the heart, is to feel and to be all that the heart is and means; for the heart is haunted ground, and there is no escaping its influences. So it is that when youth and maiden come under the grasp of the heart, or when life carries them into the sexual auricle, they are not the same people, nor have they the same names, as when under the parental roof, or in the friendly conclave; but love has located them afresh, and gives them its own new names. This is a needful memorandum in the laws of the heart and the blood. We have feeling, the fire of life, already given by fact in this organ; we need not endeavor to import it; the business is, to see it in its place.

The grasp of the right ventricle, by which it gratifies its friendship, throws the life upon the lungs, where the blood and the larger world first meet, and here the humane chaos, which is at its height in the pulmonary artery, begins to be discriminated into a new order. The spirit of consideration comes in the spaces of the lungs. In these reservoirs of the voice, the blood hears and takes part in the public murmur expressed by the breathing. It speaks forth its obstacles, and puts them off, telling its mind, and regulating its attachments. Thought and breath, as we have already seen, are the united spirit of the lungs; thought and breath only subsisting by virtue of supplies from the world. The blood raised to the height of breathing, is full of public imaginations; it has thrown aside childish things, and is polar to a new and vast ideal; the private robes and windy latitudes of its childhood are put aside in expirations, and the virile or public toga is put on, with the new airs, imaginations or liberties that belong to adolescence. It is inspired with the service of the corporate body. The transition from the veins to the arteries is an incarnation of the passing over of life or feeling from the private to the public stage; of mankind meeting the world, and flaming with a new lustre of eye when the great objects are recognized. The public air is thenceforth inserted into the feelings, whereof each goes round with an oxygen mirror that shows it the universal in the individual, and makes it shape its face and gestures into historic parallels. Every globule of blood thence conceits itself that it is a man-maker and a world-maker, and it is braced with the girdle of the public strength. The feeling in the lungs, accordingly, is one of inspiration and erection, in which the heart's proper feelings swim as in a new atmosphere of power.

The change of the blood, from venous to arterial, takes place in the lungs; a change which is ably represented in the countenance under the influence of enlarging or enlivening passions. The glow of the cheeks when life is in force is due to the same cause as the glow of the arterial blood (p. 182). Powerful healthy feeling is red and burning; and is the true blood and the only animal or living heat, to which nature, with deep architecture (p. 89), adds mineral or dead heat, itself also red, as we see in the fire. Cheeks and fires are the reasons why blood is red; inanimate countenances and cinders are the reasons why it is dark and venous. Why do we glow but because we are alive with some public spirit or motive? and why does the blood glow, but because it is alive with that public spirit the air, and bound thenceforth to act according to the pressure, and to assume the mission of the universe? We are not insensible to the pleasant weakness of this theory, but we are treating of the circulation and twisting reasons into circles. To exhibit it let us say, that the arterial blood is red because the passions are red, as witness the face while they inflame it; and on the other hand that the face is red because the impassioned blood is red. Truly a circular logic, and dear to the heart therefore. We should despair of understanding any organ if we could not feel with it and follow it; and to follow the heart, drives us into self-supporting axioms. For the heart is a proposition that never goes beyond a bare statement, but pumps through us the substance of self-evidence, which is the body of our body. The truths of blood and feeling are the *ipse dixits* of the heart.

We have now left the private or venous passions, whereof the first, the family life, ancestral house or right auricle, represents the impetus of time, causing all the movement of man and blood by the descent and tradition of generations; while the second venous passion, the friendly and philanthropic home, or right ventricle, represents the indiscriminate brotherhood of man, which gathers up the race in the second and highest of the private bonds. Both of these, as we say, are private or venous; domestic in the narrow or the wide sense: in their largest cases it is the private sphere expanding itself; for a whole clan is private and immiscible still with the rest of the community, and the largest friendly *reunion* still contemplates privacy or intimacy, and would cease as such the moment it were touched by the laws of another love, or of social rank. Both these loves are hot—black-hot, or red-hot; but neither of them can flame, because they have not met the air, which is the public in contradistinction to the private. Public feelings are therefore the arterial blood of the heart, while private feelings are the venous. A complete turn of objects exists between the two; the right side of the blood becomes the left and weak side seen in the airy mirror, and the left side becomes the right, and sits at the right hand of power. Every public object proceeds from the public to the individual—from the common air into the blood; whereas the private objects run from the individual to the public, or from the blood towards the air. The meeting-point of these two efforts, and of the feelings which run with them, is in the capillaries of the lungs, where the air converts the one into the other, and where consequently the man becomes, and feels himself, inspired.

The lungs pour the blood into the left auricle, and the first moment of arterial life is spent here; the first feeling that blushes and runs over from the redundant passion. The beginning of virility is sexual knowledge; the *mirage* of Eve at the fountain is a vision of rosy flesh seen as her own, and yet felt as other than her own. Love stands at the gate of the larger life, and revels in its flower. The arterial lungs are the puberty of the blood, coming from the dreams or imaginations, with whose hints and incentives the air is full. The left auricle is the marriage-bed where the tension of this bursting life is continually taken down, and renewed continually. Accordingly this auricle or bed is the most hidden of the chambers, decently curtained away by the rest, and least to be felt from the outside. It supports, and is supported by, the family auricle, with which it moves in step; the synchronism of parts with different functions showing the harmony of different ends, and their ultimate working for each other. The first arterial life of man is then sexual love, whose early stage, preceding conjunction, is incarnated or ensanguined in the arterial circulation of the lungs, and its consummation is in the left auricle, which grasps at the paired and mated blood, to realize sex and marriage as one of the four corner stones on which all flesh is to be founded. The amorous feelings therefore play upon the left auricle, where nature is always doing their work.

We have drawn a distinction between the private or venous, and the public or arterial sides of the heart, but it might seem as if this left auricle and the passion that lives in it were both of them of an intensely private character. Let there be a great distinction noted then between private and secret functions. A friendship that is noticed by all the world is not the less a private friendship, whilst a marriage consummated in the most secret bower, and in the bosom of midnight, is public and regular notwithstanding. The council-chambers of nations are most secret, and yet most public in their relations; they resume the state in a concentrated form. Hence the public chambers, or left side of the heart, retire from view to transact the public business of the feelings; the private business is comparatively open and forward, and is not meddled with because it is known.

But is the sexual relation arterial, aërial, or public? Philosophers discuss at great length the *me* and the *not me*, and the relation between them. There is no such instance of this as sex, which is *me* and *not me* reciprocally; each incommunicable to the other, and each fitting and enjoying the other. Man's most external world (for he has several) is woman, the living mould of himself and his faculties. In her he first knows that there must be real size in the feelings, or two beings, who may be conjoined, but cannot be identified, could not be together there. In short, sex or distinction is the beginning of breadth and body, of resisting spaces which can never sink into each other, but must lie side by side. Moreover the sexual distinction is the most manifest between human beings, and the dress which makes it secret makes it public also.

We repeat then that the left auricle is that in which the analogue of love proper is introduced into the blood, and where, by the laws of harmony, that love is physically felt in the blood which is doing its work infinitesimally. The reasons of this in brief are, that the first adolescence of the blood is passed here, and love crops existence in its bloom; that it moves with the right or family auricle, there being a co-existence of the conjugal and parental ties; that it leads into the patriotic heart or left ventricle; the state being the congregate of all the broad or spacious relations, of which marriage is the unit and the beginning; finally, though at the top of the public impulses, it is the most secret of the chambers, and is not merely the house, but the bed of the organ. Whatever difficulty there be in exploring it may be assigned to its own crimson modesty, and will furnish a fresh proof of its chaste or sexual signification.

The grasp of the auricle, which consummates this life in the blood, drives it onwards, as before, into the next chamber, or the left ventricle. The signification of this fourth heart cannot be doubtful. It is the accumulated power of the passions. The blood that it throws forth is scarlet with force—it is the systemic circulation. It is synchronous with the friendly heart, but fourfold stronger. It forms the apex of the cordial pyramid, which beats against the ribs, and aims at the world though its dearest flesh. It is public feeling in all its forms, and we have already anticipated its name, and called it the patriotic heart. Rule and empire throb for ever here, founded in the purple of the blood. The body corporeal streams from the height of the left ventricle, as the body politic from the heart of dominion, which is the instinctive architect of the state. If vicegerency of functions establishes connection between life and organism, then the love of country or empire must sit upon the throne of this ventricle, whose lordly stroke reaches the confines of the body, and seizes the central blood itself with a conqueror's grasp.

In the successive consideration of the blood or feelings in the heart, the life of the blood is an ultimate fact, which we need not endeavor to account for. It is a given truth, that the heart or blood has the feelings, and it is no part of our business now to speculate upon it. As it is with the child, so it is with the young blood—it has the capacity of going through human life, and of being and doing whatever lies in, or issues from, the heart. Our inquiry then is, mainly as to the incentives or circumstances that call forth its faculties; in short, we have chiefly to chronicle the play of its human games. These once depicted, will answer the question of the reasons of life more deeply than that question can now be asked.

But let us notice in this place that the life which the blood feels in the heart is cumulative. Each cavity seizes it with a feeling whose effects it does not lose. The family bond is assumed in the right auricle, and though rendered latent in the right ventricle, where it gives place to the friendly tie, yet throughout, the blood is the child of the heart, and remembers the household fire.

Hence life is added to life, and the four "vital principles" of the heart are simultaneously in the adult blood. It is amusing to consider how any philosophers can have sought for *the* vital principle in so many-lived a creature as man. The vulgar, in assigning to the cat "nine lives," have shown an example of common sense, which might well have been applied towards a being whose vital principle is as organic as his body, and has its very parts over again, or how could the body be alive? Hence in the right auricle, the blood-child is one-lived, or is in the parental leash; in the right ventricle, the blood-youth is double-lived, has received a second squeeze of passion, or represents friendship's hand in hand; in the left auricle, the blood-lover is treble-lived, or is lover, friend and parent in one, the three being inseparable; in the left ventricle, the blood-man is in the service of the state, which is the public ordination of all the feelings, or the carrying them out into the body. Hence the left ventricle does not so much alter the blood as receive it all, and give it high public fire. We now then see what it is that the heart confers on the body, and that it is the same set of endowments, only organic and infinitesimal, that the same heart gives to the man, and to the society.

Such is our equation between the heart of feeling and the heart of flesh, which are the same heart, only in different powers. The blood-heart $= \sqrt{}$ heart; the love heart $=$ heart^2; the heart itself being that invisible pulser which we feel under our ribs, and the knowledge of which can only be filled by a conjoint corporeal and social anatomy, or a hearty exploration of all death and life. We have indeed drawn out the parallel on only the most general grounds, and know far too little of the feelings on the one side, and of the heart on the other, to enter into details. But if the learned will bring dead hearts, and the simple their living ones, to a fair comparison, we can now no longer doubt that each will adopt the other with most specific joy and claspings.

Our hearts, we have said, grasp at their objects, and we have tacitly assumed that they get what they grasp. But how is this, if they simply drive the blood away, and pump it into circulation? In this case life would be not a substance but a mere stimulus, a perpetual alcohol of illusions. Let us remember that we are treating of body, and of a realm where everything is bodily. Does the heart then obtain nothing of that which it clutches? Is the ox muzzled that treads out the corn? And is all the real blood thrown outwards from the monarch of the feelings? Analogy forbids. If the heart is a hand that grasps, it is also a person that gets; it is a very lord of its own objects. The caverns with which it is sculptured, are so many means for retaining the blood; the little mouths with which these are studded, are eagerly absorbent and retentive; and as the heart contracts, it fills its substance with immediate blood, by the same act with which it drives the mass of the blood into circulation (p. 189). The heart, we aver, takes the central and most living blood. This it does by the love-laws and justice of physics. The best blood is the fleetest, and enters the cavities first, skirting along the walls (p. 189); the next living is another layer which comes up afterwards; and lastly, into the middle, comes up the slowest blood in the rear. The assembled blood in each cavity is grouped in a peculiar form, and contraction works upon it according to the form. The first part of the contraction is stimulated by the fastest and most feeling blood proper to that cavity, which is driven by the dead pressure of the central blood, and the counter-pressure of the solid walls, into the substance of the heart itself, where it constitutes the realized life of that one feeling or heart-beat. The racers however are different for the different cavities; the family-blood is fastest, and wins the cup, in the right auricle; the amorous blood, in the left auricle; and so forth; and centre and circumference vary as the goal is changed. We see all this well enough in life or in the play of the great heart. Each being is prompt and rapid in the working of his own relations; he who is a laggard in friendship, comes to the surface, and shines with vigor and promptitude when love catches him; showing that each man belongs to a centre, and lives eminently in and from its fires. And for the parallel of the first point, we know that the heart is not satisfied with grasping at its objects, and feeling their slipperiness and flux, but that it must have possession, or the hollowness of perpetual mockery makes it cease to grasp.

Although then the blood of the right side of the heart is private or venous, yet it is its most spirited portion that enters the living solid of the heart, and feeds the home-fires that blaze on this side of our nature. Thus this blood is more than arterial in any ordinary sense. We will here also note another point in the doctrine of the heart, viz., that the flesh is prior in dignity to the cavity; the flesh being the substance of the heart itself, while the cavity is only the general high-road of the circulation. Harvey's great discovery has had the effect of throwing this obvious truth into the back-ground for a season, but by the love we bear to whatever is solid and central in our hearts, it must be replaced, and the doctrine of the circulation subordinated to it. The body will be but a hollow shell, so long as the heart is regarded from its hollows, and not from its substances. But in order to investigate these, a new course of observations and injections will be necessary, undertaken expressly under the light which the correspondence of the feelings casts upon the physical organ.

Here we begin to see how profound is the reason for that doctrine that the heart absorbs its own blood through its walls, and does not depend upon the coronary rills for its supply, which would beggar it of what it grasps, and put an end to its motives for contraction. We must never suppose that the heart's wisdom lies in doing that which we should be fools to do in similar circumstances. If all the stores of the earth passed through our hands, should we tranship them, and then get our own sustenance back in a little skiff from a foreign country? Should we not open a retail on the spot, to secure ourselves against difficulties, winds and tides; and come as near to the wholesale stores as possible, avoiding the dearness and losses of trade? And if it were the monarch's palace through which all the goods were carried, as in the case of the sovereign heart, would he not take the pick of every object on the spot, without trusting to any honesty less than that of his own eyes? Our hearts feed more at the fountain-head than our societies; what is wise in our conscious arrangements is but the broken meat of the organic wisdom: the passions browse upon the mountains of the heart, and lap the blood of the life-covenant at first hand. We feel that they *are* our bodily life—that they *are* our blood, and that if anything in the world goes direct to its objects, it is they. For they are the exemplars of all-grasping, all-possessing, and all-holding man; and the heart is they.

But we have to notice further a point touched on in the preceding pages, viz., the mixture by emissary rills of the blood in the various cavities (p. 189). On our principles, it is a psychological necessity, and must suggest long trains of experiments. Nothing, we know, can be more mixed or miscible than the great primordial feelings, and the mixtures can never take place without mixtures between the blood of the contiguous chambers. To instance only the case of the conjugal and parental relations, how could these subsist unless by the most liberal communications? The parental love generally contains the conjugal, and *vice versâ.* The feelings, as four pure atoms, would be as barren as nature's chemistry if it consisted only of simple substances. In fact we may say that the main office of the heart consists in blendings; in the intertwinement of private and public life; in making countless binary, tertiary and other compounds of the feelings. In proportion to the combination, life arises, rich, delicate and lovely from the ground of a few simple elements, and fair kingdoms adorn what would otherwise be a stony level. On the purely moral side, the object of all existence and circumstance, is to produce single-heartedness throughout the relations; to universalize every feeling by tincturing it with the rest; to absorb the egotism and correct the weakness of each through something of a wider love inserted into each from the others; in short, to give the heart its last unity, by impregnating the private with the public, and carrying the former into developments, and the latter into exactitudes, that neither could attain on its own account. Therefore in the feeling heart, which is the true fleshly heart in contradistinction to the stony, mixture or communion are the law of life; and we arrive at the deduction that the heart is a high forum of intercommunication. The chambers are exclusive enough to constitute separate ends, but not to realize that fancied independence and unneighborliness which nature abhors. For nothing is so fluid as feeling; nothing knows so little of walls as love; of barriers as ambition; of difficulties, whether fire or water, as friendship; or of time's weary limits, as family joy. And thus we conclude on this side that the truths of communion are those which are proper to the heart; and that they are greater than the truths of trade or circulation, which are proper to the arteries.

But how are we to see this in the anatomical heart? It is true, the heart itself tells us that it is there, but how shall we make it visible? In the first place, by holding it as among the greatest of facts, and not losing heart in it. In the second place, by investigations and injections with an end and a purpose. In any case, the beginning of success must lie in the belief that the physical heart is the centre of blood communion, as the feeling heart is the centre of human communion; a fact which is very certain, though how the case stands has yet to be worked out. Something however has been probably contributed to this by Swedenborg, in his deductions respecting the intercommunications of the cavities. In the meantime we affirm, that the heart, which unites all the feelings and their blood-streams, is the bodily blood-maker; that its blood contained in its own flesh is so transcendent, that it keeps gyrating through the centres many times before it is fatigued or exhausted; that it is more constant to its objects than any of the blood of the body; thus that the circulation of the heart is not to be termed circulation but community; or if circulation, that it is many times circular—a knot of living rings; that the left side flows into the right, and *vice versâ*, without the intervention of the systemic or pulmonary circulations; the heart being a spheral thing of cycle and epicycle, a globe of golden girdles suspended in the central life. Through this passes the great arterial curve which attaches it to the earth, and the pulmonic ring by which it is hooked to the air; but the heart, so far as it is true to its place, does not swerve from its own roundness, but remains in every respect central, in its feelings alike as in its blood.

If this be so, it would appear that the ancients were engaged upon this higher problem of the circulation, but were unable to solve it, when Harvey came, and took us down to a lower field, where truth could become more definite at the time. They felt the fluctuations of nature in their bosoms, and argued that the blood went to and fro in its channels. There was a manliness and mutuality in the thought beyond the science of the time; for certainly the immediate relations of the heart are of a more spiritual power, than that large and roundabout intercourse which makes the tour of the world before it touches the centre. But the incompleteness of the latter theory, its leanness in moral significance, its comparative heartlessness, had to be seen, before the former could touch upon physics. Moreover the harmony and union between the body and common life, required to be known and stated, before ever an eye that could see the heart of man could come into the world. But a time has arrived when the feelings of the ancients can be vindicated, or when the communion, or goings and returnings, of the heart itself, can be added to the doctrine of the circulation.

It may be urged that we are reasoning all along upon preconceived opinions, which is not the method in vogue, and that we are expecting nature to conform to our ideas. We reply, that we are investigating the heart of man, in which the main evidence is the feelings. From the height and in the compass of these we would observe and experiment. The relation between them and the heart is of mathematical power, and makes prediction into a duty. We see that they cannot live in harmony with the heart, unless it be so and so; we argue therefore *towards* the fact; but leave science to find it out. That the present notions cannot be the last, is clear as day; for there is not a common feeling of mankind that attests them; not a thought of human life that connects itself with them; not a good or a brave heart among us that sees itself in their glass; and not a moral truth which reposes upon their basis, in the same way as moral relations are founded upon the human heart. They are no better than tricks of hydraulics, which are dead mummeries in our spiritual city. The living body disowns them with horror as having no souls.

But what a series of moral parallels the current physical doctrine would suggest! Hearts that grasp at their objects for lifetimes, but never catch a drop: disappointment made into an organic principle! The soul and body founded upon illusions! Hearts that are as dead as flesh can make them when sundered from our well-known life! Hearts only cognizable *post-mortem*, and without a spark of outspokenness or candor; not to be trusted till they are stiff and cold! Hearts without individuality, and which throw everything away upon other organs! Hearts that are fourfold prisons, each a solitary cell, where the felon-neighbors feel each other thumping, but have no intercourse! Hearts whose food consists in their own regurgitations!

Hearts in short which are bloodless, lifeless, sympathyless, the fools of peddling rills of circumstance, purveyors of blood to the troughs of the senses! Carbon and oxygen carriers, or coal-porters and scullions to the body at large! Truly materialism is never so ludicrous as when we see its upshot translated into the terms of life.

Our aim is the very contrary. We desire to see in the body a heart that gains its object every moment, and in so doing, ministers to the body at large. In the man, a heart that throbs with the enjoyment of his own life. We desire to take the doctrine of the heart from our noblest breasts, and to have the kingdom of the living body come over again in the sciences of the physical body. We desire to see the heart central in its relations, unspent in its operations, a communal sun, giving life, but keeping life for itself. We desire to see its separate affections open as day to the warmth or streams of their neighbor's lives and fires. We desire to see the heart built up before all things from its own immediate objects, and provoked to life by no coronary squirts, by no circuitous afterthoughts. In short, we desire that the heart should be alive and loving; that the blood should be the life; very much in physics after the same fashion that the Bible commands for our hearts in that which in these days of divorce is thought to be a different sphere. But whether or not the present doctrines of the heart may not be equivalent to the treatment of man's heart by the world, we leave to others to follow out.

In this place we will take occasion to illustrate to the reader the nature of *abstract ideas*, taking our instance from the feelings, which are so vital to us, and generally considered to be of an abstract nature when thought about. What are human feelings? In the concrete, *i. e.*, as substantial things, they are clearly the man himself alive to certain relations. Thus, a friendly feeling is no other than a particular friend through whom is passing that motion or emotion which produces certain results when occasion serves. So too friendship is the gathered maniple of all friends, quick with these emotions, and gravitating like an atmosphere to satisfy them. What then are the feelings *in the abstract?* They are the hearts of these same people with attention inwardly directed to them; the skin and ribs of the manly image are peeled away, and the naked heart is contemplated. The true process of abstraction here consists in taking away everything but the heart or centre, and regarding it alone. The philosophers have made it lie in removing *everything*, in which case it consists of words unattached to objects. Whereas its essence is, to find the core of the thing in hand, in order to see its relations centrally. Thus again in the case of the mind: in the concrete, the mind is the man; in the abstract, the mind is the brain; the difference between the two lying no whit in bodily substance, but in clothing or development. For "pure mind" is a living brain; and mind, as commonly used, is a living person. It is necessary to carry this ballast about, lest words should fly away with us. It is especially necessary in treating of the heart, and of the incarnation of the popular heart in the scientific one. But by no process of abstraction can we decompose the heart, unless indeed we find out a deeper bodily centre, which is the heart of heart: which organism will then be a more pure abstraction. We have indicated this, because the philosophical belief that feelings are not incarnate, has been a principal reason why the scientific world, duped by metaphysical terms, has never thought for a moment of seeking the feelings in the heart.

A living anatomy then gives us the abstractions of which human life presents the concrete substances; and the science of the present inner man consists in tallying the world—*le monde*—with the organic frame, as it were checking the one by the other, and translating the one into the other. It does not, however, consist in reasoning upon language, but in treating language as one among organic things, dipping it in the blood-streams, and setting it piece for piece against flesh.

Another point to be noted is, that as our plan is one of conciliation, we by no means wish to set up the heart against the head, or to deny feelings of all descriptions to our minds. We grant, on the contrary, that the whole frame or soul-house is made of nothing else, but in various degrees. The man radiates from the central parts, but reposes on the way at several stations, where he is refreshed with new names and characters. That part where the grosser consciousness begins, takes the credit of being the prime habitation, whereas the feeling is a line with several nodes, which runs down from tip to toe. Language, which proceeds upon feeling, does not attribute to the brain what is felt there only mentally, when there is another place where the same thing is felt bodily. The affections are the mind's as matters of thought, but not as matters of speech. They belong by bodily or natural correspondence to the heart, and are felt beating there, but their physical motion is not felt in the brain. In short, but for the heart, they would not be carnally felt at all, but only thought about, and it is their carnal feeling which belongs to the body. On account of this they are passions, for we *bonâ fide* suffer them in the heart, but command, or ought to command, them in the head.

We are now then prepared for attributing in the same bodily sense other things to the heart, without prejudice to the mind, which in truth is built upon the heart. Each feeling, let us remark, is surrounded by its own film of imaginations, which body forth what it is, in a *quasi*-intellectual glass. "The imagination of the heart" is the light of which feeling is the fire, and causes every emotion to shine with its own peculiar ray. Where are these imaginations localized? On the inner membrane of the organ, we reply; for the inward skin of the heart is its blood-face, in which it expresses its desires. Skins are always facial structures, and express what lies beneath them. By this membrane, the blood and the heart understand each other, or look in each other's eyes, which amounts to the same thing. Imagination then is localized all over the cavities, which are indeed the express moulds of the thoughts of the heart, or as the Scripture phrases it, the chambers of imagery. If we desire to know what these bodily imaginations are, their shape is given in the fine membranes which transmit the emotions of the blood, and the motions of the heart, reciprocally to each other, and make them acquainted.

Imagination, however, in this case the heart-house, consists of a common chamber with two doors, and here we remark that everywhere, in the mind and the body, desire opens us by means of imaginations, and thus is the dilating, opening or cavity-making faculty. We open our mouths for food under the influence of hunger working through and shaped by the imagination of a supply: we make ourselves hollow, to take in what supports and enlarges us. We open our hearts to receive new emotions; our minds, to receive new ideas. Every desire expands us, and each is accompanied by an imagination of the object, which limits the opening to something like the proper size, or constitutes its walls. And so again we are presented with the fact, that imagination is depicted upon the inner walls or front face of the heart. We notice in the lungs the expansive correspondence of imagination, where it is manifestly want (p. 131), or an elaborate vacuum produced in the man by nature every moment, to incite him to new infillings. Imagination then is the limit to which the walls stand off, the horizon of present life, or the arch of the caverns of desire. It is different for every cavity, according to its shape, objects, colors, &c.; but wherever there is a cavity it lives, and peoples the hollow with its teeming forms. It is here to be noted, that desire and imagination are one and the same thing from two points of view; the desire being the hollow or void that we feel, *per se,* while the imagination is the same regarded from the walls of the void, which imagine it full of what it wants to occupy it. Thus the heart-desires dilate its cavities before the blood is given them, and in the momentaneous void the imagination anticipates the tide which is to come; and the heart is in every stroke prepared, both by roominess and welcome, for the new life. Thus again, supply *ab extra* is the law of existence, and preparation for the supply a part of the same law.

In the heart we notice valves, which prevent the life current from running back; in the feeling heart there are *states,* or spiritual valves, which hinder the life-loves from regurgitation. To make this clear let us use an instance. The present time has its own fixed point, from which we regard the past. Manhood can only look at childhood through manhood: the experience and circumstances of the latter are the present state that flies up before our eyes in all attempts to reach the past: in short, we cannot go but only picture backwards. If the imagination is very transparent, or not greatly colored by the present, then we call it memory; if opaque, and full of existing passions, then it is merely imagination, which cannot conceive anything beyond the hour. In both cases, however, it is properly termed a state, a film detached from the present, which is applied back against the past, and constitutes a genuine valve in the feelings. Most of us are so full of these valves, that we cannot attend to anything but the present, or believe that we ever were little children, except in the sense of time, but not of state. The present in that case is made of iron, and the imagination of the past is a hatchway that shuts up, and transmits no beam from the infantine days. There are then valves in life, which prevent it from going back. And indeed this valve-function is universal. The flaming sword that turned every way was none other than the valve of Paradise. Death also is a valve, which nature keeps up during all moments of investigation, or regurgitation, and only opens from the other side when birth is to take place, and when nobody is thinking that it is the same door through which the old citizens have departed, and the new ones arrive. Yet the same it is, only shut, contracted or earthy in the one case, and open, alive, fleshly and maternal in the other.

Valves then in the feeling heart are the present by its activities shutting away the past; memories are the states of the present in which it endeavors to image the past; and they live especially upon the roof of the cavities, or upon the outspread valves.

The heart desire causes it to grasp at the object with which it is now filled, *i. e.*, the blood, and because life cannot go backwards, (for the present stands at its back, and keeps it from the past), it must go forward into the unknown, where no imagination can hinder; for imagination is our limit, but of the future we have no imagination that can dare to bound us. To-morrow is therefore always an open door, and time streams onwards. And yet there is an imagination of the future, which we term hope. Hope however is not a wall, but a hole in our advancing lives through which the firmament is seen. Its sky has an arch of definite blue, but this is, we know, not the end of things, but mercy's color for the weary point of all our sight-rays. Hope is that hole-work in our nature through which we see the heaven, and to which we stream by propulsion from behind, not less than by the unresistingness of the orifice, and the attraction of the greater space, so immense and so lovely. These hopes lead us on into the lungs or the universe of the brain, and into the system, or the universe of the body. As the past becomes impossible, the future opens, or as the blood cannot regurgitate, it is driven into circulation. Hopes then are the doors by which the little cavities of desire and imagination communicate with the great cavity where light shines in a superior degree, and into which life courses, upon the principle of all fluids, which run where there is the least resistance. This it is that gives its uniform direction to the current of time; to the little rivers of our heart's blood; to the stream of experiences from infancy onwards, and to the career of humanity from Eden until to-day; the heart being the fleshly atom that is involved or evolved in all these courses. The law is the same for them all. Wherever time or blood is to run, a present is constituted, a valvular state runs back, and hope becomes doubly open from the impossibility of retrogression.

Desire then is the heart-void or cavity, imagination is the heart-wall, our present state is the heart-valve, and our hopes of the future are the heart-orifices. Thus the heart of feeling and the heart of flesh are identical in their parts, and each is the other's.

It would be useless at present to carry the analysis and synthesis further, for the subject would run into subtleties if it were approached prematurely. Let us be content with the above generalities, and bide our time for further localization; for the soul, long estranged, will not come to live in the body all at once.

We may however here obviate an objection, viz., that in the above views, we have taken no account of the anatomical elements of the heart; but have attributed consciousness and feeling to mere flesh, when yet it is known that these are the attributes of nerves and nerve substance. We do not deny it. But we observe in reply, that although anatomically animals live in their nerves, yet the common view is deeply right, that the animals themselves, with their skin, bone, and flesh, are alive also. And we are now looking upon the heart's four bodies as four entire animals, each animated by its own life. In obedience to this point of view, we study the natural history of the heart-animals, in their habits, functions, actions, and not in their nerves; we study them alive, and know of no distinction into structures, but only into forms, as it were faces, limbs, head, and the like. This study does not deny the other, but postpones it so long as we are making a separate object of the natural history of the heart.

Nor is our attribution of consciousness to the heart without the most vast suffrages in its favor; indeed we may state broadly, that every organ is conscious, as the whole animal kingdom is conscious; and that the human consciousness is nothing more than the collective consciousness of all parts of the body. In general the whole field of consciousness is cast into the shade by the solar or central consciousness of the brain, which we call our mind; yet at all times our human form as an outline full of life, is more or less present to us, and makes us embodied men. And in particular conditions, the consciousness of any part may be raised, and as it were detached from the rest; the viscera may take on humane proportions, and open their eyes upon the scene; the magnetic animals of which we are made up, may be separately excited, and powers of an animality co-extensive with nature may spring to light in the dark chambers of the flesh. This is because life is like itself in every part, and the body is all heart, or all eye, when the proportions are varied in which ordinary wakefulness and sleep are mixed; thus when the sun of the brain goes down, the other organs may come forth, and show their ruddy lustres. For we cannot too often repeat, that the human body consists of nothing but expanses of life, all of which are ensouled, conscious or alive.

Upon the circulation itself, as contradistinguished from the communion of the heart, it only remains to repeat, that it stands in analogy with the five senses, and that its passage through the heart represents the active touch which rolls the wheel onwards. The arterial blood is the life going forth to work; the venous blood is the return circuit; and the difference between them can be best likened to the man full of resolve and affairs going forth to his calling in the morning, and to the same man who has laid care and tension aside, and is returning to his own fireside, to enjoy relaxation in the evening. The difference is in the bodily vigor and the mental polarity. Everything is distributed into arterial and venous in something of the same way; into arterial on the active side, venous on the passive: the day itself and all which it includes, is arterial from the morning to the evening, and venous from the evening to the morning again.

It is now time that we should speak of the visceral system as it is connected with life and feeling, and explain the principles upon which such connections proceed.

For the latter point we have assumed as our organon, that the physical and mental are not dissimilar, though different things; on the contrary, they are so similar that their likeness or congeniality is their bond of coherence. This similarity we express by the term, Correspondence. We find that the body and the soul do the same errands in different spheres; that the man respires objects and the mind thoughts, while the brain respires spirits and the lung, airs; which two latter are objects and thoughts, inanimate in various degrees; also that the will circulates affection and feeling, while the heart circulates blood, which is bodily affection and feeling. Again, that the areas thus equated influence each other by correspondence or similarity; the parts of each, touching and associating with the parts of the other which are likest to them. Thus respiration draws thought, and blood invites life, to incarnation. We shall recur to this law in the sequel, where it will be better understood.

In the meantime we anticipate further, that correspondence always implies some point of likeness or ground of association between things or persons; and communication to the same extent. Correspondence by letter implies a common interest of some kind, and communication within that interest. In assuming correspondence between the soul and the body, we assume intercourse also. The similarity between the two terms is the principle that we take, and the details of the similarity are the science which we desire. We see two persons—the inner and outer man—associated together; and when we find the objects that they have in common, we know the ground of their intercourse; but if they have all objects in common, though different fields to work in, we are certain that they were reciprocally made for each other, and will explain as well as supply each other's wants. We take Correspondence for granted; for the more of truth we take for granted, the more we gain; the inherited fortune of common sense sets us up in business in the sciences. Otherwise we must take for granted the dissimilarity between the body and the soul; and the non-correspondence; and then reason will not hear of intercourse between two things that have no objects in common. Success or failure, but not preliminary arguing, will be the test between the two principles, both equally assumed.

The heart is one of many viscera connected in a living chain. As the supreme organ of the body, it has a complete orbit of its own; the rest, as inferior orbits, gyrating around it, and forming an ascending stair of entrails. The brains and nerves are the mental organ; and the lungs, an intermediate field, lie between the mental and the bodily regions. But every part contains the rest; the brains are omnipresent both by substance and influences; the individuality of no part excludes them, for the nerves of an organ are its peculiar brains. So also the heart is everywhere; the bloodvessels of an organ are its private heart. The liver is everywhere; the bilious fluids are a current liver wherever they run. No feeling then is more than central in any organ; it has circumferences co-extensive with the body; for the principle on which its incarnation depends is universal. In attributing feelings to the heart, we imply that they are cordially there, but secondarily, by dilution of correspondence, they are everywhere else. The body is telegraphic with various stations; the messages are different according to the organs, but one at the fountain head.

If the heart is the centre of the bodily feelings or visceral sensibility, then on the same evidence (p. 195) the stomach and intestines have feelings of their own. The glow of content and satisfaction is incomplete until its quietude is poured over these contorting parts. The warmth of wine influences the mind by removing uneasiness from this tube. The bowels yearn with audible tenderness when the softer emotions fill them. What is the visceral inhabitant that touches their strings and feels in their quivering expanses? *Common sensibility* is its name, as inferior to the definite sentiments or feelings of the heart, and as internal to those other feelings that possess the five senses. This susceptibility is a well known phenomenon; some have it in a greater and some in a lesser degree; but it is always a magnetic or sympathetic contact with things, not on their outsides, but in their tender inward parts. Those who are most apprehensive, feel the tube in constant live motions: women especially have such experiences, and must know more than men the depth of the Scriptural phrase "bowels of mercy," and of the saying that those who are devoid of pity "have no bowels." The reason or ratio of these common words is clear. For the good services of the intestinal tube are the materiality of mercy. It stands at the bottom of the vitals, where lowly mercy also stands, for mercy must be at the bottom, or it could not have the poor and needy for its objects. It is at the top also, but beginning from the bottom; for it is the glory of the crown above the head, and the beauty of the feet of those who preach glad tidings. This mercy-tube satisfies hunger and thirst, clothes naked ribs with fatness, and lifts the starveling into the company of full men. Furthermore, it is microscopically true to its charities; it raises the food from below, taking the infant chyle from the earth, and lifting it towards the blood; and searches the unclean masses to redeem any parts that can be saved. It yearns over the new offspring of the blood as a mother over her children. Its gifts are unexpected and undeserved, for the life and spirit of man are bestowed upon fruit, flesh and herb. Long suffering is among its offices, for it makes the best of whatever is put into it, and promotes our worst meals, forgiving the abuses of our appetites seventy times seven. Granting that mercy were enclosed for a time in a prison of bowels, what could it do but what these entrails do? This is the reason why mercy, the full name for sensibilities, associates with the intestinal functions. It matters not whether the work takes place in a city, in a man, in an organ, or in a molecule. Wherever the hungry is fed, the low raised up, the prisoner let out, the erring forgiven, or desert exceeded in blessings, there Mercy lives.* But these works are everywhere, in every assimilative act (p. 166), and hence, wherever ground extends, there are processions of chyle-white sisters of mercy, carrying the world up their ladders, and passing from the dust to God. Man feels the influences of the hierarchy as it passes through

* We observed before (p. 182), that intellectual heat makes human heat; and now we observe that mercy, or the assimilation of the low to the high, makes human assimilation, or the conversion of food into man's blood. The latter can go on apart for a time, even for generations, but not for the long run, and it is the long run, or the end, that contradistinguishes man from the beasts, or separates human from animal assimilation.

his vitals; and intellect is smelted and raised, when it admits the meaning of the better longings which its own experience knows.

Hence a reason for the pleasure connected with the assimilative function. For it is a law that pleasure is given to organism according to the uses performed, and especially according to the ulterior uses associated with the material organs. Thus the parts whose functions are mercy-like, have bodily pleasures conferred upon them as like as corporeal feeling can be, to the satisfaction and blessedness of mercy. Hence we do right to thank God before and after meals for all His mercies. The organs, however, thus furnished with rivers of pleasure from this source, are capable of abuse, or stimulation without a true end. For if pleasures are assigned them for mercy's sake or for just purposes, and man finds the pleasures out, he may choose to enjoy the latter apart from the purposes; as is sufficiently well known: but in that case the first conditions are absent, and the pleasures are verging to retributions. The case of the generative organs is parallel: for the sake of the heart's love which has intercourse with them, and the rich end that they carry, they are endowed with supreme incentives of bodily delights. And in general every part is gifted with these, first according to the necessity of its functions in the body, and secondly according to its privilege of representing, or being similar to, functions of the soul which are delightful self-evidently, or in the order and justice of God.

The body itself is a group of sensories. As the eye is a bodily sensorium of light, and has intercourse with intelligence; and as the heart is a bodily sensorium of blood, and has intercourse with the hearty sentiments felt for our greater blood, which is human kind; and as the bowels are a sensory of our wants from below, and of nature's longings to be raised into blood, and as they have intercourse with gracious sensibilities—so every other part is a feeling of its own objects; and if we knew those objects in common sense, would be found to have intercourse with some principles in the soul that would explain its pleasures, and give the veriest motive of its functions. The liver, for example, the gruff king of the belly, as the stomach is the queen, is the seat and sensory of a severer process of assimilation, as of judgment added to mercy, or justice to sensibility; whence the passion of anger in extreme cases is attributed by the deep vulgar to the liver, and a bilious man by choice exercises upon his society critical and denunciatory functions, disturbing their good natures, and digesting their easy admissions with terrible stringency and power. The words of such a prepared instrument are gall to his age. These are metaphors, because the body itself is a metaphor—a flesh that "carries with it" spirit, because of the likeness it bears to the spirit. The creation is the body and pressure of all metaphors; and seen actions are done because they carry out the unseen in everlasting resemblances. Nature is a force willed from the first to sculpture the images and paint the portraits of God's attributes, in earth, plant, beast and bird, nations and peoples, wherever the one problem works, or the one end predominates. The sensorial nature of the body during emotions leads us experimentally into these realms of metaphors; for we feel that anger lies in bile, and that bile publishes anger; and out of this double marriage we cannot but infer a natural intercourse, illustrating itself by metaphors of language, and proceeding from a real ground of similarity or correspondence.

The sensory nature of the body is plain, from the fact that any part may be the seat of pain, and thus become an object of consciousness. In disease we have an indication of what also happens in the higher stages of order and content. This sign is used indeed by the sick to disprove the better condition; those laboring under indigestion long for the intervals in which they do not know that they have stomachs; and the nervous, in like manner, wish that they had no nerves. But then, on the other hand, the pleasures of the body are as much facts as the pains, but with this difference, that pain enthrals the body to self-considerations, whereas healthy pleasure either soothes to slumber, or else excites to activities and a keen relish of objects; for peace and joy are straight aims and energetic powers.

But if the body be sensorial, it is evidently capable of a liveliness of which at present we have no instances. For if the presence of objects causes so vast a rousing of the five external senses; if the eye gladdened by the light of day, stands so prominently forth, and opens our little halls of light so wide, what will be the state of the heart when it has its objects, as the eye has the range of the mighty universe of nature? And what the joy of the assimilative organs, when the communion of man plays upon them, as the warmth of the sun upon the skin, when the pleased sense opens out every wrinkle of inaction into which the cold had thrown it down? And what will be the state of the brain when it lives in the truth, and opens its sentient intellectual urns to the light that lighteth every man who comes into the world? The difference between that state and ours will be at least equal to the difference between waking and sleep; between life with and without an object; between the vigor of the hero-angels, and the drawl of history, slimy with the torpor of our own hybernation three thousand years long. Yet great as this difference is, we are bound to believe that the Christian trains for long ages have been crossing its deserts, and that a time will come when the pleasures of the head and the heart will be as sensible in the body as those of the belly; and when in consequence the body will be inwardly alive and active in its nervous and visceral depths, more than now in its limbs, or its senses.

We have further to note that the fullness of life depends upon just intercourse between the steps in the sensory ladder. Pleasure or bodily life is incomplete while its materialism only is felt, and the heart untouched by the traveling joy, which as it comes down from on high, longs to strike every cord where its music can be made. But if the heart is incontinent, or has destroyed its delicious sensorialness, the joy is spilt from it, and not felt excepting in the lower regions. First love, the comedians say, is "all-overishness;" the whole body feels it; the house is lighted and the lutes are playing from attic to kitchen, and the old neighborhood is amazed. Last love is often another thing; life in holes and corners, but the former halls dark and cold. So also the pleasures of the sensory of taste have a similar wholeness in their design. Without the conviviality which is the incentive of taste, and which supplies its heart touches, these pleasures are poor and half dead, and savors themselves are rinds with dust inside them. For as we said before, pleasure is given upon conditions and for purposes, and the fullness of the pleasures of the soul are the objects of those of the body. But the soul is all over the body, and when it is conciliated, the sensation of its approval is as it were the whole man, a choice anatomy of senses in a body of delight.

The sensorics of the body form a key-board played upon by numerous hands. The body plays upon itself through their means, and parts of it become present to other parts according to the variations of the correspondence. If the womb, in catering for the embryo, finds the existing larder insufficient, its want is present that moment to the tender-hearted intestinal tube, which uneasily yearns for the food that the nascent nature claims, making as though it were assimilating it, as we by dumb-show of eating at our mouths, show our hunger to those who do not understand our language. This sign-work is next suggested in idea to the mouth and palate, which set muscle and artifice in motion to procure the viands that the blood demands. Thus the hen, in correspondence with her eggs, picks up their shells in bits of lime that will cement and protect her unborn brood. In fact the vitals are the conductors of instincts from the depth of our nature, and the body turns first to them for whatever it wants from their respective fields. Each part is applied to in this manner as occasion requires, for every part knows the rest, and relies upon them. For want of knowing that the body is sensorial, instinct is a mystery; whereas instinct is the sight, voice and action of faculties as broad as our faces, and as luminous as thought; but these faculties happen to lie under our skins, and in our entrails, of which they are the wants, the knowledge, and the ways. The communication of each want from its starting place, through the course of organisms that lead it to its objects, takes place, as we have said, by no other means than correspondence; as supply and demand take place in human affairs. The line of sympathy thrills and trembles with the want, which becomes, as in the case cited, yearning in the bowels, expectancy in the mouth, longing in the person, suggestion in the mind, which completes the circle of the womb, gives the want both sight and voice, and then the will is stirred to pursue and grasp the object. And by the line that the want ascended as a protean instinct, the supply descends, putting off husk after husk and hardness after hardness, until the pure milk supplicated, is squeezed into the little mouth that made the cry, too tender for the mother to hear, excepting by the ministration and enlarging trumpet tones of part after part, as it were successive choirs of faëry-footed, guardian organs.

It may further be remarked, that all substances inside the body, propagate their state as interpreted by their *feel*, to the body at large. See p. 136. This we illustrated in that place by certain considerations touching the *feel* of oil and other objects in the mouth, as affording sensations of an agreeable kind, apart from the sense of taste, and which seem to be a kind of structural taste that travels beyond the tongue, and soothes the frame at large. The action of demulcents, gums and mucilages, and of cold water, are in part of this kind. And medicines act also in this double way, propagating themselves, and working out their specific problems, both by quality or taste, and by quantity or vibration. If this be true of inanimate substances, *à fortiori* the living fluids propagate their vibrations in the same way; the organic *feel* of what goes on in the heart, for example—the groupings, marriages and thrills of the blood, are propagated *per sensum* in like manner, and influence the general state. Nay more, by incessantly jogging the body, and giving it these atomic hints, the impassioned blood-populations insert into man the beginnings of an impulsion that comes out at last in full actions of a correspondent order; the still small voices whispering year after year, grow louder as the audience of vibrating organs enlarges, and ultimately play their tunes in concert with the whole world of man. The largest wheels begin to stir, to start, to move slowly, and finally to revolve, after the least; and the speed increasing, the revolution is swift enough in time to make one drop of passion into a great ring of fire, as consistent as our fate. The "biting of the maggot" that sets impulse in motion, is an ugly metaphor for these words of command that issue from our eager blood. A fairer saying is, that "all men have their hobbies," and these begin in little toys of globules, or blood-dolls, which the heart dandles, dresses, and nurses, intending that the big body shall one day do the same, when the vibration has grown up, and can be the mother of bodily action. The primordial affections of our blood, are our nature's hobbies, but after they become disanimalized or humanized, they change this name, and become good ends.

External nature also plays upon the sensorial body, and we sympathize with weather, moons and tides, because our vitals feel them as our skins feel the objects of touch. Hence come innumerable moods that vibrate towards the will, and instigate states of consciousness, and corresponding colors in our trains of action. The instincts of the day and the hour are so many, that ever-shifting nature only can produce, and God alone can regulate and know them. Sunshine and shade, moist and dry, the east wind and Zephyrus, thunder and frost, and the influences of climate, play upon us thus; some through the mind, some direct through the strings of the vitals; and hence the reactions by which we add to nature, give a new beam to her beams, or deepen her gloom by our brows. Man in this way inhabits his circumstances by a thousand-fold cunning of sensories: he palpates vapors, winds, magnetisms, and climates, with fingers finer than tact, and himself is a divining rod which points to everything, whether in earth, ocean, or air, or in the inward streams that build the crystals, and carry the messages of nature between her poles.

Man acts upon man by the same way, and the presence of persons is felt viscerally even before intercourse commences. Thus a sympathy grows up, by which those whom we meet, act upon our special organs. Do not our hearts burn within us when some men look and speak? The presence of others again is choking and oppressive. The piteous and sentimental, if we be not too steely, act upon our intestine tenderness, and call it into yearnings. The unhappy give us indigestions, and the mad confuse us, and tend to make their keepers mad. We do not often perceive these influences viscerally, unless the mind be excited and call the sensibilities into play. But from a few plain cases we may argue, that our fellows, according to their predominant character, affect us in all ways, organically as well as mentally, by their vibrating substance and stuff as well as by their actions and words; and that every man is ranged for us in a peculiar organic classification. The brain-men play upon our brains, and the heart-men on our hearts; the men of pity touch our strings, and so forth.

The influence of the mind upon the visceral man is very well known. There is not a state of mind however produced, but the body feels it, and responds to it. Our conscious pleasures, or pains,

20*

are objects of touch to the whole of the viscera; anxiety exists in the mind as a trouble and a knot, which the consistent stomach makes a knot in itself in order to feel. The viscera shape themselves upon the mental models that they may sensate the mental states, as the hand makes itself into a cube or a sphere, to feel a cube or a sphere. Thus the body is a set of sensories constantly palpating the mind, by assuming forms which answer to the mental states. It feels the ghosts of thought and passion, not by trying to grasp them, but by making itself like them, and thus experiencing the immaterial by representing it in a correspondent form.

Providence also uses the sensorialness of the body as a means to guide and shape our lives. For much arises in us without apparent cause; dictates, suggestions, feelings, calm, seeming to come from afar, and influencing us in important respects. Such vibrations arise from within, and are the passions of passions and the motions of motions. But motions in the organs, however produced, become our own, whether their causes are internal or external to our being. Within our being, work Providence and his ministers, and fate, instinct, and succession of thought, are the play of the supreme agencies, not unaccountable since we are all made of sensories, which in their veriest ground are in contact with a higher life than our own. The harp of a thousand strings is a good metaphor for this human frame touched into melody by such divers hands, and especially by Him named of David, the Chief Musician. And as the viscera are so constructed as to hear for us the whispers of the wisdom of Providence, it is not surprising that their feelings also are the source of many presentiments, or that coming events cast their shadows before, through these trembling fleshly groves; for the main actors in life are already beyond time and history, and put their fatal forethought into the piece, which then speaks, sometimes audibly, of the future as present, and of the distant as here. It was therefore by a remote application, that the augurs consulted entrails as the voices of the gods, and interpreted oracles from the sacrificial quiverings. For the intestine parts are telegraphic of what is and was and shall be, as it were wires for the electricity of ends, which runs by wisdom's way, from the future to the present. These deeply-buried natures are even as the Vala awakened by Odin for responses which daylight cannot give, and the opening of them as sensories introduces us to magnetisms and communications that the five senses will not know. The unseen world of the body of man is indeed a grave-land of innumerable prophetesses, each to be compelled to speak when the powerful enchanters come.

This sensory nature of the viscera is the complement, and the antecedent, of the motory nature which we claimed for them in our Chapter on the Lungs. It is because they so keenly feel their objects, that they breathe them, or move towards them. To enable them to do this, the respiration itself varies as the objects vary, and the parts of the machine, trebly consistent, move all together. To promote the sensory nature to a still larger correspondence with objects, the muscular system is given, which has two beginnings of ends, one external, by which it produces all human actions, properly so called; and one internal, by which it engenders all human visceral actions. As we may not have another occasion to speak of the muscular man, we shall now dwell upon him briefly, to give a little more completeness to our views.

The human body, besides containing in potency all organic sensations and motions, from visceral sensibility to intellectual perception, has moreover a will, which decides upon the objects to be sought, and muscles as servants of the will, to bring up the motories to their fields of operation, and the sensories to their stations of feeling. Apart from the will, action is molecular, and feeling, like a dream. But under the force of the will and muscles, the smallest impulses (p. 232) become translated into personal actions, and the minutest senses lead to gratifications of which the whole man is the sensorium. For the will is our sovereign pleasure with a soul added to it, and in the muscles it forms our own motions, in contradistinction to the motions of our blood and fluids. It is our continued human life, and each volition is a tick of our proper clock, without which we are not "going." We may call it the organ of progress, as the muscles are the organs of locomotion or bodily progress. It takes up the helpless viscera in its arms, and runs with them where it chooses, or where they choose, setting them in pleasant places which they could not have reached without a strong will equally supported by numerous muscular servants.

The viscera are meant in the nature of things for the will and the muscles; they are meant to be present and serviceable for every action that man performs; and *vice versâ*, all action is meant to act upon them and to modify them. By means of the muscles they are girded into one body; each has the benefit of pressure from without, and is forced to do its work in an exact space, or what is the same thing, to refine nature into skill; and all are set upright, or have their own portion of man's erectness or heaven-sightedness conferred upon them. Thus the liver, the spleen, the stomach become parallel with the eyes and hands, and action runs consistently from the soul through the skin, as well as from the will through the fingers. By means of the muscles also they receive the powers of the volitions: industry and skill are stricken into their phlegmatic tempers, and they are forced to take part in deeds, and to make blood and humors, not for molecular domestic circulation, but for human aims. The stomach, thrashed by the rods of the voluntary fibres of the abdomen, is allowed to make no sleepy chyle, but a quick and stirring milk of support, worthy to feed a workman's blood. And the liver, under the same task-masters, is not permitted to make soapy infantine bile, but sharp decisive stuff, ready in a moment to cut the maudlin of the animal secretions. By means of the muscles also the viscera have the benefit of locomotion and change of air; the will, and these his servants, are the carriage that takes the liver and its companions from London to Paris, or where we please; the muscles also walk them out, and give them exercise, which they require just as much as ourselves. The muscular system furthermore lends hardness to the tender viscera, and contains and corrects their sensibilities, just as the will makes the mind firm: thus it enables the vitals, poor jellies without it, to bear what is to be borne, and do what must be done. Where this strength is not given, liver and bowels are as it were out-lying in the world, exaggerated in size, shrinking from healthy sounds and vigorous contacts, and trodden under foot of circumstances that are the playmates of the body in its well-braced powers. In short the viscera are created to receive and appropriate the actions of our muscles, and to be humanized (p. 40), or lose their laziness and animality, by this among other means. It is a fact that they do receive these actions, and being a necessity also, we state it as a law. We therefore look upon muscularity and volition as just as much a part of the liver, as though the liver was itself muscular, which it is not, for in an indissoluble society, all things exist for each. And we notice that human inaction robs every viscus of its endowments, or leaves it forlorn among animal sensibilities and molecular movements. Under these circumstances the man and the viscera are all doing nothing; true, they breathe, but this is nothing for waking man; industry and skill, in the liver as in the person, begin the human reckoning.

The muscular system, therefore, as the organ of the progress of the rest of the body, carries the body into action, and by its backstroke carries action into the body, making us like ourselves in our tiniest humors, as well as in our works. Could we see a globule of our blood with a fine-enough eye for character, we should find that it was a chip of the old block, because it has been under the dominion of our will since it was born; aye, and inherited our will in its conception. The muscles themselves however are a sensory, or they could not have a human motory life. They are the seat of *the sense of power;* the verb *I can*, added to good pleasure, or the verb, *I will.* The power is possessed of an instinctive sense of the compass of power, by which the muscles know what they can do, according to the dose of will that they receive. This sense is a meter of the will on the one hand, and a meter of the commanded action on the other: if a leap is prescribed, the muscular man feels exactly whether he is wilful enough to attempt it, and powerful enough to carry it through. It is from this sense of power that our bodies become responsible agents, and flesh, a moral humanity; therefore we justly punish human bodies for the sins of human souls; for by sense bodies become souls whilst they are awake and here. The way in which we manage our muscles is by ideal actions shown to this sense of power; the plan of the action is held up to the moving army of the flesh, and with it a command is given to execute the action, if it be easy or known; whereas the cautious mind puts in the query, Is it possible? if the action be serious, or untried. The ideas thus exhibited are wonderful things; they are not only surface-pictures of the thing done, or of the arm at the end of the task, but organisms anatomically full of the means that lead to the end; mental skins hard with mental muscles or volitions upon the model, or rather in the archetype, of the whole muscular frame. This is of a piece with that common sense, that he who wills the end, wills the means; or physiologically speaking, when we will an action, the structural will, through the sense of power or muscular soul, wills every muscle and every fibre to take its part in the action, because the structural will is the muscle of muscle and the limb of limb. A serious view of these ends, as every one of them a human body! and yet no new view, but only a fresh mention of "the inner man." In short, the will stretches forth the arm because it is the arm of the arm, and the inner man commands the outer, because he is the man of the man. In proportion to the number of ends, ideas of action or senses of power, that are in the muscular faith and memory, is the quick vitality or presence of mind with which we move and work in the world: the weights that we lift and the tunnels that we bore; the forlorn hopes that we lead and outlive; the number of evils and errors slain that we count after the battle; the list of impossibilities flouted; is not according to the thickness or thinness of our arms, but is measured by the will first, and secondly by the sense of power which is its servant in the muscular array. Will knows nothing of the present arm, but it extemporizes muscle, building it upon its own mould, for as we said before, it is essential muscle. Furthermore, it does not wait for this, but it makes smallness do the work of largeness—a feat at which all life aims. On the other hand, the loss of these ends or senses of power is paralysis of the will and muscles, after which the flesh becomes so brittle that the consistent intellect sometimes is impressed to dictate, that the sufferer is made of glass, and will break if he is moved; for when ends become mere pictures without touchingness, nothing can stir the muscles without hurting them. These muscular senses then are the proper powers of man, or the genii of action; the maids in golden helms that attend us in the field, and bear us harmless through the hurtling showers.

The contents of the marvelous ideas which thus use our muscles and shape our purposes, lead us to another consideration respecting the external world; for these ideas take the world for granted, and assume its concurrence. It is not fair to argue, that as they govern the body, so they are designed to govern the world; and that locomotion, for example, as an end, knows no end but the utmost velocity of nature; in other words, that the little world of man is the will of which the great world of nature is the muscles. In this way, when the child says, " I want to go to that hill top," he expresses an end of which railroads are but proximate means; an end which is successively to enter upon horseflesh, steam, magnetism, and whatever other ways there are; an end which is not a picture only, but a spiritual world working upon this kosmos, and claiming it as a highroad and avenue of man. By this rule, every art is as possible as every muscular action, and the goodness and power of human wishes are the only limit of the obsequiousness of nature. The grander ends of a single head have light and magnetism for their muscles, and the stars fight for them in their courses. For observe, it is not necessary to know the order of things in order to govern or produce them; the *conscious* will in commanding a dance has no knowledge of the muscles of the limbs, or of the fibres in the muscles; yet it stirs them accurately, because the end that it proposes, contains the means, or is anatomical *by nature.* And again observe that the size of the agent has nothing to do with the power, for a little nerve puts a large muscle in motion, and the will is of no size, relatively to the mortal body. In spiritual leverage, every good action attests the Archimedean power,* and standing on the higher world, actually moves the lower. For the laws of nature, are the laws of the inner man; as witness Jesus Christ, whose disciples said, "What manner of man is this that the winds and the waves obey him?"

We cannot quit the muscular system without noticing also that the heart itself is a muscle, and that the *powers* of formation in the viscera, depend upon it (p. 98). All muscular action also depends for its continuance upon the supply of arterial blood. And as the heart is pervaded by the feelings, and the muscles are governed by the same feelings under their decisive name of wills, the correspondence between the two is regular and exact. The firmness that plants its steel in the heart, has already given iron to the muscles;

* Archimedes used to say, that given a place to stand on (δος που στω), he could move the world by his mechanism.

and a stream of blood magnetized by the unbending life, runs as a bridge between the two, over which the spirit-soldiers pass and repass. The sympathies founded upon this correspondence or human magnetism, are of all force in man. We have already noted how the muscular system is set in action by sympathy with its higher parts (p. 117); how the muscles of that high round limb, the eye, which rolls upon the beams of light, command actions in the muscles of the trunk and legs. Similarly, the instinctive muscular actions tend to be produced by the special movements of the heart. In this respect the whole of the muscular battalions, proceeding downwards in ranks from the eye muscles on the one hand, and outwards in fraternities from the heart muscle on the other, may be likened to the ends of both series, or the five fingers, which are so coördinated in action, that if one, and especially the middle finger be bent, the others instinctively bend with it, and training is needed to isolate the action of any one of the fingers. This is the motor half of that propagation of events from the heart, of which we treated in the foregoing pages, where the sensory part of the question was stated (p. 232).

We may now attempt a slight recapitulation of the psychology, as it runs parallel with the bodily parts. In the brain and nervous system we have found the mental and moral consciousness embodied; a consciousness of the whole of what goes on in the other and lower spheres, and added to this a sense of mind, or reason and will in all their forms, which through this sense strike their attributes into the body. *The sense of mind*, and the motion, is therefore the humanity or peculiarity of the cerebral spheres. Below this we find the body, which in the complex is *the sense of feeling*, or of mind, not acting, but acted upon. In the heart we have the natural feelings which connect us with other hearts; the central relations of man present by a sense in the central parts of the body of man. It is the sense of human love which is here perceived. In the lungs we have *the sense of thought*, whereby we feel bodily the fluctuations and shapings of our minds. Below the heart again, we have another but more general firmament of feeling, which lives in the bowels, and is termed *sensibility*, or by a privileged name, the sense of mercy. This forms a compacter power in the liver and chylopoietic glands, the spleen being the reverse of the liver, and testing the blood by laxity and banter, as it were by giving it its own way, and *reductio ad absurdum;* as the liver by bilious ferment and sharp examination. "Splen ridere facit, *cogit* amare hepar." Then the skin and external sensories are the abodes of the proper senses, which besides entertaining their own objects, are liable to become the fields of the other visceral and mental powers; for not only does the skin feel the objects of touch, but it feels the touch of the passions, and is a sensory of their bloods: and not only does the eye see outward objects, but it is also liable when opened to the second degree, to see mental and spiritual objects, projected through its tubes. These externals give *the sense of self.* Again, between the inside and outside, or between the mind and the senses, lies *the sense of power* and the organ of progress in the muscles, with the bones, or *the sense of stability* in progress, as their fulcrum and necessary complement. Thus man is not only sense but motion; motion in sense, and sense in motion, applied to every object in existence.

The body is simplicity itself when looked at in this living light. And now we recognize that the divine attributes, and not man's faculties, are the key of the human frame. The simple truth of God is the kingdom which has to come into the science of that structure, which is meant in God's image. Only by this means do the last terms that we use, become reliable. In man we talk of sensibility, but with no surety, of mercy. In man we find feelings, lusts, or loves, but not love itself. In man we speak of various powers, but not of power itself. In man we have many senses, but no omnipresent sense; many faculties wise in their generation and kind, but not wisdom in the full. Did we confine reasons to present human nature, we should find our ideals terminated far below reason's wants, in a mythological maze; good and evil demons would inhabit our fountains, and whisper in our groves; we should build a Babel of mesmeric oracles, each violent against the other, and stand comically aghast at our own shadows seen in the ink-pools of our palms. God is the intelligible being in these higher sciences, and common senses; man is in such snaky knots, his heart is so deceitful and desperately wicked, that there is neither scientific nor other dependence to be placed on it. To reason from it, would introduce all ungodliness and unreason by rote and rule into the sciences. And a fuddle of fetischism would then arise, compared to which the present materialism, happily dead as it is, would be but a slight evil. From this we are saved by Revelation, which gives the loadstars of every science; not indeed bestowing knowledge, but opening our eyes and speeding our aims to the ends of knowledge—wisdom, love, mercy, life, power, the Word, which are with God, and which are God.

Thus much of the visceral psychology, which as it proceeds upon sensation or experience, and the sensation is but faint and glimmering as yet, must needs be of feeble growth. Yet we have made a beginning, because the body of the mind must be added to the head of the mind,* and the arms and legs to the body, in order that the mind may be as good as the body in its own sphere, and that materialism may receive the full tilt of a personality as weighty as its own. Our views aim ultimately at a piece for piece proof that the soul is the man, and in the human form. For if the soul-doctrine has two auricles and two ventricles, and a heart; if it has stomach, guts and liver; if it has limbs, and if it has any brains, then it is in manly guise, like ourselves. And if the soul be the true body, and the only thing material to us, then it is no estranged essence, but a person with whom we can have affairs, and keep intelligible relations. Of how great moment this is in society, moral life, and religion, we leave others to say: being ourselves but a backwoodsman eager to clear away swamps and forests, and to open sane ground upon which far other workmen than we can base the truth, and carry up the spires of the pious in the impregnable cities of the just.

We now recur to our canon (p. 225), to illustrate the ground of organization, and its connection with life. And we say, that given an organic structure that will lift the juices of fruit and flesh towards human blood, choosing what is best in the world of food, and making it better by new shapes and constant reformations—then sensibility, or in the full sense, mercy, will and can associate with

* The present philosophies are decapitations, and their doctrine of immortality reminds us of Danton's consolation to his companion on the scaffold, "Our heads will meet in that sack."

this material sister, increase her tenderness, and make her alive. Thus given the bowels of man, and the sensibilities are in them. Again, given a structure that will censure the former process, and elect or reject the elements admitted by a firm afterthought and severer examination—then judgment, or just and executive anger will and can cohabit with this material judge; in other words, the bilious liver will have this vital passion with it by the necessity of the case. Again, given a structure that receives the races of our blood with open attractions, blends them into groups,* unites them in new bonds, spaces their infinitesimal lifetimes, and in their great aortic emigration pours them forth to make of one blood all the families of the body—then every sentiment of human nature will and can associate with this royal, generous and grand materialist; thus, given the heart, and love, life, will and affection must clasp

* In the foregoing pages we designated the chambers of the heart by new names, as "the family auricle," "the patriotic heart," &c.; for living terms are the proper expressions for the parts and functions of living bodies. Such terms must be borrowed by analogy from life. Already some anatomical terms are founded upon analogy. The word *auricle*, signifying "the little ear," is an analogy from the likeness of the part to the external ear. The *ventricle* means "the little belly." Such nomenclature is a precedent in favor of the use of analogies; and the roots of language, by which it assimilates things to thoughts, are analogical. But the analogies hitherto made use of in the body, proceed from the low to the high, or from the outside to the inside, and require to be corrected, and set upon their feet, by the addition of terms taken from intelligible affairs. Moreover, the dead languages have ruled our nomenclature in this country. But if truths be living, to call them dead names is unscientific. So long as this is done, universal instruction is impossible; the democracy, whose education is so vital to modern states, cannot sympathize with unknown tongues, or with sciences scaly with crusts of Latin and Greek. The body must be written out in downright English, before the horny hands can feel its beautiful weight. We inveigh against Popery, for celebrating masses in Latin services, and perhaps think it a mark of a dead church that it enshrouds its voice in the word-garments of deceased nations. Alas, in this respect we are Papists to the very ground. We put the scientific truths of God in these old clothes, as our brothers, the Roman Catholics, the religious truths. Though we are unlike them, in that our science has no fine music in it, no art or beauty, to touch the heart when common sense is away, and by a certain warmth of feeling to compensate for the misfortune of intellectual rags. But providential time, which has buried the old nations, will, we foresee, exorcise their ghosts, and free the sciences from that spectrum of dead languages by which they scare the vulgar, who, we must remember, are not always to be vulgar, but are the sons of God in potency, like the kings and queens.

it in their arms. Again, given a structure that admits nothing but the first and last essences of every nature; that knows not of the vine but through the grape, or of the grape but through the vine; that balances masses of earth by an imponderable weight; that expresses tiny illustrations of light from continents of matter, and maintains bodily heat by a solar glow that costs no fuel; that is in the parallelism of the miracles of the universe, and related to the farthest stars; that is naturally autocratic in the body, and possessed of forces that blood and humors dare not disobey—given such a structure, and in its own interest it will cease to itself and live to the mind; in other words, given the brain, and the presence of the mind in the brain, is, and is to be. Again, given the mind with its better faculties, which live with other minds, and the Divine Spirit is with it; as He himself said: "Where two or three are gathered together in my name, there am I in the midst of them:" observe—not, there "I will be," but there "I am." So again: given a structure disparted into limbs to insure material motions, or the operancy and progress of the human body; and brains and volitions, which are the loves of work and progress, will at once order it into play: whence the muscles being gathered together in the name of the will, have the will in the midst of them. Lastly, given a structure, as the human body, designed for harmony, for contiguity of parts, for opening to enlarge its harmonies, for closing to keep them secure—then pleasure and pain, or ordinary sense will live upon its self-defending plains. So much at present in explanation of our canon, which is this—that wherever in the body any organ does anything that is like the mind or the mind's doings, there, in proportion to the likeness, and according thereto, the mind, and its special deeds are present. But in the human body everything done is like the mind, nay, like wisdom working; wherefore the whole of it has the mind or the man present to it. Q. E. D.

We began with the heart, and to the heart we circulate back after this long episode, to complete the present chapter. And first we would say a few words upon the different signification of heart and lungs, or pulsation and respiration. The difference here is, that the lungs ponder and weigh the existing state, and slowly commit it to the body, or breathe it in by measured stages (pp. 108—126); whereas the heart domineers, beats with quick impatient messages at the doors of every organ, and terminates its problems by the immediacy and hot pressure of its blood. The lungs are the motives which draw us from without, but the heart is the force that actuates us from within; or in other words the lungs are the sense of understanding (p. 130), but the heart is the sense of will. Translating the organs into these known figures, and abolishing the algebraic *x y z* of matter and physics, it is easy to see the reason of the diverse rhythm between the heart and the lungs. If the breath could coincide with the pulse, the body would act out every purpose of the heart, without individuality, pondering or consideration on its own part; the man would be shot to his ends with faster than tiger pants; his eyes would gleam and glitter in the darkness of his faculties; he would become like the vision of the bloody child which rose from the cauldron before Macbeth, and would enter into the golden age of hell, and speedily into that terrible fœtal state which is called the second death. On the other hand, if the pulse coincided with the breath, caution and the slowest life would become the standards of the movements of the inner will; the candles of life would be made of ice, and burn frore; slugs of blood would crawl up and down in our veins; we should learn to walk by the science of anatomy, demonstrate the existence of God by mathematics, postpone making love until the knowledge of magnetism was complete; and in short be as hoary as Stonehenge before our first down had grown. To guard against this preponderance of either heart or lungs, will or understanding, the movements of either are different, limited, and inviolable; the heart strokes play one tune, and the breathing lungs another; the tough pericardium isolates the heart from the pulmonary engine, and on the other hand the pulse is broken by the angles of the vessels where they enter the organs: thus the heart closes its circle, and lays down its sceptre, at the cool feet of the lungs; and the lungs terminate their deliberations at the outside of the fiery palace of the heart. For will is meant to rush through us in its own way, to give life and zeal, but slow understanding is to limit the will, to adopt the part of it which consideration approves, and to let the rest go elsewhere, or back to the will, with the intelligence that it has not been received. And however harmonious the two become, still will is wilful and understanding deliberative, or to speak by nature's algebra, the heart still strikes or pulses, and the lungs still breathe, because their difference is safety and friendship for both. Thus we may say politically, that the white monarchy of the brain through the lungs is founded for ever upon the red republic of the heart.

Hitherto we have treated of the heart from the analogies of the individual feelings, but it is plain that nothing less than a human race is the prophecy of the blood. For in every two beats, less than two of our seconds, the heart runs through an entire life, and constructs relations that deployed in a line are seventy years long. It marries and has children in setting one foot down; it binds all into intimacy, and establishes the state as the other is put forward. The broad sheets of Society write out but a fraction of the momentary news of a single heart. Of all the human lives, the hearts are the least in duration and the greatest in intensity; the smallest answering to the largest sphere. The social man here is a mote and an ephemeris, most wonderful of all, since a click of time and a grain of space contain him full-limbed and perfect-lived. The heart then may justly be termed human nature, which is true to itself in its smallest parts: and it stands in the grove among the beginnings of things, in the place where the stem of the world-tree grows fine towards God. The nature of all things is their very heart. This it is that comes into the world a new essence with every mind and man, and works in restless agitations, aiming to govern the whole earth from that spring of power. The spirit from above, and the universe from without, environ, chastise and endow it, but still it remains, for ever inviolate, the fated well-head of human life. And as with individuals so with nations. Nature is at the bottom there also, inevitable nationality is there; character improvable but indestructible; destined to endure, but free for better or for worse; substantial as granite, though fluid fire of the passions; a rough and tough material, but fit to be hewn or twisted into the humanities.

The powers or issues of the heart are homogeneous throughout. Feeling upon feeling arises out of its bosom, and humanity is built up like coral continents into the red light of the suns. The ceaseless pulse which is the invisible architect, makes higher and broader dominions, and conceals itself in its works. But from its waves emerge relations and societies by the unerring laws imprinted on itself. Blood knocks against nature's ribs, and seeming never to come out, yet issues by avenues the widest, for man comes hereditary out of man, and the heart fires its progeny by a life that glows through every work, and streams through every pore of feeling. The bonds of this second heart are another coil of loves, firmer than that which begot them; more feeling and impenetrable than flesh; wound tighter around their objects, with a more terrible grasp at passing circumstance; an easier and more fiery communion of passions; fortune's greater wheel and Providence's handle; the second life of fear and of courage, of good or of evil.

But we must not tarry in this Vulcanian or Plutonic centre, which throws us forth from the mighty forge where the archetypal powers are working. It is enough to look from the safe outside at the ovens where God's slaves, the Fates, are hammering the natural sinews of the children of Adam. Everything in those dusky ruddy halls looks monstrous and delusive, and the steel of passion, lither than Völund's sword, bends in white-hot waters round our ankles. The glare also of living blood is an illumination past our bearing. The angel of the Lord must be with us in the burning fiery furnace of the passions, and even of their science, or we shall not fare forth unhurt like Shadrach, Meshach and Abednego.

We conclude the subject of the heart by tracing its principle in other spheres, for the vitals of man run through the world by permission from all the natures, and the human body, as method and light, is the *Novum Organum* of the sciences.

That which is communion and circulation in the body, is roundness and unity in the world, and the latter are the produce of an inanimate, as the former of a living, heart. The dead heart is fire, which beats in the centre of things, and is the primitive out-throw of space. Suns and systems arise like bubbles from an ocean of glowing ether that repels them into being. Fire is *the love of spheres;* first it makes all things round, and afterwards gives them revolution or moving roundness, or makes circular orbs into circulations. All things are of one color in its glow; all things, too, become tender as they touch it; the metals are plastic to its thumbs, and granite is a gristly baby and has its melting moods in those thin, dreadful arms. This chaotic love of softness and roundness is the rudest circle of the ancestry of mothers in the heart of the world, where the eldest crones of flame move purring and whirring round "first matter" in its cradles; a giant preparation for finer future tenderness in never ending circles. All force lies here in its seed; for fire is dead love, or *natura naturans.* Its pulse is the quickness of substance whereby it is self-similar in all property and action; the urgency of gravitations, chemicals, gases and stones from world's end to world's end; that gesticulation by which each thing thumps the board of existence, and lays down the law; as it were the blood of time bounding uniformly in his mundane body, and if felt at his wrist on this earth, then felt for his whole system, whether in Arcturus, Pleiades, or Aldebaran.

At the crust where earth cools, lo! a green creeping glow projecting a new earth on the ruins of the former. The first pipes of flame are its models, and the crystals, veins and rockshafts its prophecies, as they branch through the strata, or stiffen upwards in the vaporous air. Seed, plant and tree have come, full of cool, calculating sap, revolving organization; the vegetable heart, or *the love of organism,* is beating. The steaming of the terrene crater issues now through a skin of temperature, in musical flutes of a second fire, through roots into new flames, which are trees, each the candle of a milder light in which nature is seen. Brute heat ceases, and fire is cultivated. The roundness has a beginning, a middle and an end; the seed is current, and runs in gyres to other seeds; growth is the purpose of this heart, and, like a bamboo-cane, every joint puts forth a new joint, till the ground is covered with the green and gay chain-armor of the vegetable soul.

At the top of the love of growth, and running in its grooves, there is already *the love of pleasure,* or the animal fire. The round earth or tomb of the primitive heat is the ground, and organism is the scaffolding, but pleasure is the heart, of the beasts. Hence a new communion and a new circulation. This heart beats for the alliances of the animal nature; the roundness comes closer; the birds are mated and the beasts are paired; gregarious herds cluster into globes on prairie and in forest places; the community feels the thrill of its parts, and with open sensories creeps, flies or gallops under its pressure of delights. The stroke of the love of pleasure throws each creature round its world, through the length and breadth of its own climatic organs, and causes it to seek its mesmerism and shampooing from every soil, animal, fruit and circumstance, whose friction is good for food, or whose look is pleasant to the sight: grasping the lion globule, it shoots him like a meteor through a night of cruelties and destructions; softly kneading the lamb-globules, it coaxes them about over meads where dew is bright and innocence is in the grass.* The instincts are the brightness of this love—flames in which, by

* On the above principles we may offer a fresh rationale of the make of the blood and the offices of the heart. If the heart-principle be love in its genera and species, the known effects of love will answer to those of the heart. Let us take an instance to try the equation. The love of science shall be the heart that we choose to grasp us. This love lies in germ in the affinity between intellect and nature; the mind can be assimilated to natural laws; the assimilation being the stomach process preceding that of the heart. It consists in throwing down the mind before the new object as it were into a fluid state. But every science is round, or a little world in itself, and the heart functions commence by raising up our vague liquid willingness into spherical states. Thereby we turn a side to every portion of our subject, make ourselves in its image, or become soft and impressible to it. The first point in entering into any sphere is, to become ourselves that sphere: the first point in the blood life lies in making previous fluids into globules or little bodies answering to the body. This done, the next object comes, and the round mind can go the round of the field, and as it is alive and universal, can give living universality away. The equilibrium and plasticity of the spherical form enable it to pass everywhere, for one sphere is applicable to every other, without regard to difference of dimensions. Moreover the love is a force, energetic upon these round forms, and forces them to visit the parts whose affinities they feel. The blood globule is the love of the whole body, round according to its mathematics, and in the centre of the whole; hence it represents the radii of the body, and necessarily travels along each, to deposit in the circumference the virtues of the centre. The correspondence of physiological love or heart love with other love is therefore plain. For love makes its subject alive, spheral, soft, impressible, facile, cosmopolitan or universal; in short, makes it into circles, and carries it into circulations. Love is to the heart what the falling apple of Newton was to the moon; it brings self-evidence to the seemingly inaccessible half of the equation; and with self-evidence science is content.

Corollary.—There is no subject or thing that we may not explore, if we love it, and where we fail, it is that we are not loving or spherical; for if we were, we should gyrate through it as necessarily as our blood courses through our bodies.

the creative love, the animal has visions of ways and means; for instincts are animal revelations.

The love of sphere, the love of organism, and the love of pleasure or harmonic sensation, the triple hearts of nature, are included in a new heart, which circulates man through the former places, and through others that exist for man alone. The heart whose grasp is now upon us is *the love of happiness*, which carries us as men through our various day. The love of happiness must walk through the laws of happiness, and hence pleasure is bored by the augers of duty, and the happy take the course which that iron leaves free.

At the top of this, as it were a vane or guide, a heart seizes the mind, and this is, *the love of truth*, which clasps the faculties in the exactest unity, and commands them out on the most far-reaching circulations. As all that is real, is an effluence of the truth, this heart determines us everywhere; for the love of truth impels the lover of truth wherever truth lies. This insures the ubiquity of the mind, according to the declaration of the Psalmist: "If I ascend up into heaven, Thou art there: if I make my bed in hell, behold Thou art there: if I take the wings of the morning, and dwell in the uttermost parts of the sea, even there shall thy hand lead me, and thy right hand shall hold me." But in loving the truth, we love what we do not know, and hence this heart is *faith*, the force and fountain of knowledge, the muscle that commands and sends forth the true intellect since our world began.

But a new heart is upon us, not made but created, not invented but here; and this is, *the love of life*. Thereby the soul is determined to visit the avenues of the everlasting constructions; to take its architecture from all firmaments, and rear forth the body out of the marrows and backbones of the cosmic laws: to study in the schools of order and truth which are stability; to ask of all good plants the direction of the tree of life; to ponder every old story, from north or from east, that babbles of the waters of health and the fountains of youth; and to hang on the lips of revelation, which brings life and immortality to light. This is our thread of golden fire that runs through the broken moments of time. This is the sustenance of the body, garrisoned with invulnerable troops, a buckler against the hostile chemistries of earth and air. This is the safety of happiness, which with Cæsar and his fortunes on deck, rocks with treble thrill upon the waves, a life boat to which ocean danger is a game. This sends the soul to heroism, as its mark and print upon earth, and by contemplation to the starry vault, as the second glass of its endurance. In a word, the love of life, its own permanence, brings the soul to the earth, because it is God's footstool, and bids it go to heaven, because it is his throne. In these errands it runs through every other love; passes every scene, action, affection, through its fires, and gives it the immortal enamel; and crowns itself in death, for the soul that erewhile only loved life, then loves life beyond life.

And as ancient chiefs traced their lineage to the Gods, we follow their figure, and track up the heart loves until they claim parentage from the God of love. There is another heart, which is *the love of God*, and which by faith, life, and all faculty not heartless, propels us to the home of many mansions, where the father dwells. But we love God because He first loved us; and the force and pulse of his love is felt in his Commandments. His "Thou shalt" seizes our "I will," carries it forth, and commands for it the blessing, even life for evermore. The Christian heart is the Word: "Thou shalt love the Lord thy God with all thy heart, and with all thy soul, and with all thy strength; and thy neighbor as thyself." This is the first and great Commandment.

CHAPTER V.

THE HUMAN SKIN.

Of all the creatures of the natural world, man is intended as the most finished, or, in other words, the most exquisitely finite, and the skin is the instrument which first hems him in, and by limiting, completes and individualizes him. The skin is therefore his house and stronghold; the architecture of his frame; and this is the simple idea of the numerous offices that it performs. Without this boundary, the body would not contain its possessions, and would have no continence, outline and end, but be shed and dissipated in the universe. Moreover the external skin is the outermost of a series of coverings which extend to all parts of the frame, and to every particle of the part. All these are skins, connected with the general skin, and as we follow them through one depth of structure after another, we cannot but arrive at the conception that they are the firmament in which the organism is set, and that to take them away, would reduce the body to an invisible essence, or at least to a fluid, to whose volume we could no more assign a permanent shape than to the unsteady atmosphere. The skin then contributes to the idea of life the complementary functions of shape and form, which must pervade all existences before they can take up any location, perform any act, or become the organ, or object, of any faculty whatever. In a comprehensive sense, the skin, under a thousand forms, is the one vessel of the human spirit; it is all that is tangent and tangible in us. Without it, the brains would have no receiver wherein to pour their influences; the lungs would have no fulcrum whereupon to draw their breaths; and the heart, with the other parts, would redissolve into their unbounded blood: in a word, as we said before, there would be no body at all.

What is true of the man, is true of his molecules. The microscope affirms that the smallest organic atoms which are visible are little rounded vesicles or cells, very similar in appearance to some of those low forms of animal life denominated infusoria. These cells are the brick-field of the animal body, and each of them with its contents is a small skinful of liquid. Here we have the simplest form of active and passive—of skin and fluid; and if in imagination you attempt to skin the primordial vesicle, you have the simplest instance of what would happen if you could abstract the skin from the human body. For the frame in its mighty details is even such a vesicle, but contracted into chambers innumerable, and its wall folded inwards into the whole complexity of our organization.

Let us, however, concentrate attention upon the skin proper, as being the face and index of the other skins, observing first, that the general presence to which we allude in speaking of the skin, is a fact which holds, in structure or in function, of every great division of the organism. If the skin passes inwards, and seems, as we trace it, to be the whole body, the same wholeness and ubiquity may be alleged for other parts. Any separate life can be pursued to the end of the frame. The arteries and veins, issuing from the heart, so permeate the regions, organs and tissues, that when they are injected with wax or mercury, the shape of the parts seems complete, though with nothing but blood-vessels. The nerves likewise form a tree which ramifies everywhere; which has eyes, nose and mouth, body, limbs, and ends; and in short, represents the human shape. Even the viscera, as the lungs, the liver, and the like, which do not substantially extend beyond a limited space, yet reach by their motions, fluids and influences to the body at large, and draw their support from the common stock, whether far or near. The human body consists of universal principles; it is one system, or many systems, according as it is viewed; and every part is in all, and for all. Man therefore is the coalescence of many men, each fulfilling the interstices of duty that the others do not occupy, and this plenitude it is which makes him into a substance, private from the world. There is no room in his body, but only humanity completing humanity. This is nowhere better exemplified than in the skin, which is the end of the whole and the parts. For the skin-man stands beyond as well as between all his fellows, and clasps them into one. He has the rest not only in his arms, but in his legs, his loins, and his head. In short, he is all-embracing.

The skin is a threefold or fourfold clothing to the surface of the body; its layers firmly connected to each other, and consisting of the superficial layer or cuticle, also called the scarf-skin—that of which we see the polish on our hands and faces; secondly, a deeper layer, the *Rete Mucosum,* in which the coloring matter of the skin resides; and thirdly, the *Corium, Cutis,* or true skin, which itself comprises two layers; viz., a sensitive sheet of papillæ or minute eminences whose summits look surfacewards; and a firm texture of fibres underneath, supporting the papillæ and the other layers of the skin.

We will now reverse the order, and regard these skins in detail, contemplating each by the light of its function, and we may assure the reader at the beginning, that he is already acquainted intuitively with the main offices of the skin, and only requires to connect them to their anatomy, in order to gain a memorable knowledge of the present subject.

Underneath the skin, the exterior of the body is muscular and active, with considerable unevenness between the parts, some of the muscles being very prominent; and to meet the irregularity, and also to afford an unguent for comfortable motion, the surface is padded with fat, which reduces it to a level, and rounds it into convenient beauty, smoothing the harshness of the firm flesh, and gently cushioning and combining its decisive actions.

But the fat alone could by no means repress the starting muscles, or define the voluntary movements. The will would run out of the leaky vessel before it reached its jets in the hands and feet, if the good easy fat were the only hindrance. Our fat then requires to be bounded by a ring of sterner powers. This second provision is the deep layer placed immediately upon the fat, of the corium, leather, or true skin. The corium is adapted to contain the inner parts and functions. It consists of meshes of fibres inextricably felted together. These are of many kinds and many purposes. The greater part are tendinous, some muscular, some nervous, with all intermediate stages of nature; indeed they comprise the entire scale of reaction, from its passive to its active form; from the power of inertness to the power of actual resistance. Whatever is continent and tenacious in the body, sends its representatives to the assembled fibres of the corium. The compression that they jointly exercise, is facilitated, and at the same time mitigated, by the mode in which they are put together, whereby they constitute myriads of tent-like orifices, with the broad part of the tent towards the fat, and the peak towards the surface; a solid space of corium being left between these tented pores, which are filled with little bags of fat, and with serous fluid, upon which the contraction of the skin first takes effect. In this way the force is communicated almost through fluids, with the utmost power to the general surface, but with the least violence to the parts. The contractility of this envelope is a measure of the tone and strength of the body; it is conspicuous in bracing winter mornings, and in cold bathing, which make us notably tighter and more compact.

The corium, in its elastic life, not only prevents the loss of muscular force, or embanks the streams of will, but it has the material design of preventing the disorderly entrance of substances from without, and the undue alienation of fluids from within. This function supposes in the skin a discriminating power, or sense of touch —a like and dislike—to inform the brain of whatever goes on at the outposts; and also an active power, to enable the skin to obey its consequent decrees. That which feels and acts in these subtle purposes, is the next layer of the skin, the papillary cutis, more superficial than the corium. This is an encampment of small conical tents co-extensive with the surface of the body, differing however in different parts, both in the thickness with which the papillæ are set, and in their size and power. The papillary substance is a Briarean limb put forth from the corium, consisting also of fibrous elements, and quick with blood distributed in meshes of inconceivable fineness which pervade the papilla, and communicate with similar networks in the layer underneath. Our finger-ends are a type of all the papillæ, and like the fingers, the papillæ are instruments both of feeling and prehension. They are eminently movable, and being provided with nerves, and grouped together in different forms, they constitute with the corium, of whose grasp of the whole body they are the infinitesimals, the immense field of *touch and take*—the basis of the pyramid of the senses.

But the necessity that sent us from the fat to the corium, and from the corium to the papillæ, pushes us onwards to demand further instruments. The papillæ depending for their exquisite life upon the most delicate sensibility, upon pleasures and pains so fine that our senses give scarcely an idea of them, must not be exposed directly to the rude breath of the outer world, which would cause them agony, and soon blunt them to the delights of touch. The papillæ therefore cannot be the last covering. What is nature's next resource? The blunter sensations are themselves assumed by the superaddition of new layers; not however actively sensitive layers, but beds of organization that less feel themselves, than regulate the feelings of the skin. These coarser touches become the matrix and defence of the finer. The layer covering in the papillæ must of course be soft and yielding, or it would be as injurious to the parts beneath as external objects themselves; at least when contact with the outer world is taking place. This layer is the *rete mucosum*, or mucous network of the old anatomists.

It is a soft, tenacious substance, placed upon and between the conical papillæ, and furnishes a bed that groups them together, and into which they protrude. When carefully peeled off, and viewed by the microscope against the light, it assumes the appearance of a network, which is caused by the counter-sinking whereby it fits to the uneven surface of the papillæ; the thin places seeming like holes. Possibly also some of the papillæ perforate it. This layer is the seat of the color of the skin both in the European and other races, and what is remarkable, the coloring matter is chiefly abundant in its inner parts, and decreases outwards; though sometimes dark points continue to the surface, and form little streamlets of blackness as far as the exterior integument.

The unfinished softness of the rete, necessary for fostering the papillæ, unfits it for contact with tangible objects, which would lacerate it at the first touch; and consequently nature (in which utility is creative), not only compresses and hardens the rete in its superficial parts, but superadds to it a fine covering termed the cuticle, a kind of glazy platework, capable of assuming, at the bidding of the same principle of utility, whatever thickness and deadness is required. Formed out of flattened vesicles or cells, the cuticle, in its adult condition, consists in point of fact of scales; of which our finger nails are the largest and strongest type.

We have now considered a fourfold mechanism in the skin, deriving all from the general fact, that the skin gives an outline and end to the body. We will briefly recapitulate our positions, in order to cover in this part of the subject. The body requires to be levelled, both for safety to its parts, and economy of space; this is brought about by the fat, and the cellular frame-work, which round our muscular ridges, fill their clefts, and give sweep and curve to our plains. The body requires to be limited, as well in itself, as in what it takes in and gives out; this is accomplished by the elastic garment of the corium, which follows our bodily fashions, and sits upon us differently during every moment of our lives. Furthermore the corium requires to do knowingly whatever it performs; to act discreetly in its functions; to animate reaction, absorption and exhalation, with the life of the brain, and this is insured by the papillæ, which are both brains and hands to the skin. The papillæ on their part need support, encouragement and protection, which are accorded to them in the kindly rete, and in the fine, half-earthly cuticle, which latter is the direct medium between the little world and the great world, or between sense and exterior nature.

In these pages we do not treat of the sense of touch, excepting so far as it may be necessary to the general consideration of the skin, but we notice nevertheless from what has been said, that the skin enjoys a triple sensibility. First there is the sensation of contact from the three layers at once, that is to say, from the surface inwards; the papillæ make common cause with the rete and cuticle, and all feel together. This answers to cuticular sensation or gross touch, and runs in direct lines from the cuticle, through the rete, to the papillæ. In the next order, arising from the first, we have those perceptions of touch which are the consciousness of the papillæ clad with the rete alone, the cuticle not sharing in these inner feelings. This species of touch is chiefly horizontal, and spreads over the live velvet in sheets of vague sensibility. The last pinnacle of this touch, and whose activity is *tact*, belongs to the papillæ alone, with their naked summits and tiny vortical brains. It is so individual, that we can put forth almost a single papilla, as it were a finger from the finger, and play with a little agony of quest round atomical objects.* The lines of thought and purpose meet it, and set it moving in curved explorations. These three species of touch in their union, make touch itself into a substance, and give it over to the mind, which communicates with the skin, not at first hand, but through the mind of the skin.

The skin, so far as we have considered it, isolates the man, and makes him world-tight. It is, however, necessary that the world's goods should come into his house, and that his own produce, not to say refuse and wear and tear, should be carried forth: nay more, that he himself should go out and in with the common freedom that a man requires. The skin is our abode, and not our prison. It must therefore have bivalve doors and windows, opening inwards and outwards, and these, as small as the supplies that are to come from without, and collectively as large as the spirit that is to step forth from within. In short, the skin, that is to say, the body, must be porous.

The cuticle or superficial layer is a very permeable membrane, and the more closely it is examined, the more porous it appears. Three orders of perforations are visible upon it to the naked eye;

* It is not perhaps easy to realize this statement, seeing that all the layers are permanetly united to each other; but so are the parts of the nervous system, which yet may be functionally separated, and the action of the higher, or the lower, stopped off by the force of the mind. We may lay it down as a rule, that the process by which nature unites is also inversely the process by which she separates, or that all her syntheses contain analyses, and *vice versâ.* It is necessary also to take into account the existence of spheres of life as well as spheres of substances; for the life of a part extends beyond it, and can come bare to the surface, abrogating for the time the hindrance of whatever parts lie between it and its object. It must be remembered that we are not now treating of dead layers but of three coöperating lives, each of which abstracts itself from the others as use requires. Hence the prodigious sensibility of the skin, and the naked tenderness of the feelings in many cases. The difference between papillary tact and cuticular touch is but a difference of animation. In all cases the inner can be separated from the outer, and itself become the outer, under favorable circumstances. The soul can be separated from the body, viz., at death, and this is the universal of powers of abstraction or separation existing everywhere in the body itself.

namely, the large perspiratory ducts, whose orifices may be seen as little dottings on the wavy ridges at the ends of the fingers; the sebaceous follicles which anoint the skin with a thin limpid oil that prevents the excessive evaporation of watery moisture; and the passages which lodge and transmit the hairs. All these are channels of communication for influences, that is to say, influential substances, from the body to the world, and *vice versâ*. They are simple involutions of the skin upon itself, or portions of its three layers as it were pushed deeply inwards, and the end of the tube so intruded, wound upon itself into a little ball in the perspiratory pores and sebaceous follicles; whilst in the hairs on the other hand, the tube is formed as before, but the cuticle at the bottom of it grows into a prolonged cylinder that comes out again through the orifice; the hair being essentially a linear extension of the cuticle. All these organs, viz., the perspiratory glands and ducts, the sebaceous follicles, and the hairs, in the channels of which latter the sebaceous follicles frequently open, are beset with little blood-vessels; and as the matters given out and taken in, either come from, or go to, the blood, so the exterior pores are but the outer porticoes to the real pores that lie in the sides of the blood-vessels. The porosity of the outer skin is the sign of a universal porosity in the system, whereby, according to circumstances, everywhere leads to everywhere.*

The offices necessarily performed by the skin as the frontier of the body, consists in the purification of the system, in the elimination of some of its products, and in its renovation by fresh substances, all under the auspices of the sense of touch. We will now regard it by the light of this threefold function. And first, for the office of *purification*.

This is effected by the entire surface of the body, which is continually giving off an atmosphere of exhalations, differing in quantity as well as grossness at different times. It is calculated by an inge-

* The same muscles which produce action without, and carry us along our way, shape and alter the ways and channels within, determining the fluids between new embankments of substances in action; and the possibility of these constant changes in the organism, whereby every action makes its own facilities, is due to the porous coöperation of the skin man.

nious anatomist, that in a square inch of skin there is a length of tubing of 73 feet, or in the whole body, of 28 miles; this vast drain-work being intended to conduct effete materials from the system. Every organ is engaged night and day in sweating them out, through passages more or less circuitous, to the border shores of the skin. What is the nature of this large clearance of goods? or what is its common source?

The blood is the origin of the sensible perspiration, which comprises both the watery matter given off by the perspiratory glands, and the unctuous moisture which oozes from the sebaceous follicles. This is shown by the fact that the skin, and especially its glands and follicles, are beset with sheets of capillaries; and the fat itself, which furnishes the matter to the sebaceous follicles, is a deposit from the blood. The rule is, that the blood all over the system is constantly sent to the frontier, and its useless portions, averse to the life and movements of the rest, are let out through the skin, in obedience to their own unsocial tendency; the skin itself regulating their escape, and preventing the life of the body from being led away at the same time with the worn-out or ill-affected perspirations. This is shown to be the case in health by what takes place in disease, as well as during fear and other mental emotions, in which the proper constraint is not employed by the skin, and the living essences themselves run away in crowds from the irritated and incontinent blood-vessels. As in other parts of the capillary system, the nervous fibres are the intelligent agents of the above elimination, and governing, that is, opening and shutting the pores, from the minute doors of the blood-vessels to the wide avenues in the skin, they grant or refuse passports to the various applicants that incline to go out into the world. This office of the nerves is seconded by the mechanism of the parts, for the perspiratory and sebaceous ducts rise from the convoluted glands in spiral turns through the layers of the skin, and their orifices also exhibit the law of the spiral pathway, and are half-closed by the cuticle, which shuts down upon them like a lid, from under which the perspiration escapes obliquely, by overcoming its gentle force; exsinuation and insinuation being the methods of this commonwealth. The cutaneous exhalation is reckoned at nearly 30 ounces *per* day, which is supposed to be about twice the average quantity of matter breathed out during the same time by the lungs. The skin is therefore our greatest theatre of substantial change, just as the sense of touch is the most influential of the senses, impressions upon which will alter the mind when other means have failed.

The perspiration is of use up to the moment of its ejection; nay, after quitting the frame, it partially remains on, and around, the surface, fomenting the cuticle by its watery portions, anointing it by the oily, and providing it with suppleness, and power of intelligent motion or animal fluidity. Take away this natural cosmetic, and what a dried and painful expanse is the skin; as in fevers, each joint of the tesselated surface grates upon its hinges, until the dulled sense is stung with a misery that rouses the whole nervous system for its removal. Nothing can show more clearly the great though gentle power that the cuticle exercises in health, when it is soft and pliable—its power of reaction as the general bond of the skin, and of performing the functions included under this reaction—than the disadvantages that result when this thin varnish, as it is considered, is either removed, unhealthily produced, or not sufficiently fomented with its appropriate emollients and unguents. The interests at stake in the suppleness of the cuticle are known instinctively by savages, and tribes that despise dress, yet protect themselves by a garment of oil and red ochre, which serves as a supplement to the perspirations, as well as takes the place of other clothes. Our own oils, perfumes and pomades come from a similar instinct, and every one must have experienced that dryness, whether of the skin or hair, is more than a detriment to comeliness—that it amounts to discomfort, verging fast towards pain.*

Besides the sensible perspiration, another species used formerly

* The use of the cuticle, as we have seen, is to modify the sense of touch, or to produce a more external sense by reducing the papillary keenness and liveliness to earthy conditions. It is not what the cuticle seems, but what it does, that shows what it is. It produces the most general sense of all, which sets in motion the next wheels of sense according as itself is moved. Hence scaly and cuticular diseases are full of annoyance, and subvert the sense as effectually as diseases mainly affecting the other layers. Hence too the pleasant effect of cleanliness and ablutions. In considering the penetration of effects from without, it is best to think of the cuticle as not a layer but a sense, and that sense the outmost, and therefore the first determinant of the quantity and quality of feeling that comes in.

to be admitted under the head of the insensible perspiration, and which was reckoned different from the former, though now it is regarded as only the *minimum* of which the sensible sort is the *maximum*. We prefer the old classification, which marks a broad distinction both in the circumstances and form of the fluid. The one is to the other what oil, or liquid, is to gas. As a writer has said: "If we examine with a microscope the naked body, exposed during summer to the rays of a burning sun, it appears surrounded with a cloud of steam, which becomes invisible at a little distance from the surface. And if the body is placed before a white wall, it is easy to distinguish the shadow of that emanation." This emanation, as the author calls it, constitutes a large proportion of what the skin gives forth. Its functions are no doubt both purificatory and anointing. It would seem to arise from the finer pores in the skin, and from pores in the sides of the perspiratory and sebaceous ducts; for let us bear in recollection, that it is impossible to conceive anything not porous, and the same principle of permeability which is a physical fact in the external world, is an organic fact in the human body, commencing from the skin.

A ratio of size is required between the pores or passages, and the sweats. This is partly insured by the varying contractility of the former. But the difference between the sensible and insensible perspiration is too great for this case to meet it. Different sized pores are necessary. The condition of what transpires through the skin, is touch. A fine perspiration would not be felt by too large a pore, and the skin, when such perspiration passed, would know nothing about it. The story of him who cut two holes in his barn door, the one for the cocks and hens, and the other for the little chickens, is no fable in the skin. Such a provision is indispensable where the walls are feelings, and act as checktakers to whatever goes in or comes out.

But are our communications to the atmosphere limited to products which the senses can perceive, whether in small quantities or volumes, in vapors or fluids? This question, not answered anatomically, is decided in the affirmative by broad facts, fertile of consequences. What are those influences that carry contagion over great spaces from skin to skin, and which no eye, or microscope, or analysis hath seen, notwithstanding that they are irresistible ministers of disease and death? What are those odors proceeding from the skin, by which the exorbitant sense of the wild Indian tracks the footsteps of his victim over the long flight of a diversified country? What aroma shed upon the earth, and stiffer than the sweeping wind, is the guide of the bloodhound and other instinctive hunters, to the distant quarry which they seek? The footprints of the skin, have they not modified by subtle sheddings the very ground and stones where the tread has been? And how far, and how long, does the magic work from those scheming life-like centres? Nay more, we appeal to mesmerism, and we ask what it is, proceeding from the operator's body to the patient, that spell-binds every sense, or produces the play whereby the automaton sleeper is set in motion, and set in thought, at the bidding of another's brain. Clearly it is no perspiration in an ordinary sense; it takes effect through obstacles too great; yet as clearly it proceeds from the surface of one frame, and is received by the surface of another; in short, it goes from skin to skin. These facts conclude, that through certain channels occult to the microscope, quite unknown to anatomy, but assured by our babe and suckling common sense, the skin, or the nervous system through it, pours forth a subtle radiation of tremendous efficacy on other organic creatures. They render it evident that through this battery of surfaces, the animal creation, and man most of all, is constantly impressing a character upon external nature, literally magnetizing it, and producing we know not what modifications and new forms in its plastic matrices. It were foolish to suppose that emanations which engender such changes in those delicate tests, organic beings, have been powerless to alter dead things in the ages that have elapsed since first organic life, bent thenceforth upon multiplication and dominion, sprang from the seminary of the original earth. Led in this train of thought, philosophers have suspected, that the tigers, lions and snakes, and other ugly foes not seemingly of our household, were at first but the wind-cast seedlings of our passions, wicked words overheard and dramatized by nature, returning now from we know not whither to plague the inventors; though doubtless men will be slow to own such children as these.

What deduction is to be drawn from the above statement, shown to be a fact by such gross proofs? We claim it in confirmation of the existence of a manifold nervous fluid; for the subtle exhalations we have been acknowledging, pass through space, and from body to body; nay, take up a location, and keep it, which would be impossible were they not bodily themselves. We have therefore traced the sensible perspiration to the red blood, and this influential radiation to the nervous system; and grasping these extremes, we may put the mean into its place, and assign the insensible perspiration to the lymphs and colorless fluids pervading the body. Here is a trinity of exhalations, given off by the nervous fluid, the lymph or white blood, and the red blood respectively; and which are established by huge facts not to be gainsayed.

This gives new import to the transpirations of the skin, for particles constitute volumes, and volumes are groups according to laws. It flows from the previous conclusions, that there is round each man an atmosphere which has a formal existence equally with the interiors of his body. If the organs underneath the skin be different in every part, and if the blood be similarly various, then the emanations from different tracts will exhibit the variety of their sources, and the image of the body be stamped on its circumambient sphere. This spheral man environing the skin is not without a witness in ordinary sense. Observe the phenomena of sympathy and antipathy manifested when certain persons come together, and referred by common observation to the feelings, and very rightly, for the matter touches us nearly; and although sight accounts for impressions, yet a residue of influence is experienced which all will say is not seen, but felt. And mark the fear which animals know when they come for the first time into the presence of their natural enemies. Impressed by such events and appearances, we recall the doctrine of the microcosm, nor slight the influences of the sky, but recognize in the sympathetic friend some benign planet of our destiny; or in the repulsive presence of others, the malignant rays of an evil star. For as there is no vacuum in the outward world, so there is none in the human world, but all mankind touch each other, and form again a globe; the grand sphere of all individual spheres. One use therefore, and perhaps the greatest use of the emanations of the skin, is, that we may be instinctively known and knowing, and bring with us our own groundwork of sympathies and antipathies, whereby to find and maintain our places, according to the old laws, of cohesion, attraction and repulsion. For what is true in the mineral kingdom, is only more and differently true in the human.

We are accustomed indeed to think of the transpirations as not only effete, but unclean. Yet there is certainly one honorable sweat, namely, that which stands beaded on the brow of toil. Little children too are of a sweet savor to their mothers, "as the smell of a field which Jehovah has blessed." Lovers also dwell in a common atmosphere of purpureal satisfactions. The redolence of health itself is like fresh morning air to those that meet it. History moreover gives cases of persons whose presence was pleasantly aromatic. It belongs indeed to the skin to rid the system of particles outworn in its service, but these are not necessarily foul, but simply the dead forms of the same principles which are living forms in the organs. Inanimation does not imply uncleanness, but rather the consolidation of a new universe inferior to life. It is the body as influenced by perverse passions and habits from generation to generation, that makes our legitimate spheres into noisome sweats, and the radiations of our minds, actively repulsive. If the rose or the lily could be gluttonous or covetous, lewd or hating, the third generation of them would give out stenches. It is the inner world that exalts, or contaminates, the outer; and the purity of the soul in the body would add a thousand perfumes to the air, and a manifold attraction of fine sympathies to the instinctive understanding. At present the rule of tolerable scents ascends no higher than the flowers; the perfume of good and true men is an incense both prophesied and possible.

But the whole of the transpirations do not consist of effete materials. We feel on the other hand that life is surrounded by life; that a man's outposts extend beyond his skin, and convey his feelings by their emissaries to our own. Spheres touch before skins. Judging by the analogy of the saliva, which comes out of its organs alive, and returns back thither after a circuit, we deduce that the transpirations are not primarily dead, but leave the body for a space, to revert to it with fresh supplies won from the atmosphere in their journey. Man in every sense can travel out of himself with profit and interest. He drives not only trade, but commerce, with the surrounding natures. His atoms have the benefit of seeing foreign parts, and bring home new fashions. Are there not also currents or winds in the human sphere, whereby clouds drawn from one part of the surface, are drafted to other places, to descend with refreshment thereupon? for what is dead in the view of the face or chest, may be living enough for the belly and the limbs.

We therefore see reason to divide the transpirations into active and passive, or into sweats and radiations, the former of which gravitate and fall, but the latter are permanent and creative. In treating of the lungs we made a similar division, into breaths and spirits, or into bodily and purposeless respiration, and into spirited breathing. And the same remark applies to every system with greater or lesser exceptions. But this belongs to a new psychology founded on common observations, and which shows that heart touches heart with magnetic fingers, and indeed that organisms are linked to their similar parts by sympathetic columns of emanations. The reason of this is, that the soul is not only included in but also includes the body, or not only lives beyond the brain, but also beyond the skin; being the omega as it is the alpha of the microcosm.

In Chapter II. we dwelt at some length on the ensoulment of the breaths, and here we might show, did space permit, that each externization* or excretion that we make is a similar spirit-carrier. The sweats are no exceptions to the rule. In faintness whether born of fear or pain, the sweat is mortal cold, for the life has left it prematurely. In low animal passions the sweat is hircine, to suit the goat who owns it. And so forth. The range is as great as it is subtle. It extends from the shininess of oozy monk and stall-fed prelate, to the glory-rim around the saints, and again to the crucifixion before the crucifixion, the Master's agony-sweat as of great drops of blood in the garden of Gethsemane.

* The externization of things amounts to a new essence in them, whereby they become distinct objects, not only to other beings, but often to themselves. The difference between a fœtus and a child is externization acting upon a capable organism. Emanation, therefore, where it is an emanation of life, differences itself, or because a fresh object, at each successive elongation from its source. But this for the metaphysical.

Perspiration then is not only purification, but emission, and emanation or creation. We now proceed to the other half of the subject—to the renewal of the body through its cutaneous surface from the external world, or its reception of fluids and exhalations from without. And first for the fact. It is proved by the absorption of medicated substances when rubbed upon the skin; by the relief of thirst, the increase of weight, and the nutrition of the body, when plunged in baths of water, or of alimentitious fluids; by increase of weight from the air itself, during sleep, as well as at other times, and which cannot be accounted for by the food taken; also by the phenomena of sympathy and antipathy, and by the effects of mesmerism; and in short by the same evidences that attest the cutaneous exhalation. There is then a parallel of giving and taking; the man and the world cross hands; the exchanges are balanced between the skin of the body and the skin of the planet, the atmosphere being their market; the spirit, lymph, and blood modify nature by their sphere; and nature by hers constitutes their circumstances and supports their reactions. This is also shown by the organic effects of climate, and by the geographical colorings and other differences of the skin. The rule here is, that as the atmospheres, as well as the earth, environ the skin, the latter is natively adequate to avail itself of their wealth, whether ponderable or imponderable, and to feed its own system with whatever is inferiorly organic, or earthy, watery, airy, gaseous or ethereal. In short the skin man is a sponge immersed in nature's *plenum*, and soaking us through, soul-deep, with the substantial properties of every medium.

This which is established by facts, and afterwards appears self-evident, is the point whence to start in unraveling the texture of the skin; common sense being the way to uncommon sense. In this texture there are several kinds of fibres that are unaccounted for; the microscope has done its best, yet they are known but as threads, of different colors and varying elasticity. Are they not conduits of some reasonable stuff to the frame? Are the primary fibres impervious, the dams and hindrances of motion, spoiling its current by their inadmissive contractedness? Truth is shocked at the supposition, and answers that nature, the deeper and smaller she becomes, is the roomier in accommodation for her essences, which are only represented in the fluids. It is abusing the microscope to imagine that it can sound the depth of any structure: honest sight is deeper than the microscope; for we see on the broad canvas of space, effects which point to properties that no lens can verify.

But the truth is that nature is more considerate than our science, and she places somewhere in every system one or more big facts to show her designs. These are her maternal "great A and bouncing B," her picturesque alphabet, intended to be learnt by implicit rote before anything else. Too often however we begin at the end of her grammar, and never find this alphabet; and can but timidly copy her forms, without the power to spell much less to read them. Now what is the largest pore in the skin? Without hesitation, the mouth, which shows us, by an unmicroscopic exhibition, that the whole skin feeds upon terrestrial elements, solid and fluid. What are the next pores? Plainly the nostrils and ears, the purveyors of the atmosphere and its motions. And what do these cutaneous orifices declare but the universal aëration of the body by the surface in which they are constituted? But is there an opening for the ether? Undoubtedly two—in the eyes, where the mind stands face to face with the world, and the world with the mind. This is sight, which permeates the closed windows of the cornea, and sees that there are pores above the microscope, and that the universe can come in to us when the doors are shut. Those ether wells, the eyes, are the naked truths of a whole humanity of similar orifices whose lids conceal them in other parts of the skin. If these deductions be too strange, we ask then, do not we interpret other things by their leading facts? And what is our knowledge of the world but the knowledge of its faces and its heads?

But upon this subject of the appropriations of the skin, we caution ourselves that the life of each part seeks and takes the kinsmen of the part, and no other things. The lungs or air organs breathe air, gas, whatever ascends from the solid into the void, or tends to enlarge itself or to breathe. The stomach eats whatever is eatable; it fructifies fruit of every order, or puts the material finish to that process by which nature is eaten and digested until it comes up to mankind. So again, skins are the proper object of the skin; for this system skims the world, takes a film for itself from every surface, and assumes the cream of body as the matter of the sense of touch. This sense is the essence or single substance of the skin, which clothes the mind and the organism with the personable outsides of things. We are not then to think any longer of assimilation as a formula in treating of the skin from its own point of view; for we have left assimilation behind in the stomach. It is the external apposition and mutual feeling of the parts which alone belong to the skin.

We do not however dwell upon the sense of touch or the pleasures of the skin, for we are not yet treating of the senses, but of the bodily organs. Only let us remember that the armies of the papillæ, encamped in force upon the frontier wherever great sensation is wanted, and everywhere posted, spaced and grouped for watchfulness and protection, are armies of industry also, being to their adjacent pores what the hands are to the mouth. They are exquisitely quick, attentive and manipulative, and can help the skin to what it may feel, or warn it against destructions or irritations, with an agency more discriminating than our consciousness. But the mind acts with great force upon the skin through the nervous system, and depressing emotions playing upon the surface, destroy its equilibrium, damp its courage, throw it off its guard, and cause it to let in the emanations of disease. Hence during fear the papillæ desert their posts, and contagion invades us. On the other hand, the might of life is similarly transferable to the skin, which thus encouraged, walks the pest-house unharmed, and feels virtues like its own in the poisons.

Nor must we forget the motion of the skin, with which we are familiar in the breathing. Our feelings of the universality of respiration are the running accompaniments of the universal respiration of the skin. Under this, its constant movement, its other movements are performed. The sympathy of sense between it and the lungs is close: the same air and temperature bathes them both, and the winter that narrows the desires of breath, contracts the skin to fit the diminished visceral spheres. The air and earth of the microcosm both lessen together, and both expand together under kindlier seasons. The expansion of the breathing skin is the enlargement of every pore of the lung order, and causes the absorption of the various influences in which the skin is plunged. This is the active cause of the sympathy of the skin with the air and the powers that it contains. The state of the mind or the animation of the brain and nerves is in like manner concerned in the relation that the skin holds to those parts of nature which lie above the air.

Thus the skin is a natural balancer of temperature, because it is the medium between two fires—those, namely, of the body and the sun. It is a remarkable circumstance indicating the self-possession of the body, that its heat varies but little with climate, or the seasons. Standing at about 100 degrees of Fahrenheit, it repulses the equatorial fervor, and the northmost cold. Undoubtedly this is due in the last place to the mettle and temper of the skin. It sprinkles its heated envelope of atmospheres with abundant vapors, cooling as they rise; for the supply of which our thirst is commanded in the sultry hours: on the other hand, when the outside weather is inclement, and winter howls, the skin fastens up its infinite shutter-work, and the fires are fed with liberal materials, which then must be taken according to hunger, which prescribes the fuel.

The skin then suitably, or sensibly, gives out and takes in what it requires, this being the sum and substance of inhalation, exhalation, and the intermediate touch; and the sphere surrounding the skin serves as a medium to attemper, humanize, and convey to the body the forces of the external world; much as the atmospheres receive the rays of the sun, and convey them to the planet.

Whither does the skin go? or where does it end? It may be said that it is continuous with itself, and ends nowhere, as it begins nowhere. True, but it is more circular than this. For besides that it covers the whole body, it passes in along the great thoroughfares, only assuming a thinner and moist surface, and constitutes the lining of the nostrils, the windpipe and the lungs; it also enters by the mouth into the alimentary canal, and therefrom into the depths of the liver and the pancreas; and by other ways into other viscera, and into all the glands. These are its highroads; as for its private paths, they are innumerable. It passes down every hair-follicle, and by the cuticle to the tip of every hair; it runs along each perspiratory gland and oil gland, through the eight-and-twenty miles of our ingenious surveyor; and next through every pore of every sieve-like blood-vessel along the sanguineous system, and along the lymphatic system; over levels which no anatomical quadrant has taken, and through a mileage which wants another surveyor still; in short it ranges through brain and body wherever anything touches anything, and where consequently a sense of touch and a skin are indispensable.*

So much for the altitude of the skin, or its penetrancy body-deep, and its rise brain-high. But the surface of the organic miracle, its means of display, are not less remarkable. The plummet line that finds no soundings is matched in the lines of sight, which meet a constant horizon, but never a shore. Awe stands giddy over the chasms of the soul; amazement heaving with joy, finds and loses itself moment by moment, at the prospect of the landscapes of the never-ending life. The skin is as broad as it is high.

The geographical or regional consideration of the skin, to which we are now pointing, is capable of being pursued in different ways. It may be followed literally, and the diversities of skin traced in nations and races in different parts of the globe; looking on the earth itself as clothed with men as with a coat of many colors. Or it may be traced over the individual body, where it exhibits a scale of varieties, minute indeed, but apparently as exhaustless as in humanity itself. Let us briefly advert to each of these departments.

In all races of mankind the corium or true skin is white; the difference of hue residing in the other layers. Blumenbach reckons five general varieties of color; the white or Caucasian; the yellow

* The presence of skin in the body is the double of the presence of feeling, which is skin-function, in the mind; and as feeling is cointensive with life, skin is coextensive with bodily structure. Feeling however subsists in different powers and degrees so dissimilar to each other, that no boldness short of common sense dare call by the same general word the sentiment of duty, or love, and the sense of wamth, or resistance; all of which, however, are rightly designated feelings. It requires the same boldness, and no more, to denominate the membranes of the pia mater by the name skin, which signifies the general envelope of the frame.

or Mongolian; the red, copper-colored, or American; the brown or Malaic; and the black or Ethiopian. In each of these classes the general varieties are again repeated. Thus the European skin embraces the swarthy Italian and the fair Swede; and the English ranges in a lesser compass through the same shades from the blonde to the brunette. What is the cause of the great difference in the color of our skins? One cause is, that variety is the child of nature, and coextensive with existence. But why should the variety run from white to black, rather than play through the infinitude of whiteness with all coloration and enrichment? What produces hues repulsive between races of brothers, and why does nature char and begrime her offspring? Is it merely that the sun "will keep baking, broiling on?" Not altogether; for in that case the inhabitants of equal levels in the tropics, would be as uniform belts of colored tribes around the globe; which is not the fact. The reason is more complex. The true whiteness of the skin would signify the equilibrium between the light and heat of the man, and the light and heat of the sun, which were they balanced, the face would be a transparent panoply, reflecting both the spirit and the world, and self-defended against the invasive subtlety of the solar rays. If the light of life were there, the skin would bleach into more resistless whiteness in the burning noon-day, light living upon light, and energy upon energy; the brave life would stand with gleaming shields upon the extremest ramparts of the cuticle. At present, however, in the black man's skin, his darkness eats up the light, in order to effect the balance. The sable rete of the Ethiopian is an organism that by its relation to light and heat, absorbs the fierceness of the sun without making claim upon the nervous energy; and does by a fixed or unsightly circumstance what life would effect by its own witty heroism. His poor body cannot help being burnt, and kindly nature has strewn it with coal fields for the purpose. No wonder that his color is a felt degradation, for it is the police of weak-eyed and waxen virtues, that would melt under the fierceness of his skies. With respect to the other skins, red, yellow, brown and pied, all rising to their own blackness, they are to be regarded as separate organizations, or as so many intelligent resources for maintaing reasonable terms between the tyrannical strength of nature and the enfeebled life of man; and in general the rainbow of human skins is a divine signal against these floods of the sun. The power of climatization is measured very much by the ability of the skin to stand up for itself, or else to take on the restored balance that color produces between the surface and outward influences. The colored parts in each instance appear to be a distinct tribal organization, required in different shades and proportions by various races under the same sky; for there is nothing to make it probable that the English skin for instance, place it under the line for a number of generations, would assume the jettiness of the Negro, or the tawny shade of the Malay.* We may therefore see that among the causes that have led to the present distribution of races, the relative life of the skin is one; and that certain tribes would instinctively move down the sides of the planet, until they found the spot where the slanted sun-beams were dilute enough to befriend, and not injure, their undefended skins; the flaming sword behind chasing them away to more frigid destinies as their loves waxed fainter. And indeed the light and heat are continually operating in all great migrations, from the east to the west, and now from the west to the east; much as the same causes in nature produce winds and currents, which are inanimate migrations.†

* The clothing of the different races is significant, relatively to the present state and future prospects of our skins. Under the line, where the rule of the skin is blackness and also thickness, we clothe in white and light garments; in the temperate regions, in dark woolen clothes. Showing that opposite conditions to those of the skin, are sought in the clothing. Were the skin able to keep its own balance, it would be what the clothing now is, whitest at the equator, and darker where greater heat is wanted, and where less bleaching power is given.

† We see remains of the migratory principle by which the world has been peopled, and which has forced the races to quit the fairest spots, and to move onwards, often into comparatively inhospitable regions, in the fact that many persons are born out of their health's latitude, and cannot live in our English climate, but are sent away to Madeira, Italy, the South of France, or Australia, where they can continue an existence which would be early extinguished here. What was once the rule of races happens now with individuals, and that exceptionally: yet as it goes on from all countries, it amounts to the collection of new races destined to be united in the amity of new mother climates. The development of a human race is like that of a human body. At first its parts are movable, and are not born in the places which they are subsequently to occupy: emigra-

Passing from the geography to the hierarchy of the skin, from its planetary to its social developments, we find that the skins of the extreme classes of society differ as much as the classes themselves. The horny hand of labor represents a prevailing texture of the surface, which exposure to the air, coarse exercise, the sweats of toil, and the quality of the bread of toil; in a word, every circumstance connected with the lower classes, tends to produce and perpetuate. Their complexions are those of the brawny limbs of society. The present is ever a magnified time; the past and the future are compressed; and the pores and fibres are large that work upon the present, and deal with the rude exaggeration of our immediate wants. The upper classes are closer knit and finer surfaced, and breeding is unsuccessful, if it does not alter the skin. You do not see the details of their bodies, or criticize their timbers, but they come before you with a whole front at once.

The skin also exhibits the phenomena of sex with its own characteristic openness, concealing and revealing according to the eldest modesty of nature. Softness and strength are married in the male and feminine skins, and sense and sensibility are completed in that union. The woman is finer grained, to constitute a permanent aristocracy—so much of humanity as nature can keep without coarse drudging in the world. She works her feelings by fleet algebra in spaces less than our knowledge, while we swelter behind, learned porters, carrying cumbrous slatefuls of figures. It is because her skin is deeper, more vital or as it were visceral; a sweeter trembling nakedness as of a soul stripped of one earthy covering to show what its mode and doings are when the hood of flesh is gone. For this natural reason, and many other reasons, women are sometimes called angels. This sexual difference of the skin is the sign of such a difference reigning throughout the organism, for the means are carried in the extremes; and as the soul is coördinate with the body,

tion therefore is a distinct branch of embryology : and even at a late, or what we may call the historical period of the frame, some great changes occur : the testes for instance emigrate from their hot birthplace under the kidneys to the neighborhood of the pole in the scrotum. Moreover, the opening out and stretching of the infant man at birth amounts to the elongation of his feet from his head to the greatest possible degree.

the common sense of the skin proclaims the existence of sex in the soul, where the laws of diversity and union become also extended.

Again the single skin is suited in constitution and color to the part that it covers, and to its function in every place. Where great resistance is demanded, as on the heel, or in the palm, the cuticle, no longer thin, becomes a dense and almost horny pad; and in short, it sympathizes with our occupations and habits, presenting an open declaration of the statesmanship of the body; for if there be this power of wise change in the scurf of the skin, what is there not in the depth of the system and in the senate of the brain! So also the reticulum varies with the latitudes and longitudes of the person, being greatest where the sensibility is greatest, as over the hands, and upon the tongue. The papillæ also are various everywhere, and peculiarly grouped; multitudinous where feeling is quick and frequent; almost indiscernible in other parts. The corium too is thick and thin, and differently woven, as utility desires. In short the individual skin brings us back again to the social and geographical; it too runs from coarseness to fineness, from whiteness to shadow, from life to dullness; representing on one person the skins of all ranks and conditions, all seasons and zones; kingly face-skin and peasant foot-skins; swart patches of Ethiopian dusk, and tracts of open fairness, as it were temperate climes.

Turning from the general map of the body to the head, chest and abdomen, we observe that their great divisions of skin modulate through successive functions. Over the belly, receptivity is the leading office; it sustains material distention for periods short and long, and recovers itself more or less. In the skin covering the chest there is a higher elasticity, fitting it for its lifelong reciprocations; active strength is the gift of the thoracic covering, and it easily recedes before the filling breath, and then as easily runs forward after the retreating lungs. In the head, all the other functions of the skin are present, and capital functions superadded. Besides that the head-skin is mechanically yielding and elastic, it is also intellectually expressive; it runs after the changes of the brain, faster than the skin of the chest follows up the alternations of the lungs. But in robing the sovereign brain, the beauty of the organism, such as it is, dwells in its natural folds, and harmony and love come forth perforce around it. Feature or intellectual form, active fitness, and close continence, severally representing mind, motion and body, are thus the three spirits of the skin, according to the three regions; and these are the cornucopia of its uses.

We have said but little upon one important appendage of the skin, namely, the hair, which is of a threefold kind, and is universal over the body. There is the massive growth upon the scalp, the short hairs met with in other parts, and the invisible down, which appears to exist everywhere. We can only remark that its uses throughout are typified in those of the hair on the head. Warmth, clothing, conduction, protection, expression, beauty—every hair means them all. It is also observable that there is an intimate connection between the hair and the temperament, and between the hair and the nervous system, and a sympathy between the hair and the mind in health and disease. The narratives of fear turning the hair white in a single night, are so well attested, that a late clever writer on the subject has nothing to object beyond the "of course it is impossible" of his own education. Certain it is that the hair is a part of the man, and in a broad sense comes from his very nerves. In the face of anatomical limits, and heedless of learned giggling, whoever has once felt the hair of his flesh stand up, knows right well that something ran out of his brains when the fright was on his back. Let the anatomist make a scalpel out of that fact, and carve away with it at the hair follicle, and he will soon find that every hair carries a streamlet of his life.

The hair presents an elongation of the qualities of the cuticle, and analyzed by the wearer, we see in it the same round world again as in the most living organs. The moss and grass of our skulls is just like us. The head is nature's king, and nature crowns it. The essence passes into the crown, and it represents the royalty of the wearer. The manly forelocks rise with independence and openness, equally with the forehead that they overshadow. The clustered curls of the woman catch and benet with the same witchery that glitters from her eyes, or fascinates in her smile. Yea, every state and feeling makes as free with the hair as fear itself. Low-browed self abasement has a straight-haired head; its very comb puts it down by instinctive logic of correspondences. Courage and energy crisp the hair like wire. Prolific and young-blooded races like the Negro, are woolly as sheep; old and decaying tribes, such as the American Indians, are lank and drooping in their locks. Life throws the hair as a substantial shadow around the principal organs, to which it offers a native contrast, forming the colored complement of their grace. Whatever shadow is required comes by this rule: red hair is the contrast in keeping with one complexion, and black hair with another. The background is not only itself expressive, but in harmony with the portrait; the sun-faced day is set and framed in its own casket of night.

Did space allow, it would here be proper to make a brief excursion into the wild-wood nature, to note the fertility of the skin as it undulates over the plains of animal existence. What is wonderful, that demure thing, the cuticle, when applied to lower lives, breaks into a sportiveness that amuses belief. Scale in the fish, coil in the serpent, feather in the bird, many-colored coat woolly or hairy in the quadruped, impenetrable mail in the pachyderm, it shows on a mighty arena what are its delicate and silent uses in the human body. But we must not be led into the pleasant ark of zoology. Suffice it however that these properties are represented in man also, whose fine tact, and its sober instrument, terminate in arts that beat the feathered wing in flight, and outdo the fur in warmth, and the mail in protection; and his model skin stands inwards unspent the while; and all is *there* in its prolific source, which its doled out in strict though bountiful measure to the poor unalterable animals.

We recur, however, to the function of the skin as expressing the mind, and as being, with the Sovereign Artist the canvas of the beauty of the world. This beauty, they tell us truly, is only skin-deep, but none of them has told what is the depth of the skin. At all events it must gauge to profound realms, for it brings the whole man to the surface. Moreover, it is not in fact made up of two layers, but cuticle and cutis are artificial productions, conveniences of books, the work of men's hands. The human countenance especially is the painted stage and natural robing room of the soul. It is no single, double or triple dress, but wardrobes of costumes innumerable. Our seven ages have their liveries there, of every dye and cut from the cradle to the bier: ruddy cheeks, merry dimples and plump stuffing for youth; line and furrow for many-thoughted age; carnation for the bridal morning, and heavenlier paleness for the new-found mother. Masks are there indeed for every time; and tragedy, comedy, and farce are positive nature in the skin. Every shade of passion has its mantle in that *boudoir.* All the legions of desires and hopes have uniforms and badges there at hand. It is the loom where the inner man weaves on the instant the garment of his mood, to dissolve again into current life when the hour or the moment is past. There it is that loves put on its celestial rosy red, which is its proper hue; there lovely shame blushes, and mean shame looks earthy; there hatred contracts its wicked white; their jealousy picks from its own drawer its bodice of constant green; their anger clothes itself in black, and despair in the grayness of the dead; there hypocrisy plunders the rest, and takes all their dresses by turns; sorrow and penitence too have sackcloth there; and genius and inspiration in immortal hours, encinctured there with the unsought ancient halo, stand forth as present spirits in the supremacy of light. In a word, the compass of human nature, as it is seen and felt, is all to be referred to the inexhaustible representations of the skin.

Thus the skin is the kingdom of show, for whatever is seen of mankind is nothing else; a proof of the expressiveness of a covering which can be the universal face. For the skin is the fine but ample decorum in which the inward terminates; the only thing that is fit to be seen either in art, in nature, or in science. It hides the rawness of our curious fleshwork, and reveals it anew as comely and personal; it gathers up thousands of nervous fibres in the speaking countenance, and mingles the infinite colors of life into a single complexion; it groups discordant muscles in lovely or manly limbs, and braces the unseemly viscera into a statuesque humanity, toning and leveling the whole out of its own reservoirs of temperament. And so it is the official presentation, and the bright honor of the body. It is sacred too in its wholeness, and blood lies under it film-deep, to start forth red before man and nature with a divine protest against its violation.

Clothing or fallen decency comes forth from this original source;

the art and mystery of it is first given in the skin.* The materials of our garments tally with its emanations. The linen is the cool and watery envelope of the body; the cotton is the middle, or earthy; woolen and cloth are the prolongations of the hair, the unctuous and the oily; silks and satins, with their metalline glow and ocular glancing, stand for the ether and unresting fire that goes out and in, and changes forever in the eyes and countenance with illusive lustre. Jewels and ornaments also are of the brotherhood of the eyes and face, and make dress into rank and dignity. The costume of nations, sexes, times and events descends from their countenances and skins. White and scarlet and Tyrian purple are but their chromatic refractions. The Persian scarf, the Thibetan wool, England and France with their stuffs and laces, China with its cloth of gold, only continue in measured movements the infinite fabric of the weaving corium. The great first shuttle is human nature still.

This skin, like all things besides, ascends by its powers to other

* It will be seen on our principles, that clothes are natural to the human body, because art is included in our nature (p. 135), and man, the image of the Creator, cannot but surround himself with those secondary creations which we are accustomed to term artificial. Vestments however are as natural as birds' nests, or birds' feathers; but as they spring from the nature of reason and imagination, from the constructive faculties of mind instead of those of body or instinct, they are more free and various than animal clothes. But wherever the human form is, in whatever world, the principle of utility that commands the arts will reproduce the vestures to the occasion. The productions of art are in fact the comparative anatomy of the human frame, as distinct from the comparative anatomy of the animal frame, the procession of which latter is exhibited in the kingdoms of nature. And the tendency and promise of the true humanity is, a world which spontaneously follows what Bacon calls "human uses." "Houses not made with hands" are spoken of in the Bible; also garments that wax not old. In the same book we learn that the people of heaven are not naked, but clothed in shining raiment; and the Ancient of Days has "a waistcoat of white wool;" the armies here also are "clothed in fine linen, white and clean." For if manis immortal, all is immortal—sense, faculty, art, decency; and in the more plastic world of the spirit, the constructive powers realize instantaneously and organically what is here the result of the same powers working through imperfect machineries. So much for the naturalness of clothes, whose forces and forms are those of the human mind, and accompany the mind withersoever it goes. We write this note, because the Bible narratives are sometimes discredited for attributing investiture to the future man, whereas this attribution is in harmony with common sense, and the contrary is a part of the indecent doctrine, that man, when he dies, goes nowhere, and has nothing on.

spheres. Surfaces, bonds and communications belong to nature and spirit alike. Manners are a social skin, whereby our savageness is hidden and compressed, our wants are tamed into shapely occasions, and the rules that glide from man to man fold and wrap individuals into communities, and keep the warts of our eccentricity leveled down under the common tone of the time. Laws and bonds are more penetrating manners, as it were the membranous *politesse* that is binding upon the inward life. And the arts that beautify our estate, live and reside upon the new extensions of human nature formed by these amiable skins.

The universe also has the cutaneous principle throughout, as the ground of its beauty, and the substantial theatre of its changes. Landscape and ocean-scene are the planetary skin, and field, garden and grove are the expressive face that the world upturns in gratitude to its cultivators.* Atmosphere, tides and magnetic lines, varied soils and successive climates, depth and deeper depth of strata, enamelled vastness of plant and flower, mutations of all as constant as heaven is constant—these are the natural family of those minutiæ of the human frame with which we have busied ourselves here. And go where we will, we cannot transcend one pen-stroke of the everlasting order. The skin is a truth, and omnipresent.

* It would not be difficult to run the parallel between the human and planetary skins. The body with its hard muscular surface is the naked planet, to be clothed with a skin that will receive the sun by inviting but independent forms. The soil, itself chiefly of organic origin, is the fatty layer which first covers in the bare rocks and surfaces, and makes them mildly round. The network of binding grass, the interlacing kingdom of plants and trees, represents the cutis with its fibrous coverings. The animal kingdom, and man especially, is the papillary layer, quick, sentient and locomotive, living on the best and most fertile parts of the great surface. The vegetable kingdom, as it proceeds from human cultivation and through the mind of man, is the rete mucosum, a world of new growth rising to the surface, and covering in the rest. The universal mechanism of the arts, polishing everything externally, educating man, domesticating animals, pruning and raising the culture of plants, and conciliating mineral or mechanical truth with vital rotundity and flexibility, is the varnish of the cuticle, through which man and the world are put in their court dress of beauty, suitable to hold the train of light which falls from the progressive sun. Thus the skin is a circular vesture, coming up like a scarf from the earth to wrap the shoulders of man, and falling down again from him in statelier folds towards the same earth. Its ascending sweep is from the fat to the central or capital cutis; its descending, from the cutis to the cuticle.

Especially must it reign in the intellectual sphere. There it is a noble species of touch, allied to all that is great and solid—no other than our common sense—polished, inviolate, sensitive of trivialities, rejecting at once what is antiquated and useless, open as day to edification, reconsidering many things; the basis of capacity; the beauty of the emotions; the complexion of the virtues; the conversability of the understanding; the simple drapery of wise actions. This it is which fosters the man, and is the defence of an immortal vesture.

Let us, then, attend a little more severely to the correspondences of the skin considered as a function and principle, and elicit its formulas in their main departments. In fulfilling this task, we shall come to the largest generalizations, as the skin is the generalization of the body; and to self-evident positions, because the skin is ocular evidence of ourselves. This is the reason why the present chapter is descriptive, for the skin being scenery, we wander perforce around its regions, pencil in hand.

The outer skin of all is *space*, whose face robes the suns, its breast envelops the air, and its belly overlays the terrene globes; a skin adapted to its contents; space expressive to the glancing fire, free space to the atmospheres, and fixed space to the earth. At the top the name of space is light, the countenance or beaming of things; in the middle it is expanse, the chest of the same; and below it is extense, the flatness of matter, or the skin of the ground. Space limits, for nothing can exceed its space, or add a cubit to its stature. But it is porous, and keeps in none but insides of its own grossness. The light-space transpires through the air-space, which becomes luminous thereby; and spiritual things invade space with no resistance, for it is not their bound. Space has no existence without things, as neither the skin without the body. But it holds the creatures that want room, and carries them all; it is the quarry of the sculptures of shape, the canvas of the pictures of color; white and glowing in its sunny eyes, blue in its immensity, green on its seas, and verdant on its earths. As by room and show it gives free expression, it makes the outward world candid, that is to say, representative of the inner. So is space the mundane skin—all-expressive, all-elastic, and all-continent.

In the realm of growth the same skin-principle has a new name, and there we call it *nature*, the expression and limit of growth, or the vegetable term. This is the end of that kingdom, which clips it in and gives it bounds. The nature of trees is, to rise so high, and no higher; to be fast in their places; to be immiscible with other trees; to be attached to climates and fostered by seasons; to be limited by each other, and to be annual or perennial, and die when their vegetable space and time are passed. Their nature also is growth itself; the bud transpires through the bark and unfolds into the leaves, and the flowers transpire and unfold in their turn, like verdurous skies steaming out of the brown earth, and brilliant vapors collecting into petals like little suns in the cope. For their nature is to live in space and represent it, ranging minutely from its head to its feet. Furthermore, by the nature of the vegetable kingdom, the tendencies that it cannot hold, escape it, and the animal kingdom is evolved on its outside, created among its unbounded parts; and this nature is the end of development. For all is developed from growth as all lives from life. The animals are as apparitions to the trees and herbs; they come and go out of the dead spaces by no vegetable law; and by nature the science of stiff stumps avers that animals are illusions, but that birdless and beastless wildernesses are vegetable orthodoxy and truth.

We are now upon another skin-principle, a third space and a second nature, namely that of animals, and the name of this principle is *Self*, which is the limit of the love of pleasure or animal heart (p. 248); for the beasts love only those pleasures that please themselves. They assimilate what suits themselves; they go whither they will; they breathe for themselves; their instincts are for themselves; in a word, wherever they run they are contracted to themselves, which is the same as to say that self bounds them, or is their skin. Self, however, is the surface and slumber of every animal passion; for passions, or the life that "suffers itself" to be provoked, begin from the skin. Hence self-defence, in which nature seconds passion, and gives shaggy hair to one, horns and hoofs to another, beak and claws to a third. And as self is a faculty highly jealous, it incites every part for its own, that is to say, for self-defence. Again, when self is touched by pleasure, it opens and lets in the pleasure; when hurt, it shrinks away: in short, it is the texture of superficial motives, and answers every appellant with lightning-haste. And being the instigator of show, it tricks itself with the allurements of animal art, and sees its own graces in every polished surface; and while it always displays itself, it conceals by hiding one self under another, and putting the most feasible forward. Self then is the dear "whole skin" of animality, which keeps it what it is, bounds and binds it to its pleasures and pains, making it impassive to all others, and cuts it off from the universe, but connects it with similar selves. And as only a certain amount of self or animal space is allowed to any creature, an equilibrium or competition arises, which is the congregate self of animal tribes or the selfishness of animal societies. Yet through self transpires more than self, and from animality oozes forth that which it cannot keep in, a sphere of wisdom, self-denials and animal virtues, as it were the first gases in which moral life is respiring; —qualities whose origin we do not attribute to the animals themselves, but to the Creator passing through their shades, and leaving his prints among their dreams. It remains to be said, that self rigorously clears out whatever is incongruous, and makes each animal more and more itself the longer it lives. Anything unselfish is ghost and fantasy to self, necessarily denied by every art, science and heart-beat of the brutes.

Consciousness is the skin of human mind;* for at any given mo-

* Philosophers, busy with the investigation of consciousness, do not suspect that they are writing monographs on the cutaneous principles. The Ego and the Me, the self, the personality, and suchlike thin things of thought, cannot help being somewhat spare, and in themselves dry, for they are skinny subjects. And when consciousness sets up for itself, and assumes that it is the human mind, instead of the bag in which the mind lies, then the diseases of the mental skin begin, and the unconscious philosophers describe their own maladies as the history of the universe. Those who attach themselves to the "pure Ego" as having an independent life in itself, are investigating the hairy scalp of the mind more or less full of pediculi, which are "the life" that it has "in itself." Those who limit themselves to personality or individuality, are all "face" in their pursuits, and study impudence in its universals. The drier studies of consciousness are busied with the mental scarf-skin, and when the organic history of philosophy is written, the times of barren logic and method will form a chapter, headed "The Age of Dandruff." For logic is an armor of little shiny scales, infinitesimal flints of thought; in a word, cuticle; and when it is separated from practical ends, scurf; though when legitimate it is the proper varnish of the

ment our consciousness is ourself, the boundary of our apprehension. *Cogito ergo sum* is this reflex power, and means that we see our outline in our states of consciousness. These states are elastic robes to our faculties, clothing the placid eye of thought equally with the long arms of will; and where they are not, the faculties considered as human possessions are not. But consciousness is an ocular tunic, or a man of eye; it sees self, nature and space without it, or permeates all the inferior extenses.* It also looks inwards, and feels as if it saw the vitals of the mind. It has moreover different layers whereof the one is conscious of the other; and it is luminous in various degrees, from glimmers of perception to complete thought, in which it forms a tight envelope of self-knowledge that insures

mind. Again, when consciousness is sought for its own sake, it forms the mental itch, which is a general and excessive self-consciousness. And when it is taken for the world, as by the Berkleyans, it constitutes *elephantiasis scroti mentis*, in which the intellect dwells in the crannies of the warts of its own imagination. We might easily range the diseases of consciousness parallel to those of the skin, and classify the philosophies of consciousness under vesicles, papules, pustules, exanthems, and the like; but time fails, and the work would be both unseemly and irritating. It is enough, however, to indicate that the idols of modern philosophy, the Ego, and the Me, the personality, the self, the individuality, are as necessary as our skins, but that they are our surfaces and lowest parts, and require brushing and combing, washing and tending; but will by no means bear deifying; for whatever of dirt and ugliness, whatever private parts there are in us, or come upon us, these skins catch. When we look at the philosophies through the organon of the human body, we see that the skin of every subject, the sensual part, has alone been studied; and that speaking bodily, neither brain, heart, bowels, lungs nor muscles have entered into philosophy, which accordingly has been the dead plane and flatness of the human mind, the coarsest texture of all, as if it were the dishcloth of the fates.

* We may notice cursorily that the various skin principles are in their own way co-extensive. But the law of extension is different for each. Consciousness reaches the stars, not by being put on the stretcher of space, but by enjoying the faculty of representation. Conscience extends similarly, because duty is co-equal with consciousness. The perfection and relative infiniteness of things lies in their having the greatest amount of presence in the smallest space. Thus it is the pre-eminence of the eye over the skin that it touches things without contact, and that there is no known proportion between its size and that of its objects. And thus it is the perfection of motive force to move bodies without exerting any push like their own resistance. In thinking, therefore, of the higher spaces and forms, we must regard each successive perfection as equivalent to all the matter and extension that preceded it, and look within the human skin, and not in the cosmos, for the image of infinity and power.

our collectedness or presence of mind. The inner organs within this skin, each of which has its own covering likewise, are memory, imagination, thought, will, wisdom, and the parts of the mental body which find their stay in consciousness, and from consciousness run back into their deeps. For surface makes things doubly large, and adds the height and show of the outside to the gauge and secret of the inside. Consciousness too is porous, and much escapes it: within it there are humanities that it does not hold, and without it, minds and persons that it does not see; but as its grain gets closer, and the rays of its woof finer and more feeling, it girdles more of mind, and brings new invisibles to light; thus ever and anon it puts its knowing films over new worlds, folds in sciences, and gains fresh wisdom for its organs. For whatever it embraces is straightway in its body. Consciousness also defends the mind, for what is self-evident or known is our invulnerable part, in which we travel through the unknown; it holds against harm our tender and ungrown wisdom and thought. This is the *me* of which the philosophers inorganically speak, and which has the *not me* inside it for its vitals, and outside it for its society; for our substantive faculties are *not me*, and are wise in proportion as the skin of *me* is thin and pellucid. Consciousness also is an organic sieve, and clears the mind; being the theatre of mental elimination; for the mind tends to consciousness or show,* and no sooner are we conscious of any-

* From the ground of these generalizations we may give a clearer rationale of some functions of the skin. For in tracking any subject into the mind, we come to self-evidence, the mind being the arithmetical quantity of which other things are algebraic symbols. Or, to use another figure, the mind is a banker that cashes the notes of physics into its light and life, the gold currency of the sciences. But then we must know the equivalence between the notes and the gold, or we cannot check our receipts.

Now the skin, as the love of show and self-knowledge, is an attraction towards the surface. This is exerted upon the body, and calculates upon a love of manifestation inherent in the organs. The love of show is the heart of the living skins: seemliness is their body: when the former propels any particles into the latter, they are at once judged in this court, and excreted, purified, or reabsorbed. This magnetism, of the love of show, or of coming to light, explains the determination of fluids to the skin; and as the organic light judges them, we have here a self-evident account of the phenomena of perspiration and excretion. If you ask further, Why? we answer, Why do you like seemly and beautiful things? Because they are pleasant. If then it be said, Why do you

thing, than adoption, or rejection is involved. Consciousness, for this purpose as well as others, is largely porous, or full of blanks and oblivions for what is not agreeable. Sleep is its great hole or mouth through which stuff is put away, and a thousand slumbers to one thing and wakings to another, vacancies, forgetfulnesses, and assumed ignorances, are the pores by which it rids itself of thoughts and memories, and refuses them a longer footing in the mind. By consciousness also we stand in the universe of mind, and feel it in all forms and structures, as though mind were matter, and nature the caprice and building of a grand idealism.

In the state the skin-principle is *the sense of right*, which implies the former principles, space, nature, self and mind, and is the citizen's motive of self-defence, for we are jealous of our rights as of our skins. These rights are our social selves, our honors and shining parts, which make the faces of freemen lustrous, and are plain manliness in the state. Courtesy and manners are the dyes of this social skin. Moreover, according to his rights the citizen is measured, for they are his stature; where they are small he is small, and where they are absent no citizen is seen. At any given period, there is only a certain quantity of social space or room, whence each right is compressed by its fellows, yet the world of rights is we know not how elastic, and the state can expand, or make new rights, as fresh worlds introduce new spaces, and as self engenders other selves. This sense of right is the limit, defence, and reformer of the citizen; it sifts the state continually to eliminate old laws and habitudes, and opens it to allow for new: and moreover it is surrounded unwittingly by the rights of new eras, which it is one day to embrace; for its nature is the gradual assumption of all virile togas from the wardrobes of liberty in the state.

like them because they are pleasant? This, we reply, is not a question, but a question and answer, and wants no further answer. But if it be said, Why are they pleasant? that is a different matter, but involves only your account of your own pleasure. The ground of the soul's doings is, because it likes to do: and the ground of the body's doings is, because it *quasi*-likes, or acts as a material soul: this is its intelligible magnetism.

Let us then bear in mind, that the self-knowledge of consciousness, and the self-examination of conscience, are the lights of the eliminatory and purificatory offices of the skin. When the figures are thus obtained, we revert to the dead algebra, and find plain numbers underlying it throughout.

The brain of the sense of right or social man is the skin-principle of the moral man, that is to say, *the conscience*, the surface upon which the moral sense is felt. For conscience is the jealousy of the virtues, as self is the jealous principle of animal defence. The individual is an intense society in this spaceless space; a tender feeling outraged there, is a wrong to the poor and the needy; a hatred is a murder; on the other hand the virtues exerted in this secret chamber toward the least of the little ones, are done in the state of the state, and to him who lives within the conscience. The moral faculties are defended by this sense, whose monition rouses them to self-preservation; its agonies, and its happiness, are the exterior motives of the virtues. Conscience limits us to its compass; our right and wrong are according to it, for our morality is no larger than our conscience. It individualizes and spaces the virtues, for it gives to each its own grounds of action, and makes moral difference among men. It expresses them, for the beauty of no goodness can shine nakedly, but through the face of the conscience. It purifies, for it is the essential organ of self-examination,* and the supreme area of our approbations and rejections. But it is also elastic, or is wisdom living in circumstances. Finally it is a point of conscience, that there are degrees of its sense which it does not include, and hence it is open to new experiences. And as is the case with all skins, where they are not, the organs underneath them are not; but where they are, the man comes solid within them. So true it is, that in giving a new conscience, a new man also is given in the old.†

In the soul, the skin principle is *identity*, whose germ is the *I* that runs along all our time, governs each human verb, and stands as a focus over our actions; but its adult body is our sense of immortality, which binds us in all things to be and to do for ever. World

* The body is put into a sieve, which is the skin, and shaken about by the mind and muscles, to clear it of dust and debris; and the sieve itself knows the dust from the gold, and keeps and rejects accordingly. The skin is therefore the bodily principle of self-examination, the sieve of our daily life and death.

† Any one may be without some of his bodies if he is without their skins; for he does not possess until he comprehends them, or has their sense or skin. What is not manifested or revealed, *is not*, so far as man's faculties are concerned.

and man feel permanence when they touch this surface, which we may term, the human organ of eternity; for it is a sense without almanacks or clocks, which apprehends likeness with God, who is the same yesterday, to-day, and for ever. It is sanity in its castle, as on the other hand a lost sense of identity is the top of confusions. It holds us together in a oneness that is the model of coherence; and in its going forth, it is the apostle of the unity of our knowledge. The soul is protected in nature, because its very surface is a sense of living for ever; the flux of years falls down like water from the oil of such a virtue. It is limited or separated from nature by the same unviolated sense. Its beauty also is represented on this sense, for the beauty of the soul is virtue seen under its own immortal skies. This sense also purifies the soul, and causes it to throw out perishable motives from its constitution. And furthermore makes it elastic, empowering it to change without losing the identity that engirds it. For the soul becomes more and more immortal. In the body it is immortal for a time; out of the body, for other states; and so on with increment. But the deeper immortalities escape it until they are at hand, and hence faith comes in.

But our senses of duty and immortality are inconclusive unless the fact be added to the sense, or unless that which is given in love and feeling be true to the letter; that is to say, unless there be a letter of revelation to enfranchise conscience in a real universe of God. *Divine Truth in the letter* is this reality, apart from which our faculties would be baseless visions; for humanity can as little exist without the spiritual world, as animals can exist without space. But the office of the latter is, to contain and manifest the spirit. And in order that we may dwell within it, it is like us in its form; plain also in its Gospel face, dark with brightness round its Apocalyptic head, vocal to all souls in its Psalms, clothed with mystery in its Prophecies; and its feet being immovable, we do not see the ground on which it stands.

CHAPTER VI.

THE HUMAN FORM.

HITHERTO we have contemplated the human body in parcels and fractions, and have come gradually down into the world from the brain to the skin, or from the centre to the circumference. We have seen that the animation of the body depends upon the brain, and comprises both a living motion, and a living eradiation of spirit to every part of the system. We have seen that the force of the body depends upon the lungs; that the play of life is brought about by the breathing; that alternation by this way enters the frame, gives it motion and rest, and makes it the instrument of an ever-varying progress. We have seen that the renovation of the body with substances made in its own image, comes from the heart, which gives the organs a population of new individuals as the old elements die out. We have seen that the field of preparation whereby aspiring substances are fitted for citizenship, that is to say, whereby food is assimilated, and converted into blood, is represented in the stomach. Finally, that this motion and busy life, constant infusion of new blood, and insatiable appropriation, has a government of its own, yea, that it is a common government; and this is represented in the skin, which limits everything to its uses, and connects it to its fellows.

In stating thus much we have the position that *the human body is a living, moving, substantial, enduring, inviolable subject.* The brain gives it life or ends, which are the lords and masters of organization. The lungs give it motion, without which life would be futile. The heart gives it substance, without which motion and work would be impossible. The stomach gives it supplies, without which, moving substance, subject to wear and tear, could not last. And the skin gives new ends, or individuality, without which the whole would evaporate. Furthermore, the mind and muscles take this visceral man into their ranks, give him wholeness of progress, make every grain of his organization human, and stamp it with the spiritual seals.

We may now go a step further, to a closer truism, namely, that *the human body is alive.* This is what we have been endeavoring to prove by elaborate argumentations! For science does not know this fact with which its neighbors all round are acquainted, being scarcely aware that there is any difference between a corpse and a gentleman, or even that "all flesh is not the same flesh, but there is one kind of flesh of men, another flesh of beasts." For to say that the human body is alive, excludes animal life or beast body from the subject, and in this light our truism is not recognized in physiology.

We must, however, proceed from one truism to another, and further affirm, that *the living body is a man, with nothing in him but humanity.* We may illustrate this in many ways, for all things preach it. For example, in crystallization, a salt that is broken according to its cleavage, is still a perfect salt of that species; the smallest crystal is as properly the salt as the largest. What is the corresponding cleavage of the human body? We do not here enter on the question of substantial separations, but of those of thought, in which the cleavage is the representation of the whole in the parts, of the laws of the body in its members, of humanity in the organs and their elements. Take for instance a papilla of the skin—in this we see a delicate nerve—there is its brain; a miniature system of blood-vessels—there is its heart; contact with the air and movements communicated from the breast—there is its lung; imbibition of stuff from, and discharge of matter into the atmosphere—there is its stomach: finally, it is part of the covering of the body, and is itself covered with a particular vesture—there is its skin. In this way the parts are integers of the whole, and have both complete minds and bodies, and hence the whole is the microscope that reveals them. Wholes, however, are relative to greater integers of which they are the parts: the individual is a little whole compared to the mind or progressive individual, whose whole is a lifetime; and this too is small compared with society, which covers continents, and lives for epochs. There are then in the mind's optics vitreous substances of different degrees, which magnify more and more. For example, the businesses of the great world, the organs of the social man, are one power of lens with which we peer into those of the individual man. Anthropology, or the knowledge of races, with their large types, characters and functions, are a still higher power, and the body under this glass shows like a world of man, and a point in universal history. Under the moral glass again the body is governed by miracle and prophecy, and we look through the flesh as a gallery of new forms, the plastic statues of the virtues. According to our eyes, we see the human frame, either as a congeries of dots and molecules, as children see the stars, or as an infinitesimal mankind, constituting the smallest complete humanity, which is a single human body.

We are now on another landing on the stair of truisms, and we affirm that *humanity is the greatest human body, and the likest of all things to the least or individual body.* It has its brains in those who are the presiding influences of the social universe; its lungs are in those who are the practical intellect of the ages, and the voice of truth to the world; its heart is the thousandfold love that carries the races to their goals, the poets whose bold blood licks up toward heaven, and talks in tongue-like alphabets to our own; its belly is the whole schooling of mankind, all the men and means that have ever grown up, that hunger and thirst to receive infants and savages, and convert them into angels. Its skin is in those who are the bonds of their state, which are nothing more than the lines and currents of the social freedom, but looked at from without as decorous beauty, or inviolable law. In this organization, each molecule is an individual man, the account of whose functions depends upon the constitution of the whole.

So far it will be perceived, that life is the main element of our wants in physiology; which accords with common sense, for who does not desire to be alive, and more alive, in whatever he does and thinks? Yet nothing has so much agitated our science, as that which ought to be its greatest blessing—Life; nothing has seemed so much beyond it as that life which is the first perception of infantine reason, and the joy of even the beasts and birds; nothing has thrown it into such peevish atheism, as that great plain gift which we accept every moment from the Author of our being and the Founder of our faith. They cruelly say that life is mysterious, but learned ignorance alone is mysterious: how can life be mysterious, when light is truth, and light and life are one?

Let us look at our predicament physiologically, that we may see where we stand, and what we have to do. The living body is the field which we are to explore. What course are we obliged to take? We are compelled to dissect the dead: but does this give the information which we seek? Evidently not, for dead organs are the antipodes of our quest. What is to be done, to raise the dead man, and unbind his grave clothes? What divine voice shall cry to physiology, "Lazarus, come forth?" Life must be brought from the living, from the quick body and mind; also from the great forum of men, which we call life *par excellence.* Tell us whither else can we carry it into the sciences than from the mighty reservoirs of history and humanity? Beyond a doubt it can only come from where it is, that is to say, from the land of the living.

Here the question of life loses its mystery, and shines with understanding by its own inherent effulgence. There are not two lives, but one; not a human life here, and a physiological life totally different there. That which is life in humanity, that which is life in society, that which is life in persons, and in the moral soul, is also life, and the only life, in the organs of the frame. Ends, motives, passions, affections, likings, loves, virtues, are human vitality, and there is none other. For what is our life, and the measure of it? What is our experience thereof? What is it that sets us in motion, and opens us for sensation? Why do we do anything, or think anything, or keep ourselves awake to feel anything? Most surely because our being lies in cherished ends, in which success is delight, and delight the flaming of our lives. Remove these ends, and we stare without seeing, and sit in corners with hideous apathy and indecorum, miserably disheveled and vegetalized; for life has nothing to do, and is taking its departure: as in the Metamorphosis we are growing into trees, and the needy soil shall swallow us.

Apply this to the body and its parts, and we find that the ends which it subserves in the order of things, are its animating principles. They are not abstractions, but spirits embodied in works. They are not fluids, or solids, but human uses incarnate. Above or beyond these there is no life in us; but whatever is useful or endful, God is with it, and that is its only life. If you look for the life of friend or brother, it is the stature of himself working in his daily calling: no one part of him, but all parts and the whole, kindling as he moves round his orbit of ends, into more than the whole. Life is man and whatever is manly; humanity itself in the fire and work of development; it is incapable of dissection, for it never can be seen but in play; whence the end of anatomy is, to show that man cannot be anatomized. The desideratum is, to see each portion of the body united with the whole in its uses, when the life of the whole will come to the parts, and summon them to live. After which, that thing so often mentioned, and so little conceived, namely, human anatomy as distinct from cadaverous, will begin; for the life of our lungs and liver is as inalienably human, as the life of Shakspeare, or of the English nation. And then, moreover, the science of the body will lose its grossness, for the moral and spiritual powers will have their analogues in every chamber of the organs, and cleanness and chastity will be busy maids among the useful furnitures of the flesh.

If life consists in ends, there is a soul to entertain them; and wants and desires imply mankind beforehand. For we assume the soul, as also the existence of an imperishable humanity. It is a venerable creed, like a dawn on the peaks of thought, reddening their snows from the light of another sun—the substance of immemorial religions, the comfort of brave simplicity, but the doubt of today, and the abyss of terrified science. Let it come, however, scientifically, in the ghastliness of hypothesis, and let us work with it, and see what it is worth; and treating the question of life under that formula, let us proceed to the connection between the whole and the part, or between the body and the soul.

This is another branch of the same problem, which bewilders speculation, because we go away from the meaning of human life, thinking to attain to some indescribable life separate from our nature. As we said when speaking of life, so we repeat in this new field, that there are not two kinds of connections, one physiological, the other human, but the connections that we form in our daily walks are the types of those of the sciences.

The question may be divided for convenience sake into two heads; namely, 1. Why is the connection between the soul and the body effected? And 2. How does it take place?

First, for the Why? We answer to this, that the soul is connected with the body for the same reason as we are connected with the persons, objects, and circumstances that surround us, and which answer to our wants and interests. In a similar manner the body answers to the wants of the soul (p. 242—244), being the soul's wife, the soul's friend, the soul's house, the soul's office, the soul's universe. It is engaged to the service of the soul; shapen into usefulness by the soul's ministrations. As the hand shapes the pen, and then writes with it, so the soul forms the body, and then makes use of the properties resulting from the form. The connection between the soul and the body is not more mysterious than the connection between the pen-maker and the pen, excepting that our knowledge of the pen is so much more complete than our knowledge of the body. A science of the body, had we such, that displayed its uses, or its specific fitness to serve the soul, would as evidently give the motives of the attachment of the soul to the body, as the capabilities of the pen account for its connection with the fingers of the ready writer. In both cases it is the bond of service, of liking, of utility; for to intelligent life what other connecting principle is possible? If this is too simple for philosophers, still it is the ground of every connection they themselves form with man or thing.

For the purpose of breaking abstruseness from the argument, let us look upon the natural body as the well furnished house, the admirable circumstance and worldly fortune of the soul. Then, steadily regarding the soul as the man, something like the following analogical discourse may result from this point of view, in which we take our stand inwards, to gain distance for the object.

The soul being the man or real body, the natural body represents the appliances and arts of life, whether economic or æsthetic. The eye is its window, telescope, microscope, and answers to the series of means that transparent substance lends to vision, and which are as curious and exquisite for their appearance as they are excellent for use: for the eye receives the finest impressions from things, and gives the finest expressions from the soul. So likewise the ear is the hearing-trumpet of the real body, which would otherwise be deaf to the music of nature; it embraces all the means of reverberation, whether in the free air, or of cheerful voices from household ceiling and walls, or of stately sounds from the long-drawn aisle and fretted vault: in short, both the whole instrumentality and the whole architecture of sound. But the nose is to the real body the prophecy of devices that have not yet entered into arts; full as it is of membranous parterres and vacant aviaries for odors; for hitherto, aromas are but casual visitants; they come and go in brief seasons with the fitful winds, and where is the vessel that can hold them; hence the nose of flesh is deficient in circumstance, and we can only identify it somewhat barbarously as the scent bottle of the real nose. To pass over the other senses, we find that the legs are the outward art of locomotion, from passive to active; from the nails of the toes to the wheel of the knee and the globe of the hip; in short from the walking-stick to the railroad; the real body uses them in nature, whether as the staff of its lowliness, or the means of its swiftness, or the equipage of its pride; they are the columns of movement; the rich soul's carriage, and the poor soul's crutches. But the arms and hands are all the finer machineries or inventions that are wielded directly by the arms and hands of the soul; they are the pen and the sword; the instrument of many strings; strength and manipulation in their bearings; in short, the mechanics of intelligence, whereby nice conveniences of truth are gathered in the dwelling of the soul. Then the abdomen is its kitchen, preparing from all things in its indefinite stores one universal dish—even the blood of life, to be served in repasts for the spiritual man; the viand of viands, varying from hour to hour, and suited with more than mathematic truth to the appetite and constitution of the eater. Then again the chest distributes with a power of wisdom dictated from the halls above, this blood, the daily bread and wine of the body of the soul, and the wisdom that ordained, enters the feast, and it becomes a living entertainment. And the brain is the steward and keeper of the animated house, receiving order and law from the soul or brain-man, and transplanting them into its mundane economy. Yea, and the brain is its natural universe, its wide spread landscapes, its illimitable ocean, its royal library, studio, theatre, church, and whatever else is a place of universal light and contemplation. And lastly, the skin is the dress of the soul in every kind, convenient, beautiful, official; and it is also the very mansion itself; for our houses are but the largest suits, admitting our domestic movements.

By this artifice of holding out our bodies before us, we illustrate in a plain way, the connection or correspondence between the soul and the body; and though there be other motives of connection, it is sufficient to remark for the present, that by the foregoing signs, it is because the body is so replete with exquisite convenience, that it is the domestic establishment of the soul. Given a tenement of the kind, so royal with apparatus, and it is impossible that the soul to whose wants it answers, should not live in it, and use, that is to say, animate it. If the soul were not a tenant on such invitation, it would be stupider than the birds and beasts, which are drawn by far lesser affinities to their own convenient lairs.

Let us now reverse the picture, and suppose for the argument's sake, that a savage is introduced for the first time into one of our convenient mansions, and knows the use of neither table or chair, knife or fork, bed or carriage, washing or lodging; but his naked body and unarmed hand have been accustomed to rude fellowship or direct fight with nature. Can he account for the connection of the civilized man with his house? By no means. Unhoused body that he is, we see in him a type of those who cannot conceive the bond between spirit and nature, because they know nothing of the wants of spirit, or of the uses of nature to spirit. At first, then, the savage cannot divine why his civilized brother limits himself to a house, because he is uninformed of the good of a house, and not prepared for information. As, however, his wants grow, and he learns the uses of the furniture, and the proper mode of employing it, the motives and points of connection come forth one by one; and when all the uses are understood, then for the first time he understands both the reason and mode of the permanent act of inhabitation.

So it is with the body and the soul. The physiological savage (we beg his pardon) knows nothing of the body as a correspondent or furnished abode, but only as a strange looking exception to wild nature, an affront to the wilderness, and how then should he see its connection with an owner or a soul? For the uses of things are the reasons why they are used. And hence the perception of the connection of nature with spirit, is the exact counterpart of the perception of the spiritual uses of nature. To see the one is to see the other, as to miss the one is to miss the other also.

A word now respecting the second point, or the mode in which the connection is effected.

When we speak of the connection between the body and the soul, we are apt at first to think, that it is a single link or act; but this is an insufficient conception. There are as many different modes of connection as there are wants in the soul, and organs, parts and particles in the body. There are as many different modes as there are possible species of contact in the great and the little creation. The soul is connected in one way with the brain, in another with the lungs, in another with the heart, in another with the belly, again in another with the skin. To make this clear, recur to the house and its furniture. The inhabitant owns everything contained in it. Upon one piece of furniture he reclines, upon another he sits, at another he writes, upon others he treads; some contain his viands, some delight him with harmonious sounds, and some look down from his walls, and gratify him by arts and proportions; and with all these, and many more, he is connected. Now in each case it is the shape, make, form, or properties whereby the thing serves its purpose, that is the means of his connection with it. If he sits at his desk, it is because it is such or such a structure, and serves him for reading and writing; he never makes a mistake of sitting for these purposes at his coal-skuttle. Apply this to the body, and we find that its dweller uses every implement there also according to its form. The reverent soul kneels in the knees, because they are natural kneelers. The inquiring soul peers through the eyes, for they are born windows. The make of the organ is the handle whereby the inner man grasps and uses it. Our business therefore is, to show the motives for which the soul affects the body, by demonstrating the deeds that the body performs for the soul; and it is our further province to show the manner in which the soul lays hold of the body in different parts, by explaining the precise mechanism whereby the useful deeds are done. The whole is a matter of facts, too full for speculation.

This connection of soul with body is no chaining of the living to the dead, like the horrid punishments of old times, but it is the live man freely working with the finest tools of nature, the chief musician in continual play upon the choicest instrument of music. Moreover as the soul goes beyond the circumference, and returns into itself, so it includes the body, and takes it back with it alive; the immortal confers his own life upon his mortal bride: there is a reciprocal connection of the body and the soul, and that wondrous house is the model faëry-land: the bells ring of their own accord at the time of the master's volition; the chairs dance into their places for the guests; the harp and the player understand each other, and sweet touches of no violence are simultaneous with delicious sounds rising upwards to the listening soul. There is not only connection but consent. As to freedom, the relation is permanent, but liberal. There is ample space between soul and body, as between friend and friend, and yet the ever-varying bond is all the closer, being founded upon the interests of each, which have room for play in the wide interval between them.

This same problem of connection occurs wherever a higher system is united to a lower. The connection of nerve with muscle, or with vessel, is as inscrutable to the physiologist, as that of the soul with the body, to the philosopher. A sight immersed in the muscle would as little find the nerve, as our incarnate eyes can see the spirit land. Here, as in the larger case, it is the dependence of uses and forms that makes the connection: the body is a whole, because its parts are chains of correspondencies. There is no glue between the organs, but they are always loving each other, and always using each other. No wonder they cohere. Birds of a feather flock together, and likeness and liking are the gold wedding ring of the universe.

The cosmical relations of the body proceed upon the law, that the body is the form of the soul, according to the order of the universe. That is to say, as the plant grows by the unfolding of the seed, according to the climate and soil, so the body grows from its own ensouled seed, according to its soil and clime, which are the mundane system. The soul commences with the commencement of nature, and flies down, weaving and constructing, through all her kingdoms. The brain is the home that it makes on its own plan, of the solar fires, and the brain is the soul of the body anew, and the sun of the microcosm; the principles of the world are the beams of its chambers, and it lives like the eagle unharmed in the glory of its native pavilion. The nerves are the next construction of the dramatic soul, where it brings the imponderable forces and fluids into a beaten highroad of human life, making ether and magnetism into pathways for the processions of thought and will; for the nerves are the horses of the sun, which carry out the decrees of the revolving brain from the east to the west, and from the morning to the evening, of the microcosm. Next the soul is victor of the air, and poised amid the whirling winds, she knits their filmy stuffs into obedience; and lo! a fabric huge rises like an exhalation—it is the lungs, the temple of Æolus, the shrine of the powers of the air. There is the first house of liberty, the free place of elections, the Campus Martius of the understanding. Accordingly the breathing lungs are the barometer that indicate the peace, or the power, or the storm, of the soul: all the passions swing and swim, and find their golden liberty, in that aërial sea: there they rock, and waft, and sail, in gilded barge, or pompous balloon, or happy skiff, or dark boats of Erebus. But the progress continues. The work of ensoulment is carried on through the animal, the vegetable and the mineral. The heart and viscera are the stations of the soul in these uttermost regions. The abdominal organs are the garden with its trees and inhabitants, where bread and salt, fruit and flesh, are presented to the soul; and the heart is the animal man himself, the sovereign of the tribes, appropriating their tributes, and making the body an assemblage of the universals of the earth. Then, as nature moves by immense attractions from space to space, or is always making new spaces, and worlds to populate them, so the soul will do the like; and out of the moving mechanism of things she builds the limbs of human kind, which are as a new liberty; and the legs are the beginning of all progress, and the arms, of all secondary creations or arts. But from the extreme of nature—from the boundaries of death, there is something to come that nature can never bestow, and that something is sensation—the vital hoop which girdles the body, and separates it for a season from the universe. Hence at the outer gate of nature life again breaks forth in joyful resurrection; unabsorbed, unextinguished, the Maker is not exhausted, but is greater for his work. For the soul running downwards from the brain to the skin, never ending in the end begins anew, and reattaches death to life by the standing wonder of the fivefold senses. So where the formative soul ends, the conscious soul or the man begins—viz., with the senses, as the basis, clothing and incentive of the will and understanding. And as the soul descended, the senses reascend, the same grand staircase of kingdoms. Touch, taste and smell are the living mineral, vegetable and animal; and hearing is a new-born palace of the air, whose shakes are music, and its winds are speech. And the eye, round like the world, and rolling on its axis, communes afresh with the whole possessions of light, and sees all from the sun to the landscape in the gloss of that glory which is an image of the truth.

We now come directly to the human form, the stature of our souls, and the vessel of our lives; and here we quit the anatomical, and enter upon the integral sciences; nay, we also quit the body, so far as it is material, and additional to the form.

The human form is the image of God; for "God made man in his own image, in the image of God created He him," and man is a living human form. And the Creator Himself, as the archetype of the creature, is a divine man, or a divinely human form. Accordingly God's revealed faces are such as man apprehends by his own mind and likeness; as it is said, the Maker of all is "King of kings, and Lord of lords;" and again in the new covenant, God is "our Father in the heavens." And this way of speaking, accommodated to our nature, involves no compromise, but the divine truth that man, the image of God, can know God by his constitution.

The Word, in which God comes to man, or makes human revelation of himself, reveals man to his real body, or shows the manhood of the soul. For angels and spirits, who are souls, are brought to us by the Bible in the human form. Heaven also, the holy city, which lieth four-square, has the same stamp, being "according to the measure of a man, that is, of the angel." By this apocalypse we are among brethren in every world, and we know that their humanity is distinct according to the eminence of their form.

Here occurs another field of subsidiary revelations; for the world (not the skeptical but the real world) is complete in cases wherein parted souls reappear to us, and show us that the human form is not meet for death: and even when intangible to the grosser man, still the form stands before his eyes, because its very colors are immortal.

To turn from what is revealed to what is perceived, the eye opened by revelation is the eye also of reason, which now sees the same omniprevalence of the human form divine. Our thoughts of God are thoughts of an infinite humanity: the love and wisdom counted for all manliness, are attributed to our Creator by the very brightness of our minds. Whenever reason shines and becomes half angelic it talks humanities of the living God. The mind is constructed for revelation, and its functions of loving and thinking, once opened, flow according to the truth of divine things, as the body when launched into the world in breathing, falls in with the laws of the world in rearing its own constitution. It is not easy to discriminate between the gift and the giver, but in the better faculty, perception is revealed, and revelation is perceived, and our powers are properly our own, when we own that they are of God; for in the soul the claim of private property abrogates the possibility of possession. The result is that thought is necessarily directed to the God of Revelation, whom understanding and affection meet; *i. e.*, to a divine humanity: in the case of our fellows, to a soul of which the body is the counterpart, *i. e.*, to a soul in the human form: and with those who have left our world, still to human souls and shapes, to whom "he" and "she," and the like pronouns of our love apply.

Let us register here, that the body of man is the anchor of these perceptions, and attaches to every contemplation of the human form. For body is the exigence of spirit, the partner of its love and the throne of its power. And though God be not as we, it is not our form that we cast to view him, but the imperfections of our form, which as we put them off, the form becomes more sincerely human, and likens nearer to the truth. Hence the way to a divine perception is the rejection of our inhumanity. Moreover by a connection that none can repeal, the man of the senses is married to him of the soul, for better, for worse, for richer, for poorer. In thought also, every idea has a form, every form has a shape, every shape has a body, every body is a substance; and to sum up all by the same fact, human truth is incarnate. Thus the human form implies the shape of the body, but ninefold flexible, or drawn into perfections of which flesh and blood are but the footstool.

On this subject we say further, that disembodiment is hateful to man, and the fear of death itself, apart from the love of life, arises from our ignorance that the dead are men. No wonder that we shudder with all our life beside a brink where philosophers teach us that the human form is wrecked. It were weakness not to shrink from the loss of that which is the instrument of all our power. For whatever is strong tends to incarnation. An energetic purpose is at once in the arms and hands, and emboldens their play: whoso crosses it, finds that its missionary is never such tough matter as when his spirit is behind him and within him. A lively thought calls up image after image, and is embodied at once; if the images will not come, there is an end of the thought. A will not only rushes into the muscular system, which is a battalion of arms and limbs, but it rifles resistance for hardness, and plucks out the teeth from steel and stone, to make fresh bellies of strength. A high contemplation in the soul, begins to see heaven only with the upturned eyes, and to pray in the body as soon as the contemplation is full. A soul is no sooner projected than it begins to build a body for itself. Thus strength clasps and presses body, to beget works as children. Impotence on the other hand engenders with abstraction, and prating philosophies come.

In returning from this digression, we next meet that activity of mind which begins to work upon the two worlds of spirit and matter with native conceptions of its own. This is the poesy of knowledge, the sense of nature and things which is born of human religions. And it affirms the personality of its objects, not only of men and women, but of flowers and horses, stars and streams: and dwells upon them fondly, as if they were in its own sense alive. The childhood of races as well as individuals is so vital and full-blooded that it looks on the world and all it contains as a land of the living. With no permission of afterthought, it deems and imagines structurally, and its organization is the mould in which every conception runs. It believes that it is enfranchized, not limited by brains; and it knows that departure from these leads to nothing but folly and confusion; space and matter, irrespective of man, are silly monsters which it does not heed; so flimsy that thought goes through them as if there was nothing there. For the rest, the prime method of humanity is, trust to the topness of the human head. Those who belong to this method, send forth words that feed on time, which feeds on all other things: words which already play with everlasting flowers over the grave of many philosophies that have been impertinent to the human form.

We note then, among the first natures of our minds, the conception that there is an architectural human life among the beams and rafters of the world; which conception is the ancester of science; the natural loins of truth out of which come all the families of the adult understanding: and we hold, that to repulse this nature is to be brutal to the mighty mother, to bastardize knowledge, to begin the sciences nowhere, to expect to find by starting without looking, and, in fine, to select impossibility as the arena of our struggles. The attempt has been made in the inductive sciences, and with what success is known. Phenomena and their laws are the recognized objects of these sciences; substances and causes are reckoned spurious: hollowness and superficiality are thus become the postulates of knowledge. Yet causes are past exorcism, and science runs into chronic "goose-skin" before these, its self-constituted foes. But why has science been so successful? In the first place, it only half succeeds when it sees nothing but the matter and none of the mind of nature, for in this case, its use is culinary, and not educational. For what can a world of shadows, and rules of shadows, have to do with God who is a substance, and with man who is a substance too? There is no real instruction in formulas, however well they answer to facts, unless one knows why the facts come under the formulas:

otherwise man is a child in a despotic father's house, who never knows the wisdom or rationale of the father's actions: the actions go on with blind strokes like toothed wheels in a factory; and the only lesson is, to keep out of the way of this ponderous and remorseless nature. On the other hand, instruction, such as heart can love, and memory remember, arises in proportion as we see that the external world is nothing but humane ordinances, and that suns are central, systems revolve, plants grow and fire burns, in the interest, and as the vehicles, of a Divine Mind, which is the archetype of our own. To explore God's purposes is the first thought of man; and to find why God acts in such ways, shapes and properties, is equivalent to finding the wisdom of the world. The laws and formulas of things are but new spelling books of this deeper and earlier consideration.

But in the second place, where science succeeds, it is because it cannot throw away its heirlooms—its instincts of the humanity of nature—but applies them to the enucleation of natural truth. For law, rule, phenomena, order and succession, are cerebral humanities, which we get from things and give to things, because we have brains. Where would law be in our conceptions, if no legislator or lawyer had existed among mankind? Where would phenomena be found, if there were no eyes? Where is our order, when the mind is delirious or in disorder? And where succession, unless time be ticking in ourselves? These dim ideas as much imply humanity as the most carnal Sagas of the mythologies. We equally palm ourselves upon things in averring that they obey law, as when we imagine, in prattling childhood, that the flowers hold speech with each other, or that falling waters are moulded into veils and scarfs by the limbs of the Naiads. A Franklin, bottling electricity with the knowing air of a butler, and "calculating" how many dozen the skyey tun will run, is as little emancipated from mannerisms as the poor Norseman of fifteen centuries back, who hears Thor hammering in the thunderclap, and sees the brightness of the vengeance of the same God when lightning stripes the night with terror. The axioms of Euclid are as incurably human as the fairies and elves. So the difference between modern science and old Myth is not that the one is disembodied and absolute, and the other fettered to the flesh; but that in science some cerebral dot, according to the age, sets up for monarch, whereas in myth the whole man, all dots included, is tremblingly alive to the existence in nature of some living God or Gods, whom he approaches according to his revelations.

Let us, then, rid ourselves of the blur, that man is limited by being a man; when the reverse is true, that what cramps him is, that he is not a man. The human form is emancipation from the prison-house of the inferior creatures. Each man is indeed limited as a man, but manhood itself is a ladder of infinites.

The mind, as we have anticipated, is in the human form. For it is in every part of the body, co-extensive and co-intensive with the organism. This is only to say that we are conscious of ourselves. As the eye sees the human form of its owner, the mind sees and feels its own human form, though in its own peculiar way. We are solid statues of consciousness of just the size of the bodily frame. This is self-evident when it is not dwelt upon too much. For as there are objects that the eye is meant to see very seldom—viz., the face and person of him who carries the eye, so there are truths which the mind, unless full of the looking-glass, regards but for a moment, and then leaves them—viz., self-consciousness and its bodily form. Each cursory glance of this kind finds the mind in the human shape. Whence we know that the sciences and subjects of the mind are part and parcel of the same constitution.*

As for the passions, poetry embodies them in personal lives, but not before experience has well concluded that they are the humanities of their own order. Passions are men, women, or children. Imagine rage, if you can, without starting eye-balls; or love of power, without commanding eyes and hands. Think of eagerness without legs. Fancy the repose of a being who has nothing to sit upon; the intelligence of one without an eye; the joy of one without a face; the tenderness of one without heart or bowels. The residuum, after every conceivable amputation of the kind, is a philoso-

* If, however, the vanity of reflection be indulged in too much, philosophy at first sees nothing but its own pretty face in the mirror, and then as it is an intellectual person, it concentrates itself on its own obverse eyes, and falling (by the laws of hypnotism) into reverie, it finds plainly enough, from the ground of its own state, that the mind (*i. e.*, *its* mind) is nothing at all.

phical angel. "A footless stocking without a leg," was the Irishman's definition of nothing: it is the metaphysicians' definition of the soul. On the other hand, to common sense, every passion has all our parts. If the painter delineates love, it is by a beaming countenance, by clasping hands, by open arms, by beating heart, by flushing cheeks, by a bending and inclining form: or if the sculptor will show you duty, its white marble is a cheerful severity of brow, hand and frame, and a steadfastness as of limbs in the axis of the poles of the heavens. Moreover, the typical features which the imagination summons, run like light to fetch the rest; they have a creative call to form the remainder of the body. So that when the mind sees the sparkling eye and elevating smile of joy, it next, from these foci, sees the face as their radiant firmament; and the lifted arms and lifting legs, and the jocund body, come after more slowly, until the stature is there, and the joy, like its artist, is a man. But if any part of the body is missing, a letter is left out of the word, and the creative power has not spelt it. All this shows that the mind imagines the passions structurally, according to the form that they have when in play.*

The same applies to the intellect, when described not in its effects but in itself. To the mind's eye, it at once assumes the lofty brow and high conversing glance, and thereafter its incarnation proceeds by rapid stages. Here again the central features are first struck; but no sooner are brow and eye on the disk, than they call forth their brethren of the frame and mood, for they all belong to intelligence—the clever nose, the receptive fastidious mouth, the slanted open ear, the considering vibrative head, the awaiting arms, the balancing body, the movable feet, stationary but changeful. All these come part by part from the land of the human form, drawn thither by intelligence itself with the magnet of the eye. In this case, what we take for abstraction is the centre separated from the rest of the body, and held forcibly apart; for violence is needed to hold asunder

* On this subject see further, *An Essay upon the Ghost-Belief of Shakspeare*, by Alfred Roffe, 8vo., London, 1851. In this admirable performance we have the first beginning of a study of Shakspeare, not according to reasoning and criticism, but according to fact and nature. It cheers me to have such a mind as this courageous author's with me, in these untrodden paths.

elements which so love each other, and rush together, as do the parts of man. Pale philosophy, indeed, has attempted to draw a sharp cut of spells between the round pellucid intellect, and the face of which it is the jewel. But as it sits in its dissecting-room with platefuls of human eyes before it, we recognize in the amount of its victims the exact number of its failures.

We might pursue this thread of truth through every other faculty; and we should then record, that goodness, virtue, honor, bravery, industry, and all the names that we revere, are man and the organs of man, and that out of him and his they are nothing. But it is enough to indicate one or two cases, for these, as the eyes and forehead of the rest, bring with them the predicates entire of the human world, and marshal them into its form. We now, then, pass on to the sciences, to notice their impartial teaching in spheres where man seems to cease, and to be environed by apparently intractable, that is to say, inhuman forms.

It is not necessary to have recourse to any ideal state of the sciences; but their present condition serves to show, that humanity by its principles extends through the realms of beasts and fishes, herbs and stones, and even through winds and the fluid worlds. For the instinct of science in all these departments, drills their subjects with reference to order, the ends of which order are, mind in the one case and matter in the other—mind, the intellectual *plenum* of the human form—and matter, the stages of nature that lead from her deadness up to that life which is the perfect man. The schemes of the atheists are obedient to this roll-call, for otherwise they would not cohere into schemes; their vile scientific cunning can exist on no other terms. The difference between them and the religious world is, that they have the task of making man out of the ground by their own dust-strife, whereas the others hold that he is ready made out of that very dust by the breath of Jehovah God: but in both cases the dust is recognized as a fit material for building into the plan of the human form. There is then an admitted reference of the dust to man, and this, when put by experience through its degrees, constitutes the order of the sciences.

With regard to the organic sciences, the case is clear: their aim is, to range living beings in a scale of which man is the head, and to set down the lower animals as extended cases of the human type. "The animal kingdom," says Oken, "is man disintegrated." Comparative anatomy is the anatomy of the beasts compared to that of man. Take the animal forms one by one, and their likeness does not perhaps clearly shine inwards to their chief, but group them severally upon the order of the mind, which is science; and unlikeness travels into likeness as they approach the human goal. As in languages a word shall seem alien to its root, but track it through a number of kindreds and tongues, and the links are supplied that connect it with its source, and draw it from its parental stem. So man is the word to which this social science of comparative anatomy leads up the forms of animal life, and Adam still in this way gives all living things their names.

This is better the case with natural history, which leaves dead bones and *ghoul* sciences to become acquainted with animals at home in their gayeties and enjoyments. The portraits of man are so close here, that the naturalist associates with his pets, and a White of Selbourne can spend a long day with his swallows, Huber with his bees, Wilson with his *acarus folliculorum*, and Swammerdam with his snails. Moreover, we judge the animals by our own standards of pleasure and pain, attribute to them a part of our metaphysics (would they took them all); and assign them natural virtues and vices for which they are responsible with their lives. We also reverse the case, and feel no inappropriateness in giving our brethern animal names, making a comparative natural history out of the eccentricities of man; for we have human asses, geese, pigs, foxes, tigers, monkeys, peacocks; nay, in some cases we qualify humanity by an animal color; as when we say that an innocent person is a lamb; a gentle beauty, a dove; or a heroic man, a lion: which shows well the sliding scale from man to nature, in the easy projection of ideals into these distant but beautiful types. The symbolism of the Bible makes animals thus expressive. It is impossible not to interpret them by our own faculties, though reason knows that the faculties between the two are not the same but analogous. But the interpretation holds all the better; for it embraces the animal fact, *plus* an intellect that enlightens it. This is one function of man in his social universe—to give his life away to beast, plant and stone, and raise them into company for his mind: a function like the Creator's, who makes all *quasi*-things into substances, mere substances into minds, and unites the world and men to himself by nothing peculiar to them, but by the golden links of his own merciful attributions. Imputing the higher to the lower is therefore hearty truth running Godwards, or the process by which universes are made and kept. Hence it is our duty to give the stones a tongue, and to cover it with all the runes of our own eloquence and wisdom. . . . Furthermore, there is common companionship of men with animals, and duties from him to them, which duties have the significant name of humanity, as though the humane man recognized his species wherever life was seen. And according to the materialist, this is true to the very mire; for they deem that there is no difference between men and beasts but the want of speech in the latter, and the absence of tails in the former; which shows a strong dirty perception on their part of the oneness of principle prevading life: a principle which if sought at the head—for in the matter of the tail they admit unlikeness—resolves itself into the prevalence of man. Again, man never falls below himself, but animality receives him with open arms: "nemo unquam repente fuit turpissimus:" on the road to moral death, before the descender is as dead as a stone, he runs through the beast properties, transmuting form and feature right down the devil's ark, and at last crawls, serpentwise and toadwise, into his chasm and block of a thousand years. And lastly the animals, to conclude their manly participations, are many of them direct and even domestic servants of mankind: they share our lot, of poverty and riches, good and evil, civilization and savagery: they become honest, or sly, according to the rule of their master's house; showing that they are willing fellows. And as the domestication decreases by degrees, and embraces main generic heads, the whole animal kingdom may be grouped about us from this principle, either in the way of friends or foes; and thus approve a human reference of this close kind, coextensive with natural history.

The science of plants falls under humanity for similar but lower reasons to that of the animals. For their natural history is digested into botany, in which they assume order, classes, species, and families; and the order is subject to that of the kingdoms themselves, and included within it. But being a vegetative or growing order, and not an animal or moving order, it does not travel to man, but enters into him by his own choice. Hence the uses of the plants are their chief human relations. For the plants, like factory men, begin to feed and clothe us; they spin our cottons in vegetable Manchesters, and are the looms of looms, in which the Arkwrights of the sap and pod herald and provoke the Arkwrights of the mill. In this kingdom there are all things not made with hands, in preparation for all similars to be made with hands, and this again in earnest of a new growth in heaven, where all things not made with hands again appear. And the more good science we bring to bear, the more humane do the plants become; in other words, we find that they are altogether useful, that is to say, grow towards us; until at length we see all the trees as one tree, whose fruits are good for food, and its leaves for medicine. More remotely, plants are our anatomical brethren; they have sap like blood, and leaves like lungs, skin and limbs, and roots of stomachs also, each of their own order. They also have seed and progency, and in their green ambition, they tend to people all earths, and to build airy castles like our own, raising up the soil on their hods in wooden galleries to the clouds. Moreover they are symbolical, or humane to our sentiments, with which they easily tremble and glow; and in the same light they are biblical, or humane to the divine, leading him from paradise, the harmonious garden, to that city in the midst of whose street there was still the tree of life. And as the domesticated animals stand at the head of the animal kingdom according to one mode of classification, so the useful plants occupy the first ranks in the vegetable kingdom: and constituting the beginnings of series, summon the rest about them, and group their whole realm around man.

Again, the mineral kingdom is humane, because it is included in the scheme of order, or the mental humanity; and because it supports the vegetable, which grows up to the animal, which travels towards man. And here be it observed, that each realm has its own approach to us. The animal makes its way thither by a scale of likening forms and functions; it looks and acts as if it were human, to a certain extent. The vegetable, on the other hand, bears no resemblance to the whole man, nor does it act at all; it likens to us in being of use; and whether it be as fruit and flavor to our bodies, or as beauty and symbolism to our minds and souls, this same law holds: the law, namely, not of progression to man, but of use entering into progression. This use is the making up of substances into new forms to serve us. Support, however, is the human property of the mineral kingdom, and in all cases *terra firma* is its kindness to us. It stands under all things, and bears their burdens; and *in* all things it still stands under them, and makes them real. For it not only supports the stems of trees, and the feet of animals and men, but it supports the vegetative process in the one case, and the animal life in the other; so that without it the trees would be but plans of trees, and the animals but illusions. Its gold, unlike paper, is the earth of credit, by which solidity enters into commerce. Its salts are the earth of the blood, without which that soulful organism would be footless, or without a base. Its rocks are the surety of foundations for houses, otherwise than sand and mould. Its jewels are stern flowers that shine in spite of the seasons. Its stones make our palaces and prisons, which are the *ultimo ratio* of our arts and laws. It also supports the soul as well as the body, and give our ideas a terrene base, of resistance, hardness, permanence, steadiness, and much else which is low, moral and strong, so that to the soul the very ground is *terra firma* to justice and virtue. Nothing of this could be, unless the earth were coördinate with man, that is to say, in its own remoteness, humane.

The three kingdoms have also succeeded each other as the organa of so many hypotheses of the world; thereby attesting that they themselves belong to an affiliated scale. Thus we have the idea that the world is an animal; a conception attributed to Plato, but not, so far as we know, carried out in detail. A second idea of the world is spiral or vegetable, and works the hypothesis of development instinctively through all things: according to this conception worlds grow: it has been a favorite notion with races who have lost their revelations, and was set forth in that Scandinavian myth of the Ash Yggdrasil, which ash was in fact the universe: it sprang out of depths which no God or man knew; beside its root was the weird or Urdar fountain where the Norns or fates abode; its branches were heavens and earths, the giant-world, the man-world, and the dwarf-world: and every morning the Norns repaired its many wastings by lading over its head from the Mimer spring the dews of immortality. A third notion of the mundane order is, that the world is dead or mineral: this belongs to the old age of hypothesis, and to the youth of cold experience: accordingly, on this view, the planet is hoary with immeasurable years, the stiff-kneed laws of today presided over its cradle; if indeed it ever had an infancy, and were not senile from the beginning. The completion of this stony truth takes place in geology, which is a kind of mineralogy of larger growth. Yet as geology, or the science of the great crust, detects order and mutuality in the earth, so it begins to revert by analogy to organization; just as crystallography points its dry spicular nails towards vegetation. Again, a fourth view is, that the world is quasi-human, or indeed superhuman, in its parts and in the whole. On this score, the planet unites with man; the head races inhabit its head countries; in short, the human organs reappear on the surface of the globe, and account as well as they may for its relations and distributions, and for its housing of special nations. These hypotheses run in a strict and mutually suggesting series, and either all of them are true, or none. But we do not moot their truth, but adduce them to show that hypothesis has been consistent from the first; that has always handled the world as some analogue of man; and we will add, though it is beside our present object, that the hypotheses deeply imbedded within modern science, are the fag-ends of those which shone candidly out from the ancient mythologies. For ourselves, the truths of revelation are our hypothesis, and we hold that the worlds, dead and living, include every attribute that others have assigned to them, not as their own, but because they are full of the spirit of Providence, and under Him intend man in their courses. Thus we accept all analogies as tools to work with, but none as a rest.

To conclude this scale, the unattached natures, waters, atmospheres and the like, are our circumstantial relations, the support of our liberties and fluids as the ground is the substance to our feet. Blood and water, breath and air, light and sight, are playmates introduced from of old. The vitals also of nature, the nervous suns, enclosed in their rib-work and grate-work of planetary masses, are among outward things most kindred to us, and lend themselves to joy and exhilaration for all that lives. In their other considerations, as the mundane system, they offer us great helps of order and law; and enriching without dissipating our minds, remodel our first notions upon creation, and our faculties rightly applied to the universe, become mightily more human for the process.

The pure sciences, the abstract truths of the rest, are themselves the laws of mind transplanted into matter; they are therefore, as we said before, cerebrally human (p. 304), and indeed only true within the sphere of our faculties. There is no escape anywhere from humanity: there we fully agree with the subjective school; but the reason is, that humanity enfolds everything, and not excludes us from, but lets us into, the veriest nature of the world. But it is well to be observed, that there is no Kantian world of "things in themselves," but the world is social to the core.

In the view we have taken, we have with us the attestations of all parties. The theologians are with us, by the text and scope of their revelations. The superstitious are with us, by their crouching anthropomorphic worship; by their fear of eclipses, as though their god's eye were black upon them; by their tremble before all places and natures, lest the tapestry of their world should open, and stiffen them with some gliding terror. The idealists, or metaphysical ghosts, are with us, by their creed that man cannot travel beyond himself, but that sense and its world is still within the circle of our being. The materialists are with us, by their vaunted alliance with the dust; for are they not "spouse of the worm, and brother of the clay?" The scientific are with us by the actual order of their sciences, and because they cannot help it. And the practical men, including the rest of the world, are with us, in their own practical way, for not one of them cares a rush for nature irrelevant to man, and contributing in no sense either to goodness, enjoyment, the pocket, or the person. We now therefore require, that what all have in part, shall be corrected and completed in each, and that the truth of the all-embracing humanity shall be set in that matrix which opens and hungers to receive it.

It is further to be observed, that impersonality is human, jus like personality, the latter being the skin of which the former is the substance (pp. 285, 288); a substance however which is no substance without that skin to hold it. For the best things in us are impersonal, and the more they are, the better persons we become (p. 285). Upon this principle, physical as well as moral, the impersonal in nature communes with a proportionate impersonal in ourselves (p. 285).

We have already trenched somewhat upon the other side of this question, namely, the relation of the arts to the human form, and we have hinted that poetry, painting and sculpture always act personally and dramatically with their highest subjects, which are the human actions and passions (p. 305). A Michael Angelo uses no theory, and makes no apology, when he throngs his Sistine Chapel with glorious human forms as the attributes of God and man: the admiring sea of ages of hearts and heads underneath him, finds no fault with such impersonations. But art extends far below these regions; and as we have noticed throughout, there is an art in science itself, which blindly musters its kingdoms under the human form. The useful arts again, the sisters of the material or culinary sciences, aim to reconstitute nature into human images closer and more commodious than at first. The mission of these arts consists in making nature into the wraith and image of history. For this purpose the world underground is a mine; its iron must not lie waste in those old lockers of the veins, but it is drawn into rails, girders, ribs of ships, stalwart engines and vertebral bridges; and the nations look on their continents as worms full of this metal silk, out of whose bellies untold glancing cities shall be spun. The geologies of the arts are new strata above-ground, inhabited tier after tier by the centuries; so that nature, besides the Plutonic and diluvian, includes also the humanitary formations, and is never natural until she envelopes all the arts in marriage with all the natures. The mission of the husbandman and the cultivator of stock adds new varieties to existing species of animals and plants, and nature embraces the new with the old in a tighter maternity than at first: her grandmotherly love increases with posterity, and ascends hotter and brighter to her children's children. In a word, use and cultivation by humanity are accepted by nature as her own consummation of her plan. Hence, what expresses all these facts, there is not a place or thing in the known world but comes under the notion of property, evidently on the insight that all things are man's, and may be of use; and so he registers his claim to their reversion when he wants them. Thus the planet has an intra-social existence, and the men who denounce our pride in claiming all things for the race, hold their estates by title-deeds of a later credit, and by weaker seals than that of this ancient faith. In evidence of this, the flag of property, either individual ornational, waves over every region, from the northern to the southern ice, and gives Christian names and surnames to capes and headlands, bays and islands; *mine* and *thine* are the zenith and nadir of the willing earth in the grasp of mankind. But what cultivation and re-formation is to the ground, that is regeneration to the mind, and reform to the state; and as the mind and the state become more human as the rights of the soul and the society enter into and alter them, so does the world become more natural in proportion as the cultivator stamps his humanity upon it. There is more nature in that prophecied nature where the wilderness blossoms as the rose, than where the sand alone extends.

We should have enough to do with objections if our intention cared that way: but we will briefly consider only two. 1. The "moral" objection, that there is great pride in man, a little statue of dust, a crowd of pismires on a hillock, a mote in the sunbeam, &c., &c., conceiting that all things are his, when yet nature is so much bigger than he. But to this we say, that all men act on this belief, and the more faithfully, the better it is for the world about them. The savages believe it least, and give nature the greatest swing of *laissez faire;* the consequence is, that she is a dirty and undressed female, snaky-haired, under their feet. Humility however consists in obeisance to the truth, and if man is nature's king, he must not be too proud to wear her crown; for pride may rise on that side also. Kingship is either nothing, or the chiefest service; and if all things are for man, it is because he alone can make the best of them. Humanity through industry is the very dew that the Norns lade every day over the good ash, Yggdrasil (p. 311). For the rest, this "humble" objection is always made with a lordly and exulting voice, and in abdicating nature's crown, it treads upon pride with greater pride of its own. We suspect that it proceeds from dislike of work, for certainly the duties of the opposite view are sufficiently stupendous. The true humility then, we argue, lies in man taking his hard, high place with its responsibilities—the place that he has in history and property, as the lord of the visible earth. Let him accept the title-deeds, and obey the declarations, of his Genesis. The higher the station, the higher the humility.

But on the other hand we will not grant, that nature is so much bigger than man as to escape his handling. For if the world be providentially coördinate with him, then as he is, so will it be; or in other words the threads of the mundane Norns are managed in that end of them which constitutes our human nature. A world made by the Word is plastic to the Word, and the Word in its second projection exists in man alone. It is then a part of our faith in Providence, to believe that the magnetism as well as the handiworks of good men and good societies, are the first beginning of the natural laws; and that there is nothing so ponderous as to outlie the power of the moral and spiritual kings (pp. 238, 239). The harvests and rain-clouds of Judea were once obedient to the obedience of its nation to God: how if nature now be equally symbolic and ready; and for new springing of theocratic crops, only await a new Judea in which the whole earth shall be a Holy Land?

2. A word also to the intellectual objection, that other creatures besides man doubtless look out from their places as central points, consider themselves as foci, and the universe as their subject and servant. "See man for me, exclaimed a lordly goose." (N. B., the sly poet puts this into the neb of a goose.) Were such the fact, it would only show, that the animals are images of man in these perceptions, and that in the wide madhouse of animality, they all conceit themselves kings and potentates. But are there signs that it is a fact? Has an ass with ribs belabored, anything like our perception, that we are "lords of what we survey?" With regard to our slaves, and even our lower classes, they do not imagine that the state is for them; on the contrary, they complain

that they are out of the social scheme, which is only for their betters; and although they build up a little hovel of content in the midst of their sad inclemencies, yet they do not consider that the body politic exists for their behoof. True, it *ought* to do so; but the fact that it does not, shows a power above theirs. So it is with the animals. Those which bear our burdens either deem nothing at all; or supposing them *to deem*, that exonerates them from the charge of stupidity which we make, when we say that the cow knows nothing of the supremacy of the milkman; the dog, of the huntsman and lash; or the horse of the spurred and whip-provided rider. It is however the want of deeming that makes the want of the thing deemed; and man is at the head because he knows and feels it. *Vice versâ*, the animals are subordinate partly because they do not know their situation, and never can be taught it.

The argument on this score has been pushed by animals' friends, to invalidate the humanity of the Divine Nature. For while it is admitted that all our conceptions of God are human, it is contended that if man has a God-man, a cow, could it imagine, would frame a cow-god; a dog, a dog-god; and so forth. But *imprimis*, none of these creatures can imagine in this direction, which terminates the argument. And secondly, if they worshipped what was visibly above them, they would worship man, who is in such a sense their providence, that Bacon has wittily cited, that man is the god of the dog. The assertion then comes again to an end, and shows that in all respects the living creatures look to humanity. For ourselves, had we any experience of a being transcending the human form, whose mastery over us was undoubted, this, we grant, must give the form by which our God would reign. But no such being has intervened in history, either in the solemn night, or in the daylight of religions: on the contrary, the Highest has declared that man is made in his image, and that his Son is also the Son of man. In short, we find the human form, divinely augmented, burning with uncontrolled intensity in the thought of man, in the records of inspiration, and in the religions of nature; and this with its attributes, though often unseen, is a real presence in every temple.

There are indeed religions which do not accept the central symbol; there are animal, vegetable, and mineral religions; for men bow down to wood and stone, beast and bird, in all fantastic forms. There are also fossil religions, records obscure and terrifying of past conditions of our race; hieroglyphics too large to live in our petty present time; and which speak of warmth and fertility in regions of the mind now cold and dead; of great perceptions and mighty propagations; gigantic promises of the world's childhood, only to be justified in her second innocency. Again, above and around these solid creeds we have the philosophical religions, the sciences of the atmosphere of the religious world, or the theologic wind;* pretences to regard Deity under no form, and as "neither in body mankind resembling, neither in ideas." But as the reader now foresees, each of these creeds has the same principle, namely, the human form, which is portrayed in the atmosphere as in the earth, in the vegetable as in the mineral, and in the animal as in the vegetable. Abstract philosophy is the furthest of all from the centre: idolatry itself falls more easily under that rank and discipline of natures which leads to Christianity, or the Omnipotent Human Form.

But the skeptical ghost still haunts us, for are there not higher beings than man, whose form is consequently above the human. Truly such beings there are, known from of old under the names of angels and spirits; but then their form is the human. For the human is a traveling form, which reaches from man to God, and involves all beings as it goes. There are then men higher than man, but there is none superior to the human form, which itself is superiority advancing for ever to the supreme. Revelation declares as much as this, for if the human form is the image of God, there can be no more eminent form.

However, what we call the human form, is doubtless the least truth and manifestation of that Divine image. For minds, souls, societies, nations are themselves in this form. But time fails us to attempt a flight through these second spheres, and we must leave to other hands the tracing of manliness through polities and socialities,

* The sensuality of creeds which oppose nothing but *flatus* and criticism to the instigations of the flesh, is set forth in that hieroglyphical proverb, that "pigs see the wind," which marvelously signifies the face to face posture of stolidity and abstraction.

which are the laws of the collective human being. But in the meantime the reader easily sees that man on the greatest scale is more man than he on the least.

We seem in all this to be wide of what is called physiology and science; but on we go, for science cannot subsist without these heights at last. What we desire added to all, is a divine physiology, flowing according to the channels of the divine form into which we men are born. For there is no life in science unless it communicates with the fountains of its subjects; no resource in art or healing, except from the draw and gush of the same living waters. Therefore we have done our best to show, what the contents of the human body are, and what pressures of life it stands under. For it is the form of God; as it said in Psalms, we are His temples: and so it is the native land of hope, and the arena of miracles and providences: there is nothing which it cannot do, and nothing that cannot be done with it, according to its correspondence with the Most High. It is also the form of spirit and heaven, and the heights of Zion and the abysses of hell are within it, circumpressing and acting upon it, according to its correspondence with good and bad, and the other roots of man, his body and his mind. It is therefore a pipe which runs with every wine, angelic or demonic. Again, it is the centre of history, and the life of the world plays upon it from without, and brings forth tones which are myth and song, strife and triumph, the fullness of the eras. Moreover it is the parliament of the natural sciences, and the executive and army of the arts, into which knowledge arises and speaks, and from which power goes forth. Thus what it finally does and can is by these rules alone—that it is like a god among other things, like a heaven or a hell in its radiations, like all history in its scope, like all science in its laws, unlikest of all things to death, most inscient of impossibilities, most untrue to itself in meanness, and most immeasurable by lies and materialisms.

We will conclude this argument with two glorious scientific songs, which say as we cannot hope to do, the mere truth on this matter. George Herbert *On Man* remarks:—

"Man is all symmetry;
Full of proportions, one limb to another,
And to all the world besides.

Each part may call the farthest brother,
For head with foot hath private amity,
And both with moons and tides.

"Nothing hath got so far
But man hath caught and kept it as his prey;
His eyes dismount the highest star,
He is, in little, all the sphere.
Herbs gladly cure our flesh, because that they
Find their acquaintance there.

"For us the winds do blow,
The earth doth rest, heaven move, and fountains flow.
Nothing we see but means our good,
As our delight, or as our treasure;
The whole is either our cupboard of food,
Or cabinet of pleasure.

"The stars have us to bed;
Night draws the curtain, which the sun withdraws,
Music and light attend our head.
All things unto our flesh are kind
In their descent and being; to our mind
In their ascent and cause.

"More servants wait on man
Than he'll take notice of. In every path
He treads down that which doth befriend him
When sickness makes him pale and wan.
O! mighty love! Man is one world, and hath
Another to attend him."

William Blake on *the Human Form* also observes:—

"To mercy, pity, peace, and love,
All pray in their distress;
And to these virtues of delight,
Return their thankfulness.

"For mercy, pity, peace and love,
Is God, our Father dear;
And mercy, pity, peace, and love,
Is man, his child and care.

"For mercy has a human heart,
Pity, a human face,
And love, the human form divine,
And peace, the human dress.

"Then every man, of every clime,
That prays in his distress,
Prays to the Human Form Divine,
Love, Mercy, Pity, Peace.

"And all must love the Human Form,
In heathen, Turk, or Jew:
Where Mercy, Love, and Pity dwell,
There God is dwelling too."

But if the parts of the human form have functions, the whole has functions too. A new problem therefore occurs, and we now ask what is the physiology, not of brain, heart, or skin, but of the human form as an organ?

In order to answer this question, we must know what the human form is; for its functions cannot be told until we have it before us. In the lowest consideration then it is *life, being* or *substance.* For *in fact* it is the end of creation, and the end is also the beginning, and supports the means, or lifts them into standing. The end also justifies the means, or gives them solidity. The human form* is therefore the living Caryatides of the world; or more properly, in revelation it is the I AM, Who not was but is in all time and nature. Its function accordingly is the gift of substantial life to those whom it includes; ay! and of comparative life and substance to all its images and types. In the lap of this sustaining form, we receive the seals of fact, and are bound to be real: we have no escape into nothingness any more, but existence is our fate. Herein then we are in the essential organ of life; and as the body in the form receives life, so its function or necessity is, to make life.

In the second place the human form is *use.* For man *in fact* is the creature who can surmount himself, and elect the universal use. He has a spirit above his nature, or can breathe for the commonwealth of heaven. Born into this human form, it presses its industries upon him, and commands him to good works. Its signal before all his battles is, "The Lord of the universe expects every man to do his duty." "Thou shalt eat bread in the sweat of thy brow all

* That Indian myth, that the world rests upon a tortoise, and the tortoise upon an elephant, though deficient in ground for the elephant, involves a deeper thought than the metaphysical conception of substance, and stands many degrees nearer to a true answer than the barren pantheism of Spinosa. For elephant and tortoise have good broad backs of their own, unlike metaphysical abstractions; and moreover they are analogues in the series of humanity, and in this degree approach the true answer. The conception of substance belongs to the skin, principles of thought (p. 287), and is the epidermis of the conception of support, whose inner parts, as we have given them above, are, 1. Being; and 2. Life.

the days of thy life." Man's work is universal; in laying bricks to the hovels of his plans, he is working at other plans veiled; rearing far and future states; the basements of ages to come; the limbs and wills of his posterities; the strength, or weakness, of his own soul, and by reversion, of his societies: buildings, good and bad, high, and low, and broad. This is according to his incessant form, which is the essential organ of activity. And if he will not work but drone, his waste is universal. The ground supports him in vain, and his feet kill its purpose; herbs feed and beasts clothe him for disgrace; his frame puts costliest energy in play, to be manufactured into sloth; and his soul hovers uninhabiting over his slime. This is the final shape of unhappiness, the lot of apoplexed men and societies, whose curse it becomes that they are lashed to the halberds of use upside down, which cleaves with poison to their human forms. For the human form is the divinity either of Nemesis, or of God.

The third human form surrounding and entering into the former two, is *wisdom* or divine light, whose care it is to make us wise. And because we are shapes in which wisdom may dwell, it presses us with *ought* and *must*, and throngs us with duties through our eyes; for by the magnetism of our likeness it is at our side, to show us the way, the truth, and the life. We can therefore no more escape existence than elude the necessity to be wise, or foolish. In the latter case we founder upon the human form, or stumble blindfold through a world of posts and pitfalls: for our world is even as ourselves, and if we are wounded within, all things chase and gore us, as the herd, a bleeding deer. But the form of wisdom is, first, the understanding mind; hence its earliest pressure incites us to know; and heaven known, man known, God known, the world known, is its mission: the second form is the spiritual mind, which elects the noblest causes from the rest, and has ears from that scripture: "If ye will do the works, ye shall know of the doctrine." Thus as our bodies of sense oblige us to see, work and exist, to eat, drink and generate, or to pay the penalties of refusal, so our bodies of understanding and wisdom bind us to answerable courses, or mulct us in equal retributions. In this light we are organs of new products again, and necessity forces us to increase the wisdom, or the folly, of the world.

The fourth human form is *love*, the substance of substance, the use of use, and the wisdom of wisdom, which takes the names of the rest, and completes the circle of all in all. There is no such incumbence as this, of loving, for love is not only God's image but his likeness, and comes from the highest of the Highest. Its organ is the will, which is motive in all acts and apathies—a man who neither slumbers nor sleeps. This is he that ceaseless love solicits; and it stands around him in all the shapes that touch him, as child and parent, spouse and country; and by irresistible drawing, it welds him to his objects, and assembles each world under its banners. But union begets being, or love is the parent of substance. Hence love is increased by existence, as flame by fuel. And being born into the form of love, which is the human, we are bound not only to our adopted will, but to its increases. In this form then we stand under an almighty pressure towards union and generation; and are compelled to be at one with our companions; and by ever similar progeny from the loins of our love to augment the society we have chosen.

This organic consideration presents old facts in a new discourse, and we come to a functional root of the happiness and unhappiness of man, and of his manifold healths and diseases. For life and wisdom, love and action, are in his blood and spirit, and he must either produce and secrete them aright, or fall into maladies innumerable. The form which he bears, being the essential organ of happiness, wails over its own loss in tones which are essential pain; for joy and sorrow, strange in the world, are native here, and Omnipotence itself seconds every monition and delight of the human form. For the rest we have to observe, that to this physiological shape all science, art, morals and spirituals revert; for substance is the quest of science; use, including beauty, is the compass of art; wisdom is for moral philosophy; and love and its universe for the spiritual man: whence the human form is the future grammar of every school which gives real instructions to mankind.

Here too the problem of life receives its present last illustrations. For though already we have shown that the inner man is "the vital principle" of the outer, yet a new vitality is demanded for him again; tortoise and elephant (p. 321), by themselves, though they seem to support, only increase the weight to be supported; in short, the inner man drops into metaphysical dust as the outer man into physical, unless his parts are coherent through some self-sustaining life. That latter is evidently the living God, the end of problems and the font of certainties, who makes us not of cells and molecules, but of humanities because He is human. Therefore the last office of science is, to cease to make, and to accept a Maker.

We have frequently adverted before to the principle of Correspondence (pp. 225, 226, 242–244, &c.), and we again remark, that the above functions take place by its necessity. As spirit and soul are present with body because creation is a band of like superiors with like inferiors, so God is present to man when the same reason holds. The spiritual world also plays upon us by this law of likeness, sameness or assembling: all the angels rest on our good parts and deeds, and all the devils urge down into our mischievous fibres; so that correspondence is the union of things, and harmony has existence in its arms. Think then of what jets the passions and virtues are the springs; and what it is to be born into the human shape, with its celestial and infernal hydraulics! But this too is under direction; for unstop our souls as we may, their wells are supervised; for there is one "openeth and no man shutteth, and shutteth and no man openeth."

These also are things of which we spoke under other names when treating of the anatomical organs. And we find by the spiritual telescope, that the same fallacy has clouded our eyes here as that which resolved the stars of heaven into a doctrine of glittering dust (p. 290). For as those molecules of the universe are universes, so the molecules of man are men. That which seems little by distance is not little, but mighty; and that which seems little by nearness is, perchance, greater still. If the kingdom of God is within us, then have we roominess indeed. The little chariots of blood, seed and spirit are magnified into eternal processions; and judgments, thrones and principalities, horses, white and pale and red, move in awful cycles through our human forms. If we are of the dust, well! There is an end of us. But if we are images of the Most High, there is no end of us: but our breeding, educating, and healing are according to this pattern and rule—no less, and no other.

But lastly, to shut out unreality, and bolt its door, if we are in the image of God, Who—not what—is that God whose image we are to wear? The answer *in fact* is Christ. Of other gods we speculate, but none other do we know. It is his love, wisdom, works and body that clothe us with our second forms, and enable us to fulfil the end of our being in making our lives and acts co-human with our shapes. Christianity, it is obvious, is the only human religion: other religions are mineral, vegetable, animal, or anthropomorphic; but this is theomorphic; the revelation of no Man-God, but of the God-Man. Henceforth, therefore, there are no problems of God, but a history of Christ. Christianity accordingly works to the redemption of the physical man and human body, that it may be like to his glorious body. And among its other miracles is the connection of grace with nature, or of divinity with physiology; so that by the harmonies of this connection, Christ is to be Lord of the sciences, as well as High Priest of the churches. For it is very obvious that the laws of love, wisdom, and use, which are the health or functions of the body, are the obedience of its atoms to His new commandment, and that in this way the formulas of Christianity are all prevalent in physics. From the moment that this is seen, the modern polytheism is ended, for the Creator and Redeemer are the same, and the God of nature is also the God of revelation.* For be it remembered that the divine empire and pressure of Christ upon his subjects takes place by revelation: it is not left to straggle down through history, through journalists and scribes, but the book which is God's with us, is the continuation of the Man who is God with us. Whence again our human form is according to our correspondence with the Word of God.

All this belongs strictly to the sciences, and especially to those that relate to the body as the casket of the life of man. For the great Incarnation is the model of the rest. And moreover it is not this or that man that concerns physiology, but the problems of corporeal humanity in its scope. But this matter of religion, in which Christ's incarnation is part of experience, necessarily takes rank in

* See these points considered in my "Address on a late work on the Philosophy of Religion." 1849.

its way as the leading instance of all, and can never be banished from the cases upon which fair induction works. It is not the average man, but the archetypal and typical man, who is the subject of anthropological science, according to that plain rule, that cardinal and representative instances, and not extrinsic and adventitious ones, are those in which knowledge seeks for principles. Indeed there is no doctrine of man, who is a variable and fluid term, unless it be in the doctrine of Christ. The average man is unattainable; every fresh birth alters his number and figure: it is only the God-Man who can be known (p. 241). Hence the necessity of turning the telescopes of physiology and anthropology in this direction, that around the variable we may see the constant, or that science may become divine by admitting him who is the light of the world. This will give both dignity and facility to our studies; for the revealed divine perfections are the tacit axioms of all truth; and consequently in the working of these remote problems of physiology, we find here the principles that lead us on by imperceptible stages, each easy as the first, to ends where the first in all its brightness reappears. In nature's dark mines, this is a circle from light to light, in which light is the way.

We may here obviate an objection to these general views; for it may be urged that we have forgotten throughout that the original human form is defaced, and cannot now be called the image of God. To this we reply that we so designate it, in reference to what it may be, not to what it is; and claim that thought shall live upon our future health, and not upon our present diseases. For man still retains his prerogative so far, that he is capable of reconversion into the primeval image, as all religion knows. The human form has never been abolished: it has a divine strength of conservation (p. 288); so that even the poor Bushmen and Australians, nay, all criminals, are still men, and we dare not cease to say that they are in the image of God; for that same emancipative form which is the immensity of righteousness, is also the dungeon of sin. By this conservation it is that the children of those who are bending to death under all guilts and plagues, are born infants as the rest; they too begin from heaven, by whose elasticity the liberated seed recovers itself with a spring from the pressure and abuse of generations. An earnest that evil cannot triumph, or build through the world, because there is this almighty arm nursing babes in the wrestle. For the like cause every day is born afresh, and the load of life is lightened in sleep, which lets out an ever new man through the merciful gates of the morning. Thus though the divine image be defaced, yet as its features may be won again, we still recognize in man the remoteness of his archetype; just as we call reason, reason, despite of unreason; and leave to love its ideal meaning, though that substance, as such, is seldom patent upon earth.

We conclude here by reverting one stage back, to speak of the meaning and cause of death; because this background will perchance throw forth into relief the image and emblems of life. For as life comes by correspondence, so death comes by non-correspondence between the soul and the body; just as deadness in these sciences comes from non-perception of correspondence. Thus when the heart has no longer any heartiness towards the man, and no more likeness to the soul's heart, which is love (pp. 194—213, 247—251), when the blood grows callous to the formation of relations, and feels no warmth and joy in begetting, educating, marrying or advancing the children of the body, then the soul ceases to care for it, and begins by slow degrees to leave it; in other words, disease and disability prey upon it; organic vice besets it; and the inner man ceases to keep company with it; for it is a hard heart, already dead to suggestion and obedience. So when the lungs begin no more to play for thought, or to draw animation and faculty forth, the soul and they have to separate, as persons estranged from each other; for breath exists that soul may breathe, and when this is no longer possible, mortal breath stops. And when the brain no longer elects light and virtue from matter, and communes no more by vibratory states with the genii of the cosmos, then the soul quits this effete spiritualist, now a materialist; and the brain dies from the separation, just as it dies spiritually from the denial of the soul. So also, when the muscles have lost their imaginations of power, the will of the soul quits them as broken tools. Then the body falls to pieces by its own degrees of decay. After this it belongs to no soul, but to the world, whose first artisans with it are the chemistries of nature, and the anatomical worms.

But the natural decay and death of the body image those of the mind; for this it is that, after its culmination, is dying by degrees. The loves which come by nature are gradually extinct as age proceeds: the intelligence also ceases, and the sense. Thus we care much for objects at first, but gradually less, and nothing at last: that care or love is then dead. The child's dolls, and the man's, follow this curricle. So with intelligence: its objects are brightest at first perceiving; in their adult state they are active ideas; after this they become repetitions, and sit in easy chairs of twaddle; and in the end the mind dies to them, and they to it. And so with the senses also; the eye has at first much pleasure in seeing, and at last none: the whole of the five are ultimately tired to death. This is the course of nature in the natural or perishable mind for the objects of nature; and the body follows it *paribus passibus.* For the natural is a feeder to the spiritual mind—the stomach which the soul has for nature; and after a sufficiency of her ways and informations has been assimilated, and the spirit is of age, its old guardian mind and memory are deciduous, and the body drops off with them. The motives of separation between the soul and spiritual mind, and the body and natural mind, are therefore motives of common sense: and are parallel with those for which we cement, or dissever, our connections with our fellow men. For we cannot too often repeat, that the soul is a vast society, and the body a society, and that all our relations and utilities are in their focal union there.

The materialists are therefore right, when they speak of the death of the mind; for die it must, or entrance to spiritual life would be impossible. The case is like that of the proper fœtal organs, the ductus arteriosus, ductus venosus,* and the like, which unless they died to

* With regard to apparitions, or physiological re-appearances of the soul in nature (which by a law easily expounded are numerous in proportion as the state of man is low, and exercise a compensatory function for the defect of angelic communion, and faith in revelation), they are according to the analogies of the change from fœtal to personal life. For as birth may be incomplete so far as that the bodily changes which it implies have not properly occurred, so death may be imperfect, and the proper spiritual changes may not have ensued; and as "the blue disease" is the oscillation after birth between embryonic and infant life, so haunting is the oscillation between the mundane and supermundane states.

their embryonic functions, the life after birth could not proceed. And so, unless our old fixed ways of wanting, deeming and cogitating, the channels of nature's thoughts and loves, were dried up, and rendered inert, the life after death would be impossible: and in that case, having given up one life by inevitable stages, and yet not attained to the next, we should stick in death, and be lost to progress; which last, however, is provided by the successive dying of our aged minds. We therefore claim materialism as not only the cause of death, but by poetical justice, as the undertaker and burier of the dead; and we leave it with sadness to its "black work," of caring for that for which no soul can care. And as we leave it shoveling away—ashes to ashes and dust to dust—we meet our brother, the Man, again, with all his virilities and appointments intact; an eye which is mind, and a mind which is spirit, the risen person and new born infant of the second life.

Having now dwelt upon the functions of the human form, and seen that they repose upon the functions of the body, and that the one is never without the other, we conclude the present Chapter with a few words on some of *the functions of the doctrine of the human form*, which doctrine is to knowledge what the human form is to existence.

With regard to the sciences of the body, we have found that this doctrine opens up the sluices of all life, and determines its tide into physiology. We can no longer think of our vitals, or molecules,*

There is the same God on both sides of the grave, yet the state is according to the materials.

* In stating this, we know that we disagree with the prevailing views, which powder the human body into cells, and regard them as the elements of mankind. But, as we before stated (pp. 174—177), only that is essential to any subject, which is peculiar to it, and contradistinguishes it from others, whereas nucleated cells and cell-germs are common to all organization as the chaos out of which it arises. To think of them in connection with anything human, is to think of man in and from chaos. But it has required a divine hand to win the human form out of matter, and science has no prospect of gaining it, unless by accepting it as it is, without attempting to construct it out of cells. For our own part we are conscious that we can make nothing of cells, and we bless God every day that his works come to us ready made, and can neither be taken to pieces, nor put together—save by his own hand. Our cells are not the pieces, but the fragments

as objects which have ever been seen by microscope or anatomy; for the more we attribute to their life and integrity, the less are their dead bodies themselves. The effect of this upon the eyes with which the body and its prospects are seen, is hope and expansion. Blood, nerve and bone are the walls of the present sciences, and

and ruins of his works. The pieces of humanity are men, women, and children, and by the deepest analysis there are no lesser pieces than these. It is the synthesis of these, and not of cells, that constitutes fresh men and women. Every particle of the body involves the whole, or it is dead, and not within our subject: thus as a living thing it is the entire human form.

While upon this subject, we will say a few words on the *vexata quæstio* of cell-germs, the first of which is supposed to contain the whole forces of the organism in potency. For the primordial cell appears as the beginning of the body. But it is thought absurd that within this delicate vesicle forces should reside that build the frame, and continue it while life endures. In order to see how the case is, let us apply the human lens to it, and look at it from man, who is the atom or cell of the fabric of Society. Under this glass we see that the cell is a man big with a new idea. Let us call that primal cell by the name of Mahomet. This man is the cell-germ of a new religion and polity in the world. What is the case with him? He receives his mission complete as a germ in his own person; whatever Mahometanism in its purity will be, is involved in the idea of that molecular person—Mahomet. He does not know it, for who knows the contents and consequences of himself? But there it is—a logic of great events folded into that deceptively minute form. By degrees he infects new matter, that is to say, disciples, with his idea, whose life they receive, and to which they add their own life; expanding the original idea, as it was capable of being expanded from the beginning. And one such idea might range upon it the populations of all continents, and be less exhausted than when its assimilations began; for in conquering ideas, it is not extension, but want and failure of extension, that is exhaustive. It is therefore clear, that as a spiritual architectonic thing, all Mahometanism lay in the cell-germ—Mahomet; but as a physical force each new molecule or disciple contributed to it his body and powers. And had he not, from the beginning, been such a cell as Mahomet, only wanting conception or the active idea, no force could have made him an efficient part in the body of Mahometanism.

Just so, there is a spiritual architectonic force lying in the cell-germs of organization, and a material, obeisant, correspondential, bricklaying force in the materials which lie ready to be infected by the cell-germs.

A word also with regard to the doctrine of controvertibility of forces, now popular among the learned, and according to which heat has only to pass through a cell-germ to be converted into vitality. This doctrine ends in fire-worshiping, for it makes the light and heat of the material sun to be the fountains of the force of organization: and deems that these pass through vegetables, and become vegetable life; through animals, and become animal life; through brains, and become mind, and so forth. Therefore a fine day, poured into its vessel, man, their poor languid life is the rescource of the corresponding arts; but Man, the hierarchy, is the presence before whom these limits disappear, and the reservoir out of which wholeness and vitality well from perennial springs. For as the parts rise to the stature of the whole, our knowledge of them is the physiological double of our knowledge of man; and the arts of their cure and conservation, are the religions and virtues of the human race, represented under their healing forms.

All this means, that under the pressure of these views, the body begins to have a spirit. But also by the same doctrine, the spirit begins to have a body. For as the present world, like a battle field in which the armies both of truth and error have been slain, is strewed with the dead corses of the sciences, so there is a melancholy array of spectres called philosophies, fighting over again in the air the same old cause which was hopeless on the earth. But the sciences

becomes transmogrified into virtues; dark nights are converted into felonies; dull November days into suicides; and hot suns into love. This is materialism with spiritualism in its pocket. The creed is an old one, and not untrue of animality; for the beasts love in summer because the sun bids them, and their hearts are cold in winter because winter is cold. But it is false of man, and false of his very atoms, excepting in so far as he degrades himself into an animal. As it has been said of the true man,

> "The sun to others writeth laws, and is their virtue;
> Virtue is his sun."

For though he requires heat to love, yet he requires to love at all times, and that is the root of the art by which he procures a constant temperature that makes his body capable. The prior fact is his human determination to beat winter and summer; and though he must have climates, to pick and choose them at his will.

There is no convertibility of forces between life and nature; there are no cells by which heat can be filtered into vitality. The doctrine of Correspondence must be substituted for that of convertibility. It is because love and life themselves are live fire (p. 89), that they come forth by the law of invitation, when the organism which they affect by the same principle of correspondence (pp. 225, 226, 242—244) is immersed in the dead heat of the sun. All circumstance is necessarily haunted, because there are spirits akin to the circumstance, which cohabit with it, as the soul cohabits with the body. But the soul is not porous to the body, though the body is porous to the soul; and by no art can camels pass through needles' eyes; death get into life; gross heat penetrate living heat; or dead doctrines of convertibility procure admission to the distinct and fastidious truths which conserve the empire of the world.

once upon their legs again, the philosophies get into them as their proper souls. The first corse to be re-animated is physiology; psychology is the first of the ghosts which will be incarnate in a human habitation. The doctrine of the structural human soul is the basis of philosophy, as that of the organic body is the foundation of physiology. And this doctrine of the psychical man is threefold; first, what he does: second, what he is: and third, where he comes from. With regard to the first point, every house and every history is its answer: practical psychology is therefore the knowledge of the entire works of the soul. With regard to the second, it involves the human form in its brightest robes of consciousness, with all its actions poured back into it as powers, and seen not in effects, but in causes. And for the third point, as touching the origin of the soul, it is as the origin of the impression from the seal, or of the child from the parent; for the archetype by its terms is the account of the type. Psychology is therefore the science of assumptions with regard to the real soul; for it is bound to make nothing, but to assume or take up what is given in experience; and fact shows what our souls are by what they do and feel, and completes the substance of the knowledge by embracing the facts of revelation. In short psychology accepts the inner man, with all the motives, senses, and form of the outer; and its problems refer either to his associates, or if they are deeper than that world, to inner men again in a new manly orbit; and so on through benches and choirs of manliness until the answer comes from the human mouth somewhere. The knowledge of the world is therefore the analogue, as it is the beginning, of all the knowledge of the soul. And men, women and children are our only faculties, the actors in all, and the accountants of all, up to the very throne of God. Psychology is thus not abstract, but dramatic and allegorical.

Mental philosophy under the same aspect becomes equally tangible, for mind impersonated (and there is no other) is its object. Mental philosophy is always some man endeavoring to see his image either in himself, or in his fellows. The disciples of the schools look for it in the masters, and the masters either in their own consciousness, or in the works of their fellow men. That is to say, it is men of mind who are the study of the philosopher. Abstractions have indeed an empire here, because words, which are airy men, like other men beget children, which are words of words: but the first words of the mind fit to things, which are either human, or the parts of humanity. Hence in its centres every idea is uttered according to the human form.

We have shown in the foregoing pages that the faculties, powers and causes in man are referable to the types of the organization, and that by correspondence these inner men inhabit the outer body; thus, for example, that consciousness is the mind of skin because consciousness itself is the skin of the mind (p. 283). At present, we do not enter further upon this general subject, which for its completion would require us to trace the whole outline of the mental body, as it runs parallel with the organs of the physical frame. But yet we will select one or two metaphysical ideas, in order to show their fitness to the human form.

And first we select that of *power*, which has caused terrible staring to eyes philosophic, for they have been trying to identify power with nothing at all, and to see it notwithstanding. From Locke to Brown, through the exhausted receiver of David Hume, the essence of power has been spilt, for want of the vessel of the form. And what is the form of power but the human arm with the will in its muscles? The metaphysicians have asked whence they got the idea of power, but the prior question is, Whether they have got it, or not? Certainly if ever they had it, they have lost it, and their chains of causes and effects are ropes of sand or sequences of weakness. On the other hand, it is plain to common sense, that power is shown by deeds, and that the idea of power comes from the showing; also that the arm of man is the central executive of experience. All other carriage of causes through effects is but the comparative anatomy of that prime organ. But the central form is the main symbol, or the essential body of the idea. And by the laws of correspondence, the idea has that body with it, either consciously, or unconsciously, whenever it is used with force. And moreover from the richness of that body it draws out imagery which is the sharpness of the occasion.

The controversy about power could not fail to lead to the conclusion, that there was no such thing; which was quite true of the systems of the disputants, but not of the industrious world. The consequence was, that power was beyond experience, and disappeared out of philosophy; though to avoid the shame of avowal, its name was kept for a cloak. Power, however, becomes again an object of study when it enters its own proper form. Let us run along the arm, from shoulder to fingers, to notice how the emblem offers a true habitation to the idea. First, man is a living *force*, a lake of energy, from which the arm pushes forth as a river of waters of works; for force is the trunk and body which urges at the back of power. Next, the shoulder is proper *power*, the place where will meets force, and where power is born. Again, the arm is *strength*, where the biceps and triceps which are the blacksmiths of action live; the men of iron know where their iron lies. The forearm is *toughness*, or the home of sinewy power. The wrist is *flexibility*, where sleight of hand begins. The palm of the hand is *possession*, or the power of having; it itches for the finger tips, which give it the power of holding. The grave thumb is *firmness* or steadiness, and it stands short and sullen among the fingers, which are of the rapid temperament of *skill*. In the closed fist again the emblem of force comes forth, and completes the circle of this series of powers. This, for the one-sided view: there are however two arms; the left for all passive powers; the right for the active; and the two are married and sometimes clasped, when the most important works are to issue. It is fortunate that the thought of power inhabits such an engine as its proper *word*, for as we said before, a wealth of suggestion arises out of the natural language of things, such as could not proceed from artificial symbols; for the etymology of this language is the whole nature and anatomy of the arm as the exponent of the idea which dwells in it: hence it introduces us into an infinite field in the study of power. But if we study it in words or conceptions first, we get no more than the word or conception involves, and we leave the nature of the case out. It must surely be evident that a child working as well as he can with divine forms, in which his mind too is native, is better circumstanced for truth, than a sage fumbling ever so cleverly with notions of his own, into which no creation, but his own laborious conceit has induced him. Experience itself declares, that a thousand ideas arise when nature is the source, for one that can come from the closet-cudgeled brain. Hence the mercy that has still incarnated our poor metaphysical minds in dresses of such everlasting suggestions.

We will try to illustrate one other philosophical idea—to embody another of the ghosts in which philosophers believe—and it shall be the idea of *progress*, of which the limbs of man are the essential emblem. Here first the feet are proper *progress*, to which the ankles are *speed*, the emblem of which is the *talare* or winged sandal; for the feet are the measure of advance, and they twinkle with swiftness as they run, by virtue of the nimble ankles. The legs are *straightness*, the shaft of direction, keeping the line of the object in the midst of the rush of joints and muscles. The thighs are *effort*, or the brawniness of progress; and the hips are *motion* itself, globe-jointed; while the trunk which they bear is weight, pressure, or *necessity*, the incumbence of all man's vitals and wants to generate his progress; as the same trunk determined to the arms is the by-play of all to produce his skill; and carried to the head, is the coronation of all in his will and intelligence. If now it be asked, What is progress? we say it is all advancement, *upon this model:* walking and running are in its definitions in every sphere. No writer has ever talked of it for many sentences together, without falling into the symbolism of the legs. Such an one "makes a stride in advance" of his age; such another is a rearward spirit, and slow in the movements of his mind: this one leaps to a conclusion, and that other goes rather backwards than forwards. If these terms are metaphors, it is because the soul inhabits a metaphor in dwelling in the human body (p. 242—244).

It may be evident from these sketches, that our organization (containing us, as it does) has the power to take up, and arrange on its plan, the various abstract terms which hover over the philosophical world. For intuitions may be put into eyes, and reason into brains; and philosophers may become seers, when they get these organs of which they have mutilated themselves so long. In common humanity let us pray for this consummation. For the cripples of thought are painful objects to us. It is sad to see the philosophers of progress leaving their own good legs behind, and hobbling along through the learned world on wooden pegs of abstractions! To see the reasoners on cause and effect ordering automaton arms of their surgeons, and stowing their own arms away in the closet underneath their libraries! To watch the Kants putting off their brains, volume by volume, that their "reasons" may be "pure," and filling their skulls with pure bran instead! These curtailments are too lamentable for laughter, for we are all maimed in our brethren of the schools. We can only hope that they will henceforth condescend to the human form, and not make damnable imitations of it, in words of "leather and prunella."

The immediate mission of the doctrine of the human form in philospphy, lies in the constitution of analogy as the method of reason in this department of experience. For when the mind is consciously in its moulds, it becomes allied to all existence; like God and like nature; and philosophy is assimilated to theology on the one hand, and to science on the other; nay, and as the middle term, to the life of man in the world. Hitherto analogical reasoning has had no proper place in thought, that is to say, knowledge has been deficient in the principle of association; each branch was afraid of foreigners, not knowing that their races have the common stem of humanity; but this once seen, the principle of analogy or friendship begets intercourse, and the mental kingdoms begin to communicate. In this case the analogy falls into no stereotyped methods, but corrected and fortified by an abundant appeal to facts, it touches that infinity which relatively to us is one of the first lessons which existence has to give.

The first object, therefore, whether in questions of mind, or body, is to find the fact, that is to say, the prime form, or shape, of the thing in hand; the second requisite is, to see its illustrations, or in other words, to view it in connection with its universe; and the third thing, or the end, is, to see its universe over again within it, but according to the facts of its own being. The method here is analogical, running from the form of the thing, through the forms of its circumstances, and back from these again to the starting point, which is thenceforth no speck, but a full-sized world. We have already illustrated this in speaking of the idea of power: in discussing which we first find the form and the body of power, which is the arm; next we move down the rails of this through comparative powers in nature, and engines and machineries in art, and complete the circumstantial empire of power; and thirdly, we see in the will itself, with its own distinct facts, the arm of arm and the mechanism of mechanism; and one limb of the inner man is rescued from nothingness or abstraction. In a similar way the analogy of the human form makes the circuit of the world in the interest of every other faculty, and brings back for each a portion of the reality which belongs to man in experience. Nor does it matter whether it be a psychological problem, or a problem of causes, or of ethics; for the mould of man embodies the latter just as the former, and accounts of the nature of things, and theories of duty, take shape and order with the same facility under the banners of the human form. Social theories also are incarnate there; for it has been known from of old of the human body, that its parts are the exemplars of a perfect commonwealth, intelligence and morality; in other words, that it is Christianity practical to very matter, built into its own shape of a man.

But if the principle of analogy be thus coextensive with just thought, it cannot fail at the top to suggest the more living principle of universal symbolism; for if the mind be like the body, and actuates the body, the latter becomes expressive through means of the likeness and the action; and in that case the series of existences in the world becomes similarly expressive, that is to say, symbolical of the mind. The face of man thus travels throughout the universe, and love and intelligence look out from things with an infinite variety according to their capacities. Nature and form are therefore full of meaning to the true philosopher; that is to say, they always signify something in the mind and soul of man, and the last account which can be given of them is, to trace out what that something is. For as the arm means power, and the eye sight, (though we could not guess as much unless our own potency and vision were therein,) and as power and sight seem different from these tools, so the horse and the ass, gold and silver, the rose and the lily, have also meanings just as apparently remote from their forms; and these meanings are inner functions from which they come and to which they return. Without such meanings these things have no communion with the soul of things, just as without power and sight the eye and arm do not partake of the animation of the body. Philosophy therefore has hard work, and a glorious mission before it, in hearing what all forms say; so much indeed to do, that we foresee a good time coming, when it will throw down its luggage of classified faculties, empty its pockets of its bad money of "objective and subjective cognitions," and anxious for labor, like blind Bartimæus, cry out exceedingly, "Lord, that I might receive my sight." Then will it find that there are not only more things in heaven and earth than it dreamt of; but that there are also more heavens and earths. For if there is an inner man, there is also his experimental circumstance, his comparative anatomy, botany, mineralogy, cosmology; in short, an inner world like this world in its form, though built of spirit and life: and to learn about this will be the last privilege of philosophy, when it becomes teachable, as a little child.

Having indicated that the symbolism of things is the soul of the comparative scale of nature, and that the knowledge of correspondence is the *punctum saliens* of the sciences, and if we may be pardoned the thought, a stream from that divine expression, the creative Word, we pass on to one or two other points connected with our subject. And first let us glance at the application of the human form to the doctrine of history.

In this field the idea has been steadily but tacitly growing up, that there is an organic connection of races and affiliation of ages, so that what seems to be a world of men, is also but One Man. The doctrine of association, admitted of the mind on the one hand, and of social interchange on the other, helps us to conceive that we are all members of one body. The method of thought which runs from the great to the small, and magnifies our parts into humanities, in its other curve from the small to the great, brings mankind together in a single human body. What may be the ultimate triumphs of this idea—what sciences may come out of its mighty loins, we dare not venture to predict; but one thing is clear, that it will purge history of much occasional matter, and aim constantly at that which is the sublime part of all annals, the history of Providence in the world. For though we may think that we shape our own courses as isolated parts, yet if an architectonic whole be admitted, the scheme is necessarily God's. Do we not foresee, that history from this point of view will unwrite itself, and proceeding by the wisdom of erasures, find that its grand task is done for it, and that Old and New Testaments are its truth. For when all is said, the complication returns at last to simple pieces of good and evil. Paradise lost, missed, and regained, are the main concerns. This is the upshot of the birth, education, career and end of the "Grand Man" upon earth; and writing has to show, that this is already written, for this among other reasons, that there is no man could write it. And so from the garrulity of old experience, and after the fag of unrollings and discoveries, we find that we are right exactly where God says to us: "I told ye so from the beginning."

But if the doctrine of the human form brings unity to our conceptions of mankind, and sees the terrene man as a globe of societies and churches, its effects upon moral philosophy are not less important, for it embraces kindreds and tongues in one fraternal whole. The human is the form of function, and function is diversity of act founded upon aptitude. If the race be indeed a man, what vast differences are needed in parts and individuals, to make up his body of such various wants! The members must consent to differ, as the head differs from the feet, or the liver from the fingers. They must also consent to agree and succor, as closely and quickly as the bodily commonwealth itself. They must further know that they are bound together in a common lot of health or disease, and that there is no wholeness until the entire system is well. In this light the doctrine of the human form is a standing policy of regeneration to man, and sends out the sound to bring in the sick, that their evil spread not to the frame. Here, in short, we have the doctrine of fraternity, sympathy, help, or the foundation of ethics.

I also deem that the mystery of Christ comes more home to the scientific mind through consideration of the same doctrine, though far be it from me to attempt to base Christianity upon inductions, when it stands on a divine rock of truth. But the mind, which does not know its claims, seems a little advanced towards them, by finding that the human form is no accident among things, but is the ruling soul of the world, and that to enter it by divine means amounts on these principles to a reconstruction of the universes of spirit and matter by virtue of such Incarnation. For if the end be changed, and a divine humanity substituted in nature for our perverted manhood, then the means leading to the end are altered also, and the kingdoms of nature and science advance thenceforth to no man but Christ, who takes all power, or is the First and the Last, who was, and is, and is to come. It were easy to say more on this theme, but the hint must now be sufficient, that a science resembling Christianity is possible, based upon the divinity, humanity, and omniprevalence of the human form.

The doctrine also of immortality is not untouched by these considerations. The belief in a personal immortality, is a belief in the never-dying perfections of the human form. The hope of a social immortality, or a state in which the spirit is no lone existence, rests upon the foundation of a permanence of true brotherhood in a grander human form. Ultimately it presupposes, that all races under the natural skies, and that majority who have passed away, are one stupendous humanity whose life is God. But we may not dwell upon the overpowering vision. Let us be content with declaring, that the more of God there dwells in society, and the more of society in the individual, and the more of the individual in the body, and the more of the body in its members, the more living is our life, and the truer the science that represents it. For of one thing we may be very sure, that our proudest knowledge is either nothing, or a rill from the cataracts of heaven.

We have now, kind reader, endeavored to travel together over some portions of that garden of the Divine Wisdom and Love, the human body, where the tender paternity of God is so very manifest to all his children, and I trust we have found as we proceeded, that what He does is actual as well as perfect; that His works like Himself are positive substance, and meant for our positive knowledge. In the course of our journey we have conversed repeatedly on our common sense, regarding it as the foundation of our discourses. And once more, what is this common sense? It is the active light which follows whatever is largely experienced and rightly done; the mundane version of that scripture, "If ye will do the works, ye shall know of the doctrine." It is not common opinion or thought, both of which may be wrong, or vague; nay, it is not opinion or thought at all; but it is the whole living present hour, standing on the pier of the whole past, and about to embark with cheerful courage upon the unknown welcome future. This is the fortune of life and action, and the price of the sciences; nay, the reaper of which science is the gleaner. This is the bond of times and the principle of toleration. This makes learning humble, and simple natures great.

ADDITIONAL NOTES.

p. 300.

The reverse of the biblical revelation of God as a Divine Man, is the philosopher's notion of God from physical immensity, as an indistinct being whose sensorium is space. "*Jupiter est quodcunque vides*," &c., &c. The modern astronomical sublime, in which the greatness of the Creator is deduced from the mileage of the universe, and God is increased by naughts according to quadrillions, quintillions, and sextillions, is like the child's notion of a great man or a great book, which amounts of course to a giant, or an imperial folio. It is not, however, by piling Ossa upon Pelion, that divine size is gained: the way to it is by common sense following Scripture, and recognizing the intenseness of those human qualities which are the greatness of man, and to which matter is a servant, for they say to it, "Go, and it goeth; Come, and it cometh." So likewise omnipresence is purely divine, and has not to do with space, but with wisdom; in which latter space itself is but a nook. The biblical or human view is then according to the uprightness of reason; whereas the notion of the Godhead from space is an infinite sprawl, reason down wallowing on its knees and nose.

p. 301.

We here take occasion to warn the reader against the fallacy of the notion that man is an infinite who has been degraded to the human form, when the fact is that he is dust, which is ever being raised by God through and in the human form. For as we show in the text, the human form is mind, life, and infinity.

p. 314.

The humanizing function of the arts is indeed a phrase which admits and succinctly conveys the whole of this argument.

p. 319.

We may regard this view of the human frame as the doctrine of the freedom of man, for it opens avenues into all expanses, and sinks its shaft into liberty, which is the principle of sufficient space. The dead microscopic view, on the other hand, is the doctrine of cells or prisons, not indeed unfit for a limit below, but which ought to be kept below, lest the mind, tied to what it contemplates, should become a scientific bottle imp in the cases of its own museum.

p. 338.

By virtue of the human form all the ages of thought reappear upon the scene together; the mythologic and the scientific; both of them contained in that which is the first and the last, or the theologic epoch: but by inhumanity and atheism man is a perpetual bastard, always vilifying father and mother; like an enemy burning the bridge of time behind him as he runs; and those who come after him being of the like mind, of course his days are short in the land which his own conceit gives him.

p. 339.

The human is the form of originality, or of special gifts and activities to each member of the body; for its perfection lies in the unlikeness, yet accordance, of the uses which each part ministers to every other. Agreeably to this form, every man has a genius of his own (p. 80), or he would have no function in his commonwealth. The animal form is the reverse of this, for animality limited to self, works for no commonwealth, but aims to put itself in the room of others, and thus to engender universal sameness of functions, each animal striving to have all advantage, and to be everything in its sphere. There are, then, two opposite forms in man, and out of him, viz., the human form divine, and the monkey form infernal. And there are two classifications of nature corresponding with these unlike principles. See our tract published some years since, on *The Grouping of Animals*.

CHAPTER VII.

HEALTH.

AMONG the functions of human life, the presentation of ideals in mind, body and estate, is one which influences all motion and sensation, and gives faculty its direction and play. It seems as if inertia were so tied to motion, pain to pleasure, and imperfection to perfection, that they tend to run out of themselves, and to seek something beyond them to which we give the general name of health. How we come to think that we have a right to be healthy—at least that this is our proper nature, is a problem admitting different solutions; but that our bodies feel disease as a grievance, admits of no doubt whatever. Shall we say, according to the last Chapter, that it is one of the functions of the human form (p. 322), to subject us to ideals, and to put us under this air-pump of wholeness and happiness, that we may gasp to fill its limits? However this may be, the fact remains, that through all disease we look wistfully after soundness; from the depth of incompetency aspire to strength, and long to be *well* through our little day, that duty and pleasure may not be stunted in our hands, but enjoy their legitimate proportions.

We seem to know sentimentally what that general health would be which we desire. To fill our places in the world, and to love to fill them, are the best ends of our aspirations: to be so organized, or so minded, which you will, as to be spontaneously able and cheerful in our labors, at the same time that those labors are not only our choice, but the wants of the time. This includes the rapid direction of every muscle to the private in the public service; the bending of sense straight to the objects in hand; the limitation of sensibilities to the occasion, or the running of life in the pipes of duty; and finally the control of the all-controlling mind under a genius which is called felicity when its works come forth with complete adaptation to the time and space which they are to fill.

Health, in short, by the old definition (and we know of no better), is harmony in its most considerable meaning—harmony of the parts of the body with themselves—harmony of the mind with the body—and harmony of both with the circumstances and ordinances into which we are born : harmony also of the human frame with the climate that it inhabits, and with external nature in its variety. The science of health, then, is ideal physiology and psychology, and the art of healing embraces the means that may conduct us from the present or any state of unhealth, to that picture in the clouds which we cannot give up if we cannot reach it—the means which may gradually make some part of our ideals real.

There is, of course, a dark side observe to health, in the existence of innumerable maladies and diseases, which beset human nature ; otherwise health would not have been heard of, but instead of it existence full of the play of power, and of the power of play: but upon our present experience as a background, pain, which is the writhing and restiveness of the human form away from, and against disease, sketches out with the pencils of hope and desire the lineaments of a bright possibility. In this light we look at health from disease, which is perhaps its only point of view. For, as we said before, if that ground be left, the name of health becomes too negative, and perishes; and in its place other substances arise, such as joy, love, activity, and all those states which are blessings irrespective of bans. In that case we do not think of state but contemplate action, and valetudinarianism ceasing out of mind and body, leaves us free and fearless for our business.

Health is therefore only the beginning of a just consideration of the human being, as it were the birth of man into the realms of humanity; and yet as he is ever beginning afresh, it pursues him with new exigencies along the stages of his journey. To be well with the world of this hour, and equal to the existing situation, is a demand which is always changing, and health must be flexible to meet it; especially so, in a being, who is a child to-day, a youth to-morrow, and an old man in time, and who has no experience of these, his eras, until they come upon him. For health implies a perpetual self-adjustment of a new needle to a new pole. And thus, however high we rise, the problem of health, or some difficulty of being well, may be expected to recur to us. The beasts are better off, and worse. They are acclimated from the first, or if they need change of air or season, they are naturally bi-climatic, and run along the magnetic streams to light and warmth, caring little for earth, but very constant to summer. Then again the cup of their heart's blood is measured to objects, and they are drunk with no desires but those which nature prospers. Their muscles also are full of spirit, and do not tear from contrariety of minds. And even if instinct be wrong on one track, and disorder them, animal magnetism is their physician, and like clairvoyants they run to grass and herb, and nibble their healing leaves out of the pharmacy of things. In a word, their potent life burns up sickliness, and makes medicine of little avail; excepting indeed in those cases where a false domestication denaturalizes them. So far they have a superior lot, and earth is their heaven. But on the other hand, brute life and health are not enviable for us. The beasts are nature's simpletons who are pleased with a little, and that little, of the lowest order. They are well with their world, because it is so single and small. Could they have another shown them, by those other eyes which we possess, they would pant and struggle as we do to the ever new adjustment. Instead of living on the bare surface, they would dig in the mines, and build up their palaces of sanity. Such is undoubtedly the cause and object of human diseases—to carry us deeper and higher than brute health can go; or to make the health of soul, mind and body inseparable and coördinate. For this reason, there is no joyous inhabitation of the earth for man, unless the inner man also be right with his world, and the social with his; or unless wholeness be fulfilled. Our maladies therefore are warnings and signs of a lost integrity, which is to be sought, and found again; and where cure does not come, it is an evidence that the problem has been stated and worked on some partial ground, and that a further view and a higher sacrifice are asked. We may terminate the comparison between man and beasts by saying, that the health of the latter is already complete and natural, while the human can never be completed, but is always integral and progressive. If reason shows unwholeness, the pain of the proof is contemporaneous with the journeying of a star from the east, which stands at last as a heraldic sky-point over a new born health.

In the few remarks which we have to offer in this Chapter upon the subject of health, we shall follow that division which is generally admitted, and consider the individual from two points of view; firstly, as containing within himself the grounds of a certain stock of health or disease, which falls under the ordinary department of medicine: and secondly, as being surrounded by circumstances favorable, or otherwise, to health; which constitutes the field of what is termed public health, or as we might name it, social health. It is however no part of our design to contribute directly to either of the branches of practice which are busied with these two walks; but rather to complete and conclude the physiological and psychological subjects that have already occupied us in the foregoing Chapters. A popular education on these themes, is very near our aim; and general notions only can therefore enter into our pages: yet we feel that if the eye can be opened to the compass of the subject, we shall not fail in the long run to have brought to practice those first helps which depend upon the insights and expectations of the learner. There are abundance of good books, and zealous men, engaged upon details and sciences: be it our endeavor to elicit some little of the order and light in which they are working, and to present it to memory under an organic form. In the first place, therefore, to proceed from the great to the small, we shall consider the subjects and method of social health; and in the next place, we shall speak, not of individual diseases, but of the various systems of medical treatment which are applied to disease, and follow its fashions and moulds. I do not know that our design can be likened to anything more aptly than to a map, in which the fewness of places, the smallness of size, and the poorness of outline, are themselves valuable for giving a first view to that public which does not travel. There will be tracts of *terra incognita* also, for which the excuse shall be, that they are left blank, and not completed for deception.

And first a few words on the difference between the private, and the public or social health.

The science of private health is of individual concern, and lies in making the best of our own circumstances, for the strength, improvement and enjoyment of the organism. It chooses a healthful place to live in; keeps clean the person and the house; superintends diet and clothing, and all that belongs to cheer; and aims also to keep the mind easy. In short, it is the analysis and perfection of bodykeeping and housekeeping. But it stops for the most part with the front door. It gives you the best of everything, but without insuring the goodness of the best. You can have excellent meat and wine on this principle, if the town supplies them; good air, if the neighborhood be favorable; good drainage, if there is a natural outfall, and the sea washes up conveniently to carry away your refuse. This private health is the property of the strong, the vigorous, the wealthy and the fortunate, who have the pick of circumstances, and are the favorites of the hour; but even with them it is casual and impure, not the *maximum* of the public health, but the *minimum* of the public inconvenience and disease. It is like a high hill at whose base the fever-vapors curl and steam, and whose top they threaten to invade, and by subtle fears do invade, and some one or other of its inhabitants drops down ever and anon, shot by invisible arrows from beneath. That superior vigor of mankind which seems to need no tending, and to burn like fine wax without scientific trimming, is the subject of this private health, which is no system or doctrine, but a resource of carnal virtue and goodness above, and in spite of, the elements. Nature has done what she can in producing the robust individuals who belong to this class, but it is committed to ourselves to enlarge the class until it embraces everybody.

The science of public health undertakes this task, and aims to do for everybody what it seems nobody's vocation to do for himself. Private weakness and impotence is its field of operations; the want of virtue in persons is what it has to compensate. It knows of houses only as little dots in streets, and streets only as fine lines in towns. In short it looks from the community at individuals, and is necessarily tyrannous until its work is done, after which freedom of a new kind breathes everywhere. It washes the foulest faces first, strikes at the Stygian neighborhoods, keeps company with publicans and sinners, and always begins where it left off, with the remaining dirtiest man. Soap and towels from the toes upwards; "he who would be clean, needs only to wash his feet." Yet the problem grows up street after street, until we find that it is the whole metropolis that is stated. In good faith, there is no such thing as private health; health is the Saxon for wholeness, and wholeness is the public health. We shall further illustrate this presently.

Public health is either an autocrat, or nothing. Independence is its aversion, for it has to trace and cleanse the dependence of man upon his circumstances and his fellows. When it has driven its ploughshare through a foul neighborhood, sown with salt the foundations of sin, and carried rivers of water under the new streets, it then knocks at the house-doors, from the worst to the best, and rummages privacy with a curiosity most detestable and proper. It insists that the public fountain shall have a squirt in every room; that the pure air outside shall widen the windows and space the rooms; that the underground kitchen shall be plucked up, and set in the sunshine; that the chimneys shall burn their smoke; and the sewers have their decorum thought of in their beginnings in the chamber. Thus it ordains that the anatomy of the house shall spring by a fibril of propriety from that of the street: and this, by a fibre from that of the town; and that every one shall know what his neighbor's income of health is, that pretension may die, and the easy manners of a common stock of sanity and consequence arise. Such is public in contradistinction to private health; the latter is the *vis medicatrix naturæ*, which takes stock of the existing health, and sums it up, for life or death, by a balance of figures: the former is the *vis medicatrix hominis*, which enlarges the stock, and promises to make it all-sufficient. The business of public health is prevention, but that of private health is cure.

We shall now speak a little more in detail of the exacting nature of this new duty, and for this purpose we shall group the various branches of it upon the bodily organs. For it will be evident to the reader of the former Chapters, that our method of considering the world and the society as the procession of the principles of the human frame, offers facilities for the treatment of this question, and throws it into an order parrallel with that of the body itself. (pp. 107, 155—157, 165, 229, 230, 289—300.)

Like babies on the subject, let us begin with the stomach, which is the most interesting thing in all our early considerations. And what are the demands of the stomach and the alimentary tube as regards the public service? Evidently they require that cultivated nature shall be conducted into them, and their own rejections be cleanly drawn off into nature again: that is to say, that the world shall be an alimentary tube leading to them from the one side, and a hidden intestine or *rectum* of drains passing from them on the other. Health immediately, and existence soon, demand that the stomach shall not be isolated, but that the earth shall be cut into ways that correspond to, and converge into, it. Wherever there is a break in the tube of functions that conducts from the earth to the mouth, a corresponding inanition takes place in the latter; and wherever there is a stoppage in that other tube that should run from the belly to the ground, a constipation is created somewhere in the upper parts. Our towns are in a state of permanent inanition and constipation, owing to the fact that markets, and drainage, are not laid down to, and from, every mouth or inhabitant. The violation of the calls of the frame, which is satisfied with nothing but the conquest of the external world, leads to loss of health, public and therefore private. And here we observe a second time, that public and private health are inseparable, and only exist apart in exceptions that prove the rule of their union.

But how much is involved in that single reconstruction of circumstances of which the alimentary tube is the missionary: how much wise industry in the creation of food, and honesty in delivering it from hand to hand; and on the other part, how complete a belief in the interest of the commonwealth in cleanness and order. This lowest department of sanitary duties, begins to demand virtues and faculties which are not yet real; in fact, the universal kindness of a higher polity than exists on earth. But what then? shall we give up effort; or shall we not rather be grateful, that past our powers to attain, the wants of the body are prophetic?

There is no organ, however, which stops with the material question, but all flesh has psychical inhabitants, who make their own fearful demands. The public peace, prosperity and ease constitute an atmosphere of circumstance around the stomach, which allows our food to do us good, or causes the reverse effect. Health is like the funds, and digestion and indigestion have their daily quotations, if we could but read them. Low anxieties and love of money for its own sake, neglect of the Divine truth, "sufficient unto the day is the evil thereof," make whole ages dyspeptic; and neglect of the republic of other men than ourselves, creates stoppage in that which should be a unanimous society. Hence the public health of the stomach embraces even these considerations; and indeed as the lower parts of its duties are fulfilled, these higher ones come out only the more prominently in their claims.

On our lips, the alimentary tube changes to the skin, and the next realm of public health which we mention, is that which concerns the latter organ. The skin is as exacting upon us as the rest of the body. A polity of healthy skins can be maintained by only the most vast demands upon our industry and sciences. Everything about us must be clean-skinned, or half our personal washing is wasted. The skin leads outwards by forceful channels, and will not be stopped even at great distances, without its emanations recoiling upon the health; it allures surrounding influences inwards from equal lengths, and will not be deprived of them, or supplied with them in a malignant form, without withering, or diseasing, the organization.

The effects of climate and circumstances upon the skin, are not less remarkable than obvious: for it sympathizes directly with the places and spaces around it, and takes its complexion from them. The inhabitants of the regions of gusty winds have weather-beaten faces, and lines as of the tempests blown howling into their skins. Mountain races have stony or granitic features, as of rocks abandoned to the barren air. The people of moist and marshy places look watery and lymphatic. Those where extremes of temperature prevail for long periods, are leathern and shriveled, as though their skins had given up the contest with nature, and died upon their faces. And so forth. These events show how much the skin is influenced by the circumstances about it.

It is equally certain that the surfaces of the body, represented by the skin, are the medium of contagion, which is the railway of the public disease. For this organ, which compasses all our parts until they form one, true to itself, offers a sympathetic plane on which the health and disease of the community also tend to universal oneness or diffusion. There is no breach of continuity on the surface of mankind, but the skin of the poor joins to that of the rich, and epidemics run without ceremony from the one to the other; only more sparse as they spot the palaces, because cleanness is more studied there.

After the stomach has taken care of our nourishment, and the lungs have looked to our breath, the skin has to provide for both in a kind of infinitesimal sense. For it supplies us with food, and disburdens us of excretions; though both its aliments and rejections are for the most part invisible: it also washes itself in air, and keeps itself in motion; the former, by itself; the latter, under the superintendence of the lungs. The skin is the theatre of influences; the other organs we have mentioned deal with more palpable stuff. There is a corresponding delicacy in the question of the public and private health of the skin.

Great, however, is the plainness, and equally great the mystery of cleanliness. It is one of those terms that will hardly be chained to a physical sense; we no sooner begin to treat it, than it buds like Aaron's rod, and blossoms into morals. Frequent ablutions wash away the *sordes* of our bodies, open our pores, enable us to emanate with freedom, and with freedom to take in what the atmosphere can yield us. The model and mirror of these effects is presented in our daily washings, which make us *feel clean.** This clean feeling

* In speaking of the saliva we took occasion to say a few words on the emotional nature of the secretions and excretions (pp. 152—154). The field is a wide one, and embraces all the living processes. Feelings are changing in our bodies with every transport of the animated fluids from place to place. The current of blood above is the signal of one set of feelings, and its rush below produces another; for it carries soul with it, and its sentiments are according to the parts from and to which it is sent. The same is the case with the perspirations; they carry away old feelings, the perspirations of the soul, and remove them from the person. So also do the dejections of a grosser kind. Hence the exhilaration consequent upon the latter functions when satisfactorily performed. In washing, especially, an effect is produced which we may term material-psychical, and which every one must have experienced; a sentiment of new vigor, as though the mind itself were washed in the skin. We do not know "how we are" in the mornings, or what is the promised complexion of our day, until our ablutions have taken

is the basis of correct perceptions. It gives self-respect, which marks us out from the things about us (pp. 282—285); makes us judicial among our associates; establishes a ring of healthy sentiment around us, and between us and other things; and enables us to discriminate between clean and unclean in whatever seeks to enter our feelings, or aspires to stay there. In short, it places a cordon of pure life around our bodies, as a troop of angels around the bed and before the path of the faithful. Between the life, thus whitely washen, and its objects, nothing intervenes to hinder immediate judgment and action so far as the surface is concerned. The light of the sky and the vigor of the man, kiss upon his skin, and cement a covenant of justice, in which every predominance is conceded to the lordly organization.

On the other hand, dirt upon the skin is not merely dirt but dirty feeling; and the latter is no sooner set up than it travels soulwards. The skin is given, among other ends, as a vivacious sentinel to prevent the entrance into us of whatever is alien and impure. The purity of the sentinel is of the greatest value to this exercise of his functions. Dirty feeling does not know dirt when it comes, but is bribed by it, and lets it pass the barrier. Hence an unclean skin, besides adulterating the feelings, admits a material adulteration to the organs. Furthermore, by clogging the pores, it prevents the beloved dirt from escaping outwards; until at length the body, crusted over with itself, abrogates the skin functions, and finds another and violent eruption in disease. For nobody can stop long in himself; he must go forth as a messenger of life, or death, to those about him. And when he ceases to transpire health, specific sickness is conceived in the struggle; the system makes new

place. The result is often quite different from what we feared before these processes were undergone.

The generative process, with its sheddings, is the head of the material-psychical acts, and as it secretes the germ of man himself, it affords a light as to the force and function of the other living secretions. What a quantity of imaginations the seed at once carries out of the mind and body; and what an architectural effect it produces on its matrix! Now, likewise, in their places and degrees, the other excretions export emotion from the organism; and when they have come out, and reached a suitable nidus, they show that they are seeds, and after their own fashion germinate. In this way the world is sown broadcast by the natures of all men, animals and plants. See our Chapter on the Skin, pp. 261—268.

terms between itself and nature; a part of the privileges of life is ceded, and the various maladies appear. This is the history of one class of physical evils, engendered by the neglected cleanliness, not of years alone, but of generations.

The private health of the skin subsists in the public health, private cleanliness also in public, as a man in his society. There may be excellent citizens in a debased community, and cleanly persons in a dirty town; but the surrounding influences are against them; and they are good and clean in spite of example, by mere manhood and as it were miracle. The labors of cleanliness, though cheerfully undertaken, are Herculean and incessant. Often too they are unsuccessful, for the laws of nature work in masses, and public neglect is visited not unfrequently upon the just as well as upon the unjust. Hence the necessity of treating these questions from the public side. A clean house in a sooty town; a well ventilated room with an adjacent swamp or churchyard; a chastened appetite with unwholesome provisions—these are the impossibilities which the prudent ones are laboring to establish in the city and in the country. It is plain, however, that things move all at once; that house and town, room and sky, dietetics and food, are the same essence in different quantities; that the large is the continuation of the little, and *vice versâ*—the little the pensioner of the large; in short, that health has two ends—the health of the man, and the health of the people; which must be treated as one by doctors and clergy, because they are tied into one by the Great Physician.

But if a clean skin supposes in postscript a clean habitation, and this a cleanly location; and if the skin functions preach to us to cultivate purity in every field; they also enjoin another lesson, of warmth and clothing. The skin, as we have observed, is the garment of the organs (p. 278.) It has wonderful powers of keeping us warm, and of moderating temperature (p. 270). Nakedly, however, it cannot stand against the extremes of heat and cold, but it beseeches to be fenced with coats of other skins, derived by textile skill from plant, worm, and fleecy animal. These too must continue the skin, have pores like the skin's, and be clean as the skin. The duties of cleanliness recur in the clothes; the necessity of warmth—in a word other clothes, is also broadly stated. The clothes must be clad by the walls of a comfortable house, and the hearth fire artificially elongate the brightness which the life fire commences. Only thus can warmth become solid. For it is here as in space and logic: there must be three terms or dimensions; viz., the man in a whole skin, the skin in a decency or wholeness of garments, and the man in a house of comfort, in order to make the substance of a citizen. And again the skin of the circumjacent earth must be washed and dressed, as the double of our own, in order that the reservoirs of outward cleanness may be filled; and lastly, our minds in their skins (pp. 281—288) must be clean and whole, lest eruptions worse than can come from without, should break forth from within. Such is the logic of duties, easier to say than do, which deduces itself by sanitary necessity from the skin.

The lungs cry aloud for still another public science; and as they provide us with air and motion, they claim circumstances around us convenient for the supply of these two demands. Had we no lungs, or lungs that had not begun to breathe, we might then, so far as these organs are concerned, live like embryos, closely surrounded by walls. But the lungs require both space and atmosphere. These are easily procured as a raw material; we need only walk abroad to breathe freely enough. Other agencies however intervene; the fields are cold and the ground is damp, and we need shelter as much as space, and warmth as much as air. Thus the problem arises; for the world is too big and too windy for our constant lungs. The first size and air which we have to manage is that of the house. For we soon find that we cannot live in nature, but only in art; and are constrained to carry out the lungs in the mansion; to raise to the second power the air which they contain, and to have an airy room without us, answering to and supplying the air chamber within us. So too with the free motion of the lungs; it requires to be taken up and continued by a liberality in the dress, and this, by a space in the house, which admits the corresponding free motion of the body. Our apartments must be large enough to enable us to do with our whole frames the duties by which we live, as the body by its chest is large enough for the play of its organs (pp. 100—104).

Nay, but the lungs are bigger missionaries still, and they not only supervise chambers and chimneys, and love ventilation, but also stand on the Acropolis of all towns, and preach their sermons there. The rooms are but air-cells, and the chimney shafts bronchial twigs, but the streets are the branches, and the main streets the trachea, of the outward social lung. Breathing comes down from the larger into the lesser, and all the windpipes depend upon the greatest tube. Hence the tyranny of this organic doctrine, which detests individual smallness when public size is concerned. It is plain that building itself has its diseases—its *phthisis pulmonalis*, and the rest—which ought to be cured by direct prescriptions from the state. In the vastness of nature, when our windpipes lead thither, impurity dies by extreme dilution; our *sordes* perishes in the great sea and the fleckless ether. Nay, take dirt far enough away from its generators, and it becomes the pabulum of some other set of natures, as the noxious breath of animals is transmuted into life, glossy green, for the plants. We have, therefore, to make the lungs continuous with the grand atmosphere. The management of streets and cities,* and the cultivation of the earth's surface, so far

* The providing of places of exercise, where pure air and pleasant sights can be found in or around large towns, has naturally claimed attention. London, Manchester, Liverpool, &c., &c., have their "people's parks," formed either by private or public munificence. This is well; but a park, at best, is somewhat formal and valetudinarian. It does not invite walking so much as strolling, which is a very different tension of mind and muscle. The country adjacent to the town is more inspiring generally than the park. We cannot but think that the subject of *footpaths* belongs to the care of the sanitary men, and if properly considered would open new energies of health to the people. Your footpath has a refreshment distinct from either the park or the highway. But alas! the footpaths of the kingdom, and even those around the metropolis, are gradually vanishing under the encroachments of the proprietors. We propose to the reformers to have a statutory registration and map constructed at once of all those still extant; and to make them an inalienable possession of the people, which each parish shall at all times be bound to claim with fines on the mere showing of their registration. And we would also have them maintained, within a given number of miles of the towns, in a passable state during all seasons, winter as well as summer. This would be easily done by means of asphalt laid upon a good foundation; whereby also trespassing would be discouraged. We claim no new right, for the paths belong to the people; nor could the first expense of winning them from the seasons be considerable. On the other hand, this is one of the most ready means of drawing the population out into circumstances of vigorous exercise, and enjoyment of nature. The whole neighborhood of towns is a public park ready made, if only the paths were duly administered. We

improves the expirations of all, that they grow lighter, and sail away into the blue places where impurity is at a constant *minimum*. After which, no doubt, the prayers of congregations go up more readily to heaven, because they have no longer to fight a passage through a putrid roof of vapors overhanging the church, and of which the steeple is the pillar. They are not skunked by vested interests, reminding them of the dead, or tied like coffins to their flowing robes.

But the psychical office of the lungs is the true and the major problem. How to carry out every breath through the society, so that man shall inspire, expire and conspire! It is the problem of the production of a common understanding of what is to be thought and done—of how all the men at the same windlass shall heave in tune (pp. 121–122). Truth here is the sky in which all breaths must be received and purified, and successive aspirations thither are these second social windpipes. Moreover, the due sectarizing of mankind is required; unanimity or one-breathingness in masses; no skepticism from one man straitening the breasts of his fellows; for skepticism is social asthma, in which the age forgets how to breathe. The conspiring of the efforts of many to one end, is then a desideratum for the public health, enjoyment or integrity of the lungs; and when it has place, each effort pervades a society, and self-breathing is the most attenuated possible. But we do not know the length of the public vibration of the lungs; the size of any chest when it has the truth of Man in it; or the delight of expanding and contracting for a commonweal. Still, this is the problem.

Our exactions rise as we enter more deeply into the human frame, to ask it what it wants for a world. We are almost afraid to interrogate the heart upon the point; if we open its mouth, and encourage it, the organ may prove too voluble for any good purpose. But remembering that the brain has to speak after it, we shall allow it a voice, trusting to correction some day from that sequel.

commit the subject to Prince Albert, as worthy of his patronage, and as one that would add another lustre to that diadem of national services which he is determined to wear. The people will never forget the framer of an Albert Charter, if it make the scenes of our fair islands accessible to the poor as to the rich.

What then is the public health of the heart? We do not now speak of that food which allows us to make blood, further than to say, that the public health of the stomach lies at the foundation of the rest, and that bread and wine require to be heartily conceded by the community to its members, on the principle that the laborer is worthy of his hire. It does our hearts good to eat the bread of toil, because it comes charged with the votes of God and man. But we now canvass the heart in its popular and living sense, in which there are three points to be noticed, all bearing upon the question of happiness, that river on which health with its white streamers floats. A man is healthy in this sense, 1. When his heart is in his work. 2. When the relations of the heart are carried out for him. 3. When there is an atmosphere of cordiality about him, supplying the individual affections from the social, as the lungs are supplied from the air, or the thought of the writer from his age. These three things are one in the proposition, that the man shall exist *con amore.* It is indeed useless to treat of this matter, excepting in so far as it may show, that the highest relations belong to the human frame, and come as prophecies out of its study. This however is motive sufficient, and therefore we dare go on.

I. A man's heart is the muscle of his muscles, the lion of his strength. But muscles work together by balances and coöperations; in a dance, for instance, there is a marvelous association and change of powers to make the rhythm which answers to the music and unity of the soul. If a muscle or a fibre in one leg be out of tune, it will either be torn, or make a limp of the dance. And when the main muscle of all stands out, and will not enter the quadrille, as in ill-assorted tasks, the fire of industry expires, and legs and arms move languidly enough. The joint of joints is out of joint, and the inferior limbs are but crutches on which painful duty carries the cripple about. On the contrary, in happy moments, when the man and his work are at one, each muscle comes parallel with the heart, true to its rank, file and moment, and the strokes of the man are constant and imaginative as his heart beats. Heart and hand then grasp the same thing, and are working in united pulsations. So much for the first requisite of carrying out the bosom, namely, that the man's heart shall be in his work.

II. The happy carrying out of the heart's relations. Home, friends, children, country, are the immediate world of the heart; and when its love can reckon them over as its own, its every beat against the breast is answered from without, and the heart eddies through the society in widening concentric circulations. The blood is made under the auspices of feelings which are the sweetest enjoyments and the dearest bonds, and the body is tinctured with a stately fire, larger than its individual life. On the contrary, where there is no issue for the feelings, or no proper objects to love, the breast is shut, and probably the senses absorb the soul, and carry it out to death through their vicious doors. The disappointments of the heart may either break it, or wither up countenance and frame, showing the picture of a man whose blood carries no live motives in its current. Moral freedom is the formula of this kind of health, which imports that the walls of nature and circumstance—of heart, ribs, manners and laws, shall be no hindrances to the structural affections of mankind. We have got now to labor *con amore*, *plus* life *con amore*. What next?

III. The supply of heartiness to the heart from without. Man is a being who lives forever upon grounds, forever breathes atmospheres, sees suns, is gladdened by light and heat, chafed by electricities, and pulled by magnetisms. No wonder; for all forms are Pan-anthropal. But the planet which is ordered to accompany him forever, puts off its exuviæ at every stage, and shows a fresh core or surface. The heart-man does not live on mineral, but on social grounds, breathes not airs but thoughts, is warmed by blood heat, or affection, and drawn by living magnetism, which is love. And this set of circumstances is a true universe which environs us, and whence we get life *ab extra*, as we get nature from the world. According to the constitution then of the social world, is the supply of the air, sunshine and waters of our existence, and we can no more live out of the one world than out of the other. As we have all from nature being nature's subjects, so we have all from life as we are the subjects of life; ourselves alone being free, a germ of manhood plunged through and into all things, to grow through, and to outgrow, the more limited planets. When man is all in all in the secondary sense, he will be in the image and likeness of God, who is all in all in the primary. The fitness of society to every man, is the condition of this last demand of the heart for health. It is a mad claim: worse perhaps than that of those sanitary reformers who would grasp the winds and wash them clean, sweeten Leeds and purify Sierra Leone, abolish ague or typhus from the rural districts, or drunkenness from all classes. We believe however that it is only the same problem as theirs, but stated for a very exacting organ, the heart.

How shall any of this be secured? If we cannot manage common nature, and lead it as health into our dwellings, how can we sweeten this vast and terrible life, which gives and takes our moral diseases, and whistles its comedies through our ruined affections; which accepts the pollutions of bad hearts, and the wail of broken ones, and mixes them in its columns to press us more heavily into the kingdom of pain? All that we can say in reply is, that the heart must deserve its universe before it gets it; the present world is the fatal logic of its diseases, the other, which we have described, must be the logic of its healths. It is an untieable knot, excepting on the principle of a religious alteration of man, or a *Deus intersit.* For if already the community depends on the individual, and *vice versâ,* then they are equated, and without the intervention of some third power, there is no hope of improvement. But still, under the guidance of God, both the man and the society step out of the fatal circle, and acquire new duties again: and hence we cannot doubt that in the public cause the community must be worked upon by communal means, and the individual by private means, and that both these will be reformatory, entering into the spirit of Providence.

But let us reassert our claim to be map-makers, and not travelers or colonizers. It is allowable to exhibit an organic geography, like the physical, though on a glance at the result, it is seen to be rather a mighty blank of desiderata than a habitable globe. Here, we say, are ten thousand square leagues on which no grass grows, no rain falls, whence no rivers issue, where the oases have dangers of their own, and where a few sparse lions haunt with horrible hunger the white bones of old camels and travelers. Here are regions where man's greatest works and his smallest pests, fleas and pyramids, come together. Here are the verdurous shores of pestilence, inviting as sin, apples of Sodom whose rind is half a continent. Here are the chilblains of the frozen zones, which cool down the love of the sun to a senile moonlight and night-heat for the poles as they creak and grind in their ice. Here are the islands where elemental anger begins, and infant hurricanes spin their tops. So much for the inhuman map, and then for the human. Here are the savages, and here the barbarians; here man is in the fulness of his present stature, in the manufacturing heaven of Manchester, in the Arcady of Ireland, or in the holy cities of Paris and London. If the map has no other purpose, it shows that the world is still round, vast and habitable; capable of intercommunication; it enlarges our ideas of the field that duty has before it. In defect of maps founded on the spherical whole, the ancients regarded the tropics as mystical fiery lands where man could not live; and in defect of the psychical map, we are become afraid, not without a show of prudence, of the richest heats of the heart, which we regard as the realm of phantasmagoria and monsters. Yet if tigers and snakes abound there, men also may be found; moreover the sugars and spices and gold of both kinds, moral as well as physical, can come from no other than those sunniest rings.

But in the meantime we are sure of one thing—that man must work at his post, and not desert it, if he is to find health for his heart. Want fools us, unless we drive it through our daily works. Present circumstance is the top of desert, and the means of happiness; and it is in the reformation of our own fields, and no others, that our future is to come. The statesman does not emigrate to carry out his plans. And so, in like manner, it is not new but old industry that is to become attractive, by a new heart given by heaven during fair toil: it is not altered social relations that will make us contented, but a better love in those which exist; and it is not a new world of men that we expect, but a conversion of the old to the types of the commandments. The heart, as we have stated it, prophesies how newly, under these circumstances, all things will appear.

We now pass on to the public health of the brain, which has two parts; *firstly*, the mental and spiritual culture of the time, and *secondly*, the national education.

With respect to the first point, or culture, it embraces the progressive state of religious and other knowledge, but under a form established and recognized by government for the time being. The beginning of it is an established church, under which comes an established or sanctioned literature, to give the tone to the secular thought of the year. Private churches and literatures may exist in any numbers, and they belong to the department of the freedom or private health of the brain; but over them, in the order of things, there ought to be public encouragement of the admitted best among them, both to gain a temporary rule of thinking, and to stimulate the individual to the greatest possible degree. A government loses its chief handle when it ceases to dispense the prizes of thought; it cedes the courage with which dissoluteness of mind and morals is to be repressed, when it no longer grasps the chief literature of an age. The thing to be guarded against is the supposition, that the highest reach of thought can ever come under patronage at the time of its attainment; that is not the point sought; it is the best average which alone any state machinery can be expected to select. The prizes of a state will be the leading points by which the next ages decide upon its own condition; and progress will be marked chiefly by the successive adoption of works and thoughts which have been rejected, but are at length received. Nevertheless, in this department, religion must ever claim the first place; that is to say, the Christian religion through the Bible; and we regard it as a fundamental to take no thought of infidelities in the council-chambers of culture, but to proceed as if they did not exist. Let them have their private range, according to the freedom of the age; still, they can enter into no general plan, but must be discouraged, though not violated, by the state. By means of this laudable supervision, the governments of free nations will begin to operate upon the general thought, and with various success, according to their own wisdom, will stimulate knowledge, both natural and divine. Much, indeed, is done in this respect already, by numerous corporations addicted to special walks of art and science; but it remains to crown the heights of the possibility by a State adoption of what, from time to time, seems the best.

It will be borne in mind that we recognize two modes in all these fields, the private and the public. We do not desire that the one should usurp the place of the other, or that freedom should be lost by any direct action of government upon matters not strictly required by the commonwealth. No censorships of the press are needed: the laurel given from above is a more effectual means of command than any leg-locks which absolutism has devised. And where public opinion presses, as in England, the average best men of thought and art will be the prize-men of their year or decade; and more could not be expected. Those who are before the age, can wait until their age arrives.

We must, however, guard against the supposition, that one establishment will have anything to do with the choice of another, or that the clergy of the church will nominate the clergy of literature. On the contrary, all these public organs must spring directly from the head of the commonwealth, or they will not possess their own souls, or be taken as they are found. In the contrariety between the best products of an age, lies often the balance and safety of the state; and the church and literature will never be more mutually corrective than when they meet with the rights of an equal origin around the person of the supreme magistrate. Hence in treating of the second point, or national education, we are in nowise tied to the order of the first, or culture; but we begin our course afresh with existing possibilities. It is as in cultivating an estate, we offer prizes for the best fruits, vegetables and stock; and by that means, we hope to stimulate all the labors of the various tenants; but when we come to feed our people, we can only give to each the best that can be had; and if their stomachs are full of fancies and indigestions, we must take them at their word, and supply them with such food as they desire. The point however is, that the state is bound to give them mental food, and this, by a system, not of coercions, but of encouragements. And if like Hindoo castes, they cannot eat anything that has been touched by those of another sect, this, too, must be respected; yet not so as to render the main object impossible. The first thing is to make the knowledge so general and public, that no complaint shall arise from the jealous persons—that the waters shall not come through our neighbors' houses, but in pipes down the middle of the streets. The people will then feel that this is knowledge direct from the reservoir of the age, and that if it is not good, they cannot complain of it. We therefore regard secular education as the only public education which is possible; and the public schoolmasters consequently as a clericy of which the Queen is the head, and which must emanate, coequally with the church itself, from *the progressive State.*

But the great brain-builder is action, or industry directed by knowledge; for real doctrine is the brilliance of good works. The encouragement of action is therefore a department of public health, and indeed the chiefest of all. This occurs by the stimulation of examples, great pieces of art, or model actions; and the prizes of the workers have no proper source, lower than the hand which grasps the sceptre. For if the cradles of the land hold the dew which is ascending into private existence, the throne, or mountain of the State, is the altitude whence the public rivers must flow. The clergy of industry, which is God's church in the muscles, in England, in 1851, receive their first ordination. It is fortunate that we write in a year when the throne is the centre of the arts: when a great prince, faithful to the height where Providence sets him, dispenses the provocatives of new achievements to bless the nations of the earth; when private selfishness and *laissez faire* have been brought to their knees, and old inactions have yielded under a royal lash of shame. It is mere desert to say, that ALBERT is the victor in this social battle, and that the standard which he has planted in his crystal camp will not be plucked down again, because it belongs to the nature of man to take the fire of his industry from his chiefs.

The conception of public health implies the reconstruction of all the circumstances with which the organism is surrounded, upon the model of its natural and spiritual wants, and the presumption is, that many diseases and vices will die which circumstances, and not the choice of individuals, have engendered. This result itself, however, can only run *pari passu* with the increase of private virtue, and hence, as we said before, the throne to which the whole problem perpetually refers itself, is the regeneration of man. After any given circumstantial operation has been effected, an intractable mass of evil will still be left, which requires new circumstances of cure, originated by new physicians of good. Thus the private circulates into the public, and *vice versâ*, and to dream of any Owenite reform, is to postulate a state which is desirable, but by that means, unattainable. The "best possible circumstances" mean the best possible brain and heart, sent by God to the occasion. But again, on the other hand, without new circumstances there is no new hope for man.

Were it worth while (which it is not) we might follow out the other organs in their exactions; and group the production of man around the generative parts; his first endowments around the organs of the senses; and so forth. But enough has now been said, to show that the human frame is a natural method of thought, even in the social and political spheres. Indeed, to the eye of faith it is a great sanitary prophecy, unfilled by the mind, as the world is unfilled by the race. And in this respect also it is an urgence of immortality. For health, or the whole inhabitation of man in the body and the world, is an expansion and a deed so great, that no time can exhaust it, or work out any final equation of the correspondence: the fitting may go on for ever, not for the race alone, but for each individual. We repeat also that each end of the chain of health is contagious in a universal sense; that a renewed man and society are not only consciously, but organically also, working for the perfection of every climate, for the unfreezing of every ice-lock, for the fertilization of every soil; for the extinction of monsters and the increment of creations; and for the birth of more men, and new men, who come as cultivators in their seasons. The greatest problems (pp. 338, 239, 316) are not intangible; but it is ordained that our little frames shall give nature the signals of change, and ring the bells of new eras.

This body corporeal then, when the light is set shining through it, projects into our space new social and political bodies, one after another; the desiderata of health call about them the planets of health, and geography and astronomy are its satellites. It is none of our doing that the subject has this shape; the sanitary reformers have to answer for it; for it is they that have shown, in their first days of existence, that man, in his body equally as in his brain, is by nature a factory of Utopias.

In the meantime, the community of things is proved by evil, if not by good; and in bringing this to light, the pioneers of health have already great merits. They have shown that nothing is lost, but that disease springs from neglected social duties. The map of dirtiness is also the map of disease; the map of intemperance, ditto; the map of pauperism, ditto. They have afforded new force to the truth, that all things are in the effort to be promoted into man. Swamps and ditches, in their new rank and honors, are fevers, choleras, and the like. If we have not dwelt upon these subjects more in detail, it is because our vocation here leads us to illustrate organization, as a means of opening the mind itself by a just education. But for all present purposes, we concede the palm of usefulness to these brave sanitary inquirers.

So much for the curing of societies, which the public health-doctors have undertaken to superintend: we now come to medicine in the more ordinary sense, as it proffers aid to individuals in their maladies, according to certain rules and methods, new and old.

The earth appears to be netted over with a traditional healing power; for some knowledge of herbs and simples belongs perhaps to every race, to enable it to apply the products of the soil to the cure of its own immediate distresses. As each land has its national music, and in some countries almost every dale has its melodies, so each seems also to have its national medicine, derived from immemorial times. How the places came by the knowledge is not often asked. We cannot but think that in this fact also, we have traces of an ancient state of man of which history has not brought us the records. If in the earliest national myths there be evidence of a condition in which the faculties and imaginations were more full of tact and grasp than now; if the first bards loved, and were loved by, nature, with a fondling intimacy to which we are strangers, there is reason to suppose that this intimacy extended to every walk of life, and that there was a primitive shine of knowledge anticipating the arts and sciences, and in which medicine, and the skill of herbs, would partake. If newness of powers is still so great a gift, that infancy is a delight through the livelong day because the world is fresh and sense unworn, what must have been the newness of those first senses which opened upon the morning of time? And if delight is the bed of genius, what must have been the brightness of the insights that went penetrating forth from the eyes of the children of happiness in the beginning? The word happiness, in its human propriety, contains it all. Such races would *happen* on the needful laws of the world, and its harmonies with their frames, and when the first stings of sin tore them with their first physical pangs, they would aim at cunning evasions of consequences by arts and herbs with which they had been familiar under other auspices. The speed of their reasons would be like revelations compared to the slowness of ours. And thus if we choose to call the mythologies, revelations, we must also call medicine by the same name, so far as it has come down from the original stem. The thought of harmony corroborates this view. For primeval man was in harmony with his circumstances, and his mind had a fair start: he had his own native version of physiologies, botanies and the rest, though unlike ours; and when he swerved to our tree of knowledge, his child-genius, like all natural things, could only desert him by degrees. Doubtless it hovered over him, until it created for him the heads of that observation which is still traditive with man.

There is another version of this, but it is hardly worth our notice: criticism, as the art of picking things to pieces, is its parent. For they say that man was at first a savage (or according to learned Vestiges, an ape), who by dint of a toilette of some thousands of years before the mirror of philosophy, has "titivated" himself into civilization. This contradicts experience. The savages* are not infantile but senile races, with no loins left; they tend not to pro-creation, but extinction. They have no power of imitating the models of their betters, but where the white man comes they die out. If the Creator had planted the earth with such, he would have been like a Colonial Secretary peopling a new continent with scum. Pretty babes these to have appeared so soon after the morning stars congratulatively sung together and the sons of God shouted for joy!

* We do not reckon the negro races as savages; they are evidently still infantine, capable of propagation and education: nay, capable of passive conquest, by the very destinies which have enslaved them, and which are insuring to their posterity many spacious kingdoms on the face of the earth.

The tradition of such a crew of darkness must have been a Botany Bay and Norfolk Island reeking through all myth and song: the arts and sciences must have smelt like dogs and monkeys from these kennels. Let alone Eden—I do not see where Egypt and Babylon could have come from, if skulking aborigines, and hide-dried cannibals eating their own heads off, had been our Adams. Long before this, not a baby would have been left to hope, but the human form, like a boat launched with a hole in its bottom, would have foundered on the brink of the year 1. It is according to all analogies, to stick to the paradise of Genesis, which gives a divine and redundant youth to man where he so wanted it, that otherwise no human race could have grown up. And if we know not how to reconcile every petty circumstance to this piece of God's common sense, let us know that it is only because our winter cannot comprehend the spring in which His creations blossom : and let us wait until the second human spring enables us better to commune with the first. We can easily receive any science, without letting it meddle with the divine proprieties.

As then it is so much easier to fall than to rise, we have first to chronicle the *descent* of medicine from old times, when the race, like a woman's heart, had feeling and insight full, and when also the schools of healing originated, in the persons of those who in the well-distributed humanity had greater gifts in this kind than their fellows. These were the gods, from which medicine was said to be descended, and whose powers might well appear to be miraculous to their successors, who had lost the insight upon which the *tutum*, *citum* and *jucundum* of healing depends.

But we do not profess to sketch the history of this interesting subject; it is sufficient to know, that there are two great classes of healers, namely, 1. The medical schools, and their disciples; and, 2. The people itself as a depositary of mere traditive medicine. With respect to the schools, they also contain two elements; in the first place, their own traditional lore, the history of medicine gradually purifying itself towards modern science; and in the second, the cumulative experience and induction of that science itself. As for the first of these elements, it has grown weaker and weaker with each generation, so that one might say that each age of doctors never had a grandfather; the family likeness vanishes more and more out of the race. Orthodox medicine in this century is a substitution and not a continuation of the science of the last: it has no right to be offended with upstarts, for it is not more than fifty years since itself arose out of the crucibles and dissecting rooms. In a word, it has many experiments, but almost no traditions. Each fresh version of our Pharmacopœia carefully weeds out old simples, and fills their places with chemicals, exterminating this and that to make room for the last new compounds.

With regard to the second, or scientific element, we do not find that it has placed physic upon any basis but that of experimentation; and that not integral, but chemical experimentation. If the body were a pure acid, or alkali, housing nothing but affinities, it could hardly be more industriously tested by all the new products of the laboratory, to see what they will do with it. And when the perilous experiments have been made, no law comes out of them, excepting that mankind, "having suffered many things of many physicians, and spent all, and being rather worse than better," is wearied in the end and worn out in the process. This is to be expected, where the most acrid substances which art can wring from nature, are put in large quantities into the living frame, on nothing more than an experimental hope. It is furthermore to be expected from the course which physiology has taken in the rearing and breeding of medical men. For it teaches them to think chemically of man himself; to imagine him not as a human being with a spirit inside, but as a plaster-statue of nucleated cells, whose life is manifested in composition and decomposition. There can be no common sense in treatment, where thought wanders into this chaos of petty prisons: indeed one feels that if man is such a compost, blue pills and black draughts are fit stuff, not only to medicine, but to feed him. In the meantime, however, what becomes of the treatment of disease?

The consequences of this system, in which nearly all but the chemical side is neglected, have been seen by its orthodox adherents, and if we remember aright there is even a school calling itself "Young Physic," which recognizes the probability that medicine does more harm than good, and that the secret of the superior success of parties to be mentioned presently, depends upon their good fortune in "doing nothing." This protestantism of theirs demands a few words from us, for really the thought occurs, Whether, experimentally, we know how soon diseases would cure themselves, if they were left to nature? No observations have been made upon sick people visited by the physician, who has made up his mind to have nothing to do with them, but only to use his gray eyes.

It is a mistake to think that there is such a thing as the natural history of disease; it has none but a human history—either benignant or terrible. Man as he lies stretched on his pallet, is still both a soul and a body; he does not grow, or show, like insensate wood or bark, in any direction. If you are there to heal him, he takes good out of your frame, sympathy and skill, even though you pour bad out of your bottles; he feels the warmth of a brother, present as the envoy of one of the arts of heaven. And if you give *placebos*, you are still acting upon him by the inevitable medication of your person. But if he finds you at his bedside, to watch him for cold experiments' sake, that his life, death, or throes may figure in your statistics, do not suppose that you are innocuous, for you are a potent poisonous drug, calling forth sorrow, despair, fear, and other destroyers, and aggravating the sensibilities of his organs. The best feeling that you can cause under these circumstances, is indignation, which, as it rouses the life, may be remedial in your teeth. But there is an absurdity involved in the question, What would be the course of disease if left to itself?—because it never can be left to itself without a breach of manhood; and even if it occurred on a desert island, the mind of the sufferer would play an important part in determining the illness towards either death or recovery. The brutes know better than to resign themselves to doing nothing under natural afflictions: they lick their sores, and seek their herbs; and the domestic animals bring their sorrows of this kind to man, and assume a kind of patience under the treatment that he enjoins. The record of cases requiring care, and yet to which no care was given, would constitute, so far as it could exist, the unnatural and inhuman history of disease.

We therefore reckon that any medical practice which has had a few precedents to correct it, is better than that profession of nothing, to which some of our brethren have directed their hopes. Yet it is equally true that many cases would have recovered better by simple watching, and the application of a few obvious means mostly suggested by the patient's feelings, than where a large apparatus of drugs has been employed. Nevertheless, in every social state, the physician, in some form or other, is indispensable, whether he be called priest, magician, friend, mother, or by whatever name that has authority over our mind and matter.

But after all there is great necessity to do nothing relatively to what has been done; for we have been on a wrong track, and must change it; and by a common law of our nature, the new will seem nothing to the old, so long as the old has new eyes to see it. Buried, as the existing medicine is, among fuming acids, sharp stimuli, chemicals and caustics, it cannot believe in the power of gentleness, or the first smallness of good causes. Widely extended though it be, we must pass it by; but not without the recognition, that it is the chaotic mother of children fairer than itself, and on their account deserves to be respected. But now that these have come, it only keeps them out of their fortunes by lingering too long above ground, and cumbering the earth with its age and infirmities.

The first considerable child which it has born, is that science which HAHNEMANN delivered—we mean HOMŒOPATHY, or the treatment of "likes by likes," which was a legitimate fruit of the previous drug medication. For in the whole, the idea of medicine itself is homœopathic; it does not give health-producing agents to engender health, but poisons which would issue in disease: it is, therefore, the general application of the law, by which like is to be cured by like. It is in the particulars that medicine does not recognize the application of the Hahnemannian formula; and thence, whenever it comes into details, it is in contradiction with its own idea. It is homœopathic in theory, and allopathic in application—a house divided against itself. And in the matter of doses, it is subject to the like remarks; for no one gives physic in the same quantities as food, but a few grains of calomel, or a few fractions of a grain of arsenic, are considered sufficient even by "heroic practitioners" of the old school. Why is this, but that there is a working in these poisons, which takes them out of the category of the ordinary materials which we put into our mouths? And if a grain will produce results upon a man of fourteen stone weight, where is the absurdity to end, without experiment, which may choose to show that the millionth or decillionth of a grain will have even better results? I marvel how men who lift fourteen stone by the equipoise of a skillful grain, can sneer at other men, who do the same nice balance by incalculably lesser weights. For it is evident that all medicine is on this railway of smallness, and is more perfect and harmless for every fresh terminus that it reaches. If the allopathists were accustomed to give calomel porridges, their wrath against small doses would be consistent; but when they are themselves reduced to grains, why should they cavil at other healers, who, by experiment, have found out the value of grains of grains.

It was Hahnemann to whom all the world is indebted for the scientific deepening of medicine in both these fields. He, first of men, saw that if poison *in genere* is given to disease *in genere*, the aim will be more neatly hit if poison *in particulari* be administered to disease *in particulari*. This conception of his, involved the working of a very peculiar "science of correspondences" between the effects of drugs and the symptoms of diseases, so as to discover exactly what poisons, and what order of them, would answer to the symptoms and flux of special maladies. In the ideal of this great sportsman, each shot in the gun was cognizant of its own part of the prey, and the line of sight was the science which brought poison level with disease. May we not extend the metaphor, and say, that man in sickness is like two men, each wrestling with the other; and that the physician comes to shoot the worser man to death, without a grain of the charge touching the better: in this case the homœopathic dose will not hit the struggling health, because the shot can wound nothing but disease; whereas the allopathic bullet, having no scientific speciality in its projection, generally riddles both the men, and leaves mere death, or its antecedents on the field.

The matter of doses depends upon the fineness of the aim. In everything there is a *punctum saliens* so small, that if we could find it out, a pin's point would cover it as with a sky. What is the meaning of that invisible world which is especially versed about organization, if there be not forces and substances whose minuteness excludes them from our vision? We have not to batter the human body to pieces in order to destroy it, but an artistic prick—a bare bodkin—under the fifth rib, lets out the life entire. Nay, had we neater skill of deadliness, a word would do it. The sum of force brought to bear depends upon precision, and a single shot true to its aim, or at most a succession of a few shots, would terminate any battle that ever was fought, by picking off the chiefs. If our gunnery be unscientific, the two armies must pound each other, until chance produces the effects of science, by hitting the leaders; and in this case a prodigious expenditure of ammunition may be requisite; but when the balls are charmed, a handful will finish a war. It is not fair to count weight of metal when science is on the one side, and brute stuff on the other; or to suppose that there is any parallel of well-skilled smallness with ignorance of the most portentous size. The allopathic school is therefore wrong in supposing that our "littles" are the fractions of their "mickles;" the exactness of aim, in giving the former a new direction, takes them out of all comparison with the unwieldy stones which the orthodox throw from their catapults.

But again, there is another consideration. Fact shows that the attenuation of medicines may go on to such a point, and yet their curative properties be preserved, nay, heightened, that we are obliged to desert the hypothesis of their material action, and to presume that they take rank as dynamical things. A drop of aconite may be put into a glass of spirit, a drop of this latter into another glass of spirit, and so on, to the hundredth or the thousandth time, and still the aconite-property shall be available for cure. Here then we enter another field, and deal with the spirits of things, which are their potential forms, gradually refining massy drugs, until they are likened to those sightless agents which we know to be the roots of nature, and feel as the most powerful in ourselves. How can such delicate monitors be looked at from the old point of view, or assimilated to the violence that is exercised by materialistic physic? If the latter would stir the man, it does it by as much main force as it dares to use; whereas the former moves him by a word, through the affinities and likings of his organization.*

* There is something unfair in the manner in which the public criticises cases that do not recover under homœopathic treatment. None of our systems will

It would be curious to consider how it is that medicinal substances so attenuated are still true to themselves, and exert their properties upon the body; for the fact is beyond question. Although we cannot resolve this enigma, yet there are analogies which somewhat domesticate it in our understandings. In the body, for instance, a grain of any substance, or a spot of any feeling, is participatively present throughout; a dose of calomel influences the frame, and if we may use the expression, calomelizes it; the wound inflicted by a needle point gives a sensation to the man, or hurts the great body itself. It is not that the agent is materially everywhere; but the patient is set in the attitude of the effect, and feels it universally. We may therfore look on organisms as universes that vibrate from one end to the other with every force that assails them; and thus in their own way become the magnitudes and propagations of the force. This is plainly shown in contagions, which engrafted upon the smallest part of the body at first, in the next place run morbidly through it, and ferment the frame into a new but greater vesicle of a morbid kind; after which the social propagation begins, and a continent may be infected from that first grain of disease, thus shaken and dynamized in human body after human body, and not diminished but aggravated by its propagation. In this we have a clear image of the dynamization of medicinal powers, which received in any vehicle (*e. g.*, alcohol), and properly vibrated, seem to convert it into their own likeness, just as if it were an organism capable of transmission of effects or community of feelings. But let our explanation stand for no more than it is worth, and be considered as a word

cure every disorder. Nor is to be expected that an art which is in its infancy, can do more than greatly surpass in safety and virtue the Hippocratic medicine of 2,000 years' standing. Yet, whenever a death occurs under homœopathy, the neighborhood argues and acts as though homœopathy had invented death, which was a phenomenon unknown until Hahnemann brought it from the infernal regions! Why! the bills of mortality since Hippocrates are the bills of allopathy. And in most cases, let the worst that can occur, it is no worse, and no more, than happens daily under that practice. But if the patient dies under allopathy, he dies by precedent, and there is no responsibility; if homœopathy is at his bedside, he departs unsanctioned, and the survivors have to answer for him to public opinion. This must be borne until the battle is further fought, and those who are not prepared to endure it had better not dabble in homœopathy.

to those only who like to see, not merely that the fact is so, but that besides being true, it is also not improbable.

Whatever be the hypothesis of the properties of drugs in infinitesimal doses, the fact remains the same, and Hahnemann has the credit of testing pharmaceutical substances with a rigor of which his predecessors had no conception. The vagueness of medical practice disgusted him, offending intellect and conscience alike, and for a time he retired from a profession in which he had so little faith. His own discovery—that like is to be cured by like—then came forth, and by an easy process the whole strangeness of homœopathy developed itself. The diminution of the doses took place by degrees along a road of linked facts, in which there was little room for fallacy: there is no case in inductive science in which experiment was more minutely perfect. Cures followed, and have ever since followed, on a scale to which the orthodox medicine was a stranger: the statistics of homœopathy, taken in Government hospitals, and under military strictness, show a lessened mortality as compared with the tables of its rival. And wherever it is fairly practiced, the same average results occur; so that in spite of much opposition, it spreads from the healed to the sick, and the rumor of its beneficence is stronger with a sensible public than the diatribes of a very active and influential profession arrayed against it.

The practical blessings of the New Medicine are dependent, as we conceive, firstly upon the science of correspondence, which bringing poison and disease together with a completer fitness, poisons the disease, and kills it; and secondly, upon the smallness of the doses, or we would rather say, the use of the spirit and not the body of the drugs; which use gains its cause by no destruction of our tissues, but by giving the body an attitude that neutralizes the disease, and then itself ceases after a certain duration of effects. Drugs given in the latter way are more like ideas than material bodies, and when they have served their purpose, they either vanish of themselves, or may be countermanded by their appropriate antidotes.

I suppose it is impossible to overrate the consequences of Hahnemann's life. Even the negative results are vast for our future wellbeing. How different, for example, from the pale faces that we note in every street, will be those which belong some day to undrugged generations! What vigor may we not expect from the later posterities of those who have not hurt mind and body by supping on material poisons? How much better those childhoods will be, whose parents and grandparents have neither been bled nor salivated *secundum artem*, but who have kept their own current in their veins, and given it entire to their race? And on the positive side, what another gain it will be, when hereditary maladies begin to be displaced, and the crust that hides man drops down from his skin by degrees! What virtues may we not expect, when with all higher helps to good, the body itself seconds the monitions of the soul! What talents also, and what happiness, when the frame is set in parallelism with the order of things! For though we do not attribute everything to body, yet a sound body has consequences which make it needful to speculate upon it in all views that concern the advancement of our species.

On the theoretical side, Hahnemann has approximated drug-healing to the pure sciences; and by instituting experiments on the healthy body, he has expanded the properties of each medicine to a human form of symptoms, naturally, by that form, applicable to man. I think of medicines now as curative personalities, who take our shape upon them to battle in us with our ills. The testing of their characters is also capable of being carried to the utmost exactitude; for drugs may be "proved" upon many persons in different places and at different times, and their symptoms curtailed, sifted, and if we may use the phrase, pared and sculptured down, until only their essential and nude form is left. When we get these heroes on their feet, they, and not their discoverers, will be the great men of an ever-young physic.

It is hard to imagine how any profession can disregard the service that Hahnemann has begun, in the constitution of a rational pharmacopœia. When we look to what was known of medicinal properties before his time, and then compare it with the state in which he left the subject, the difference is like that between light and darkness. No one had imagined that each drug ran through the frame, and evoked fresh symptoms from organ after organ; nor indeed without the *similia similibus curantur* would any application come from the fact. But it is an attestation of that formula, that it leads to a knowledge of drugs infinitely special and diversified compared with the science that preceded it.

The number of superstitions also that Hahnemann slew, entitles him to the gratitude of all those who dislike to be frightened by unreal shapes which a strong man can walk through. He made the true experiment of doing relatively nothing in medicine (p. 369), and found that it was abundantly successful and humane. Purgatives were one nasty superstition which he banished. Bleeding was another of these vampires. Long before we met with homœopathy, we wondered why we bled our patients in inflammations, according to the common practice, when yet the attack struck in a moment, and there was no more blood in the body after than before it occurred; and we thought that it was but a wrong distribution which caused this rapid assault upon life, and not a plethora of blood; and that skill would lie, not in butchering the disease, but in restoring the harmony which was lost. We had seen some of our best beloved friends sacrificed to the murderous lancet, and our's was the hand which let out their life—though under the legalizing sanction of the most accredited physicians. Would that we could recall the dead; but they sleep well! Who has not had similar experiences? And who, in the long run, will not reproach himself, if he does not accede in an inquiring spirit to the New Medicine, which has availed to exorcise this host of killing superstitions?

Among the other benefits of homœopathy, we reckon this also—that it tends to make us think more worthily of our bodies. I defy any man to be a physiologist who is in the habit of bleeding, purging, and poisoning the human frame. The body abhors him, and dies rather than tell him its secrets. What idea can a man have of life, if he is accustomed to take blood, which is the soul's house, in pint basins from the frame; and to think that he is doing nothing extraordinary? What notion of living cause and effect can any one entertain, if he deems that such an abstraction of our essences can ever be recovered from so long as we are on this side the grave? What imagination can be felt of the music of man, by one who orders purgative pills *pro re nata* to play upon our intestine strings, in the delusion that their operation is temporary, and confined to the first effects. I see in the whole physiological science the large written evidence of these stupid sanguinary methods; the doctrine has followed the works with a vengeance, and the science has been purged and bled away until nothing is left but chemical dust on the one hand, or germ-cells on the other. This has gone so far that it is doubtful now, whether the medical profession has any further power of pursuing human physiology; doubtful whether that great knowledge must not pass to the laity and the gentiles, and become a non-medical science. Certainly, the hands that have least been crimsoned in the bowels of the living man, seem by nature most fit to receive his tender and amazing secrets.

Another department also is that of mental effects, in which homœopathy stands preëminent. If each drug evokes symptoms throughout the body, it also affects the mind wherever it touches the organs; and hence the new pharmacopœia groups around it mental and moral states so far as they depend upon the body. In this respect homœopathy opens a field which was untouched before, and includes the healing of moods, minds and tempers under the action of medicines. How valuable this is as an adjunct of education, will suggest itself at once to all fathers and mothers; and how new a power it is, those best know who have become converts to homœopathy after practicing the old system of medicine.

It is, however, in the eradication of chronic diseases and hereditary taints, that homœopathy promises perhaps the greatest of its benefits. On this subject the views of Hahnemann deserve the attention of philanthropists of every degree, whilst at the same time they are highly interesting to the medical philosopher. Nay, there is a touch of the sublime about them, such as only comes into the scientific spirit in its happiest moods. As Hahnemann teaches us of the trine contagions that have come down with man from early days, we seem to hear echoes of every mythos that has struck us with significance before, from the Parsee dualism of Ahriman and Ormuzd, to the blue-white Hela of Scandinavian faith; nay also we are let into the understrata of that evil which throws out sulphurs and geysers in the human and inhuman worlds: and we cease to wonder that no cure comes, when the pit of disease is so deep. What a chasteness of genius too in Hahnemann, that instead of swerving to speculation, he forced these conceptions through the outlets of his method of cure, and thought nothing sacred enough for his attention, but the recovery of the body from its ancient pests. If there be such a thing as bodily disease distinct from psychical, then he was right in his devotion, and is rewarded already in contributing to the whole sanity of his kind.

In a strictly medical point of view the Hahnemannian theory of chronic disease comports with principles which are beginning to be admitted on all hands. The multiplicity of diseases and epidemics is suspected to be the mask of a unity of which so-called distinct maladies are but symptoms: just as on a large range, different languages are but dialects of some common stem. Whether Hahnemann has hit the central forms of malady of which the rest are the procession, it would be presumptuous in us to say; but at least he has put us upon the search, and indicated that the confirmation of what he has deemed, or the suggestion of something truer, will grow, as his own views did, out of the bosom of practical healing. Moreover, his science of medicine has the advantage of springing from both the roots of the past (p. 367), for as we said before, it germinated from the scholastic side; and as it grew, it took in, and retains, the traditional medicine which is found among the people. In fact, the homœopathic law gives specific justification of the popular usage of many herbs and simples, which accordingly now reappear as parts of a scientific system, affording new evidence of the probability that should be acceded to practices which are immemorial and of world-wide acceptation. And in another respect it unites with the instincts of animals, as well as with the pharmacy of the "old wives," in prescribing simples and not compounds, in order that pure operations may ensue, and causation or cure touch the ailment with a finger-end of tact, and not with a rude indiscriminate hand of confusions. The homœopathic law also accounts for the cures that have taken place under the other practice, and shows that they are owing to a latency of homœopathy in the common sense of its predecessor.

We are indeed convinced that the law of treating like with like, is the one intellectual formula to which the healing art has attained. Nevertheless, we do not assert that at a given time any art is prepared to stake itself completely upon practice dictated by science.

The genius of man walks willingly with positive knowledge, but there come times and cases, when he falls back upon the unknown chaos, and trusts for instinctive revelations there. We would not therefore cut connection with allopathy; because there will be a certain number of instances where there is no knowledge, and where chaos is a resource. When these arise, it is a comfort that they can be committed to that respectable body, the old medical profession, which to do it justice, has its own stars in its own night. We think however that it is a mistake to call its art, *allopathy;* it should be termed *chaopathy*, because it is without a formula, and welters down time by that set of falls which are vulgarly known as good and bad luck.

Homœopathy requires many changes, and new brains of the Hahnemannian order, before it will do itself justice according to the conception of its founder. I know no set of problems which would better repay severe thought founded upon observation, than the properties of the drugs *quoad* the human body. It is not routine practice, but penetrating investigation, which will introduce the next highly necessary improvements. New views of the human frame are requisite before the science of pathogenesy can attain to any degree of perfection. Among the first of these, we reckon that natural pathogenesis which the powers intrinsic to the body daily exercise upon it: viz., the powers of the mind, soul, and the inner man. By eliciting this, we shall get at the leading idea of pathogenesis, and also obtain rules for the succession of symptoms and states as welling from a single fountain head. Otherwise, unless our eyes be thus armed by these greater knowledges, the various symptoms that drugs evoke in different parts of the frame, will seem to have no connection with each other, and the memory will be unable to retain them, at the same time that they will lie as so much incoherent dust in the way of the intellectual powers. The subject is so important that we must take leave to illustrate our remarks in a few words.

If the effects of medicines should be decided upon the healthy body, it will be conceded, that the knowledge of the healthy body itself is the canvas upon which our medicinal science must be drawn. But the body is tenanted by various lives, each of which pervades it; and hence these also come to be the subjects of any physiological investigation which seeks wholeness as its aim. We may liken the powers of man to drugs that produce symptoms all through his frame. These symptoms are the basis of the science of health, and of the corresponding art of healing. But who has ever studied so much as one of them? Where is the natural pathogenesy which is the foundation of the morbid pathogenesy in the body itself? It is not yet extant in science. But we must strive after it, or verily we do not know the human subject to whom medicine is to be applied. There is a brilliant mine to be worked here, and one which in giving a deeper basis to our art, will also constitute a knowledge of psychology such as the world will be glad to receive.

In our Chapter on the Heart we took occasion to trace the passion of Fear through some of its pathogenetic states. Let us explain our present meaning by following some of the symptoms of Grief in the same way. First, what is the index-symptom here, of which the bodily state is the sequence? Weeping from the eyes is the finger that points to all the other signs: *the falling tear* is the beginning of the propositions of grief. The voice wails, and the sentences fall out of the mouth through plaintive lowering cadences which end in sobs. The head is depressed, and the hair weeps over the face. The chin falls, and the lips melt as if they were big drops of sorrow; the saliva also trickles forth as though the mouth cried. In children the mouth does cry, and sobs, like tears of tone, roll heavily forth; the lungs weep out both words and breaths. The blood and excretions are wept away, and pallor and loss of bodily spirit are manifest; and in long grief, the body itself pines or falls away. The arms hang at the side, the knees totter as though they would trickle down, and the frame droops with willowy sadness. In extreme effects, no tears flow, but the spirit weeps itself out; just as in the last cases of fear, the man runs away not outwardly but inwardly (p. 192). Now the points to be noted in this slight outline of the pathogenesy of fear, are twofold: in the first place, the mind, the body and the organs are each affected in their own manner by the emotion: in the second place, the cardinal phenomenon gives the cue to the other signs, and we find that in grief the whole man weeps. We have, therefore, the best known term of all to interpret for us the kingdom of grief. The like is attainable with regard to the bodily train that accompanies every other passion, and in each case the head sign will run through the entire phenomena that belong to it. This is what we mean by the pathogenesy of the inner man.*

Now apply this to the action of drugs, and observe that it demands two conditions, only one of which is at present fulfilled. For not only are all the symptoms to be recorded and grouped in their places round the organs, but a head symptom must be found which is their common denominator; a principal fact from which the remainder flow. When this is done, you will have in your mind a portrait of the drug, as the painter has with him an instinctive limning of the faces of the intellects and passions; and you will then become acquainted with your pharmacy, be enabled to divine symptoms by insight without cumbersome catalogues of them impossible to remember, and to apply them with something like genius to the moving facts of each case as it arises. Until this be accomplished the heart of our drugs is unrevealed to us. First, however, we repeat, it will be better to begin with the study of the healthy body and mind, and with the effects of healthy agents, not only because this is the *præsuppositum* of a knowledge of diseased manifestations, but because the mind's actions are so much more intelligible than those of drugs, and the easiest lesson should be learned first. Afterwards, when we have accustomed ourselves to the new mode of working, we may approach the drug problems with a little more chance of success.

By this means we hope to see that difficulty of practice which is the chief opprobrium of homœopathy removed, and the handles of the instruments of cure placed in the hands of medical men. This is the more imperative upon us, because our morbid states of mind have the effects of true poisons upon the body, and must be regarded as spiritual drugs, each having a pathological history of its own. Can we doubt, for example, if *ignatia* or any other medicament be

* Perhaps our meaning will be best explained, if we say that we desiderate a new JAHR to give the symptoms of health instead of those of disease; in other words, a pathogenetic psychology and physiology, which shall stand to homœopathy, as the existing chaotic physiology stands to the old system of medicine.

homœopathic to the effects of grief, that were the pathogenesy of grief itself worked out by a law of order, the application of the drugs allied to the state, would become infinitely more precise than at present? What form the science will take under this development, we cannot prophecy; but that it will hold more light and more readiness, we very much foresee.

Here we close our remarks on homœopathy, as the basis of the medical sciences. It is the intestine system of medicine in its finest form, and cures as it were by touching the beginning of the body in taste and the mouth. The law, however, is not confined to drugs, but has an application we do not know how extensive. Some glimpses of it will still be seen as we proceed to speak of other organa of the healing art.

We have said nothing on the vexed question of the doses, for very little is known about it. The effect of a millionth differs so little from that of a trillionth, that in the existing state of the science we cannot make any available distinction. Fine points are the last to be settled in any department. In a contagious disease, which spreads from a single person to a town, we know almost nothing of the differences that are caused by the transmission, or of any altered manner in which the last occurring cases radiate their effects upon susceptible people. Nor in infecting new portions of spirit with a drug, have we the wit to see what the several dilutions* signify. It is sufficient for this time, that acute and temporary disorders are most amenable to the dilutions that are nearest to the matter of the drug, while chronic maladies are chiefly touched by those high "potencies" in which only the spirit seems to be present.

Very different from homœopathy, or the cure by specific drugs, is HYDROPATHY, or the cure by water as a general element. The former appears to contend with bodily diseases in their material strongholds, searching them, and forcing them to quit their poisonous grasp of the organization. The latter, or hydropathy, is not medicinal but hygienic, and operates by stimulating or depressing the natural processes of the system. Fluid, either in the shape of vapor or water, exists everywhere in the active organism; whatever

* In preference to dilutions of drugs, we would rather term the higher potencies, *transmissions*.

part was dry, would be also dead. And temperature accompanies the frame in the fluids as vehicles. Now the water cure is allied to the natural moisture everywhere, and applies itself specially to the universal skin-system (p. 271). Whatever relief can be obtained for suffering, by means of the perspirations, is carried to the greatest degree by those parts of the water treatment which so powerfully favor perspiration. Whatever benefit can come from the tone which cold water so instantaneously arouses, is heightened in the plunge bath, the wet sheet, and the douche. Whatever calmness and coolness can do for the irritations and heats of the body, is brought about in an extraordinary degree by the cold quietude of the sitz bath. In short, the water cure is the exaggeration of the hygienic processes, until they come to be powerful agents for changing morbid conditions. There is nothing like the action of specific medicines in this method of healing; for though the actions and reactions of the body are quickened and strengthened, they remain true to themselves, and assume no toxicological phase. Water is not a poison any more than alcohol (pp. 159, 160), though both of them stimulate and depress to an alarming degree, if their use is abused.

It is hardly necessary to enter the lists in defence of the water cure: the public has derived from it so much benefit, and withal has acquired so much knowledge of the subject, and caution in the knowledge, that the art and science of PREISSNITZ may be considered as established things. The water cure takes a capital stand among the bases of a common-sense hygiene, and while fountains bubble and rivers run, it will not be abandoned by those who love the welfare of their bodies.

Though hydropathy be not medical in the drug sense, yet it is impossible to say without experiment what number of diseases it can cure. We have no certain discrimination of those maladies which arise from stoppage of the natural processes, or from over or under stimulation of the frame. And although every loss of the balance of health depends upon defect of the whole organism, which is too weak to protect itself against circumstances, yet the symptoms of that loss may be combated by circumstantial means. Cure is another thing, if by cure be meant that restoration which provides against relapse: such cures are unhappily rare. But for the curation of present ills, the water treatment is often singularly prompt and apt; and as we said just now, experience alone can show how wide, or how limited, its powers in this kind are.

We incline to consider *energizing* as the formula under which its effects come. Of course we are now speaking of the *cold* water cure; for the formula of the hot is the very reverse. It is true that the application of cold produces by reaction, glow and heat, but this is an energetic glow, and its perspirations break out with force. On the other hand, the warm bath leaves a sense of coldness, and its sweats are languid and profuse; the chilliness that is in them proves that they trickle out of the weakness of the containing parts. The energizing of the cold treatment proves itself in two ways—firstly, by bracing the body itself; and secondly, by bracing, that is to say, increasing in power and quantity, the secretions. Both these effects depend upon reaction; whence, if the energies be not excited, depression is the result; and of this latter state, which is desirable where excitement already exists, the water-cure doctors have made great use. The drinking of cold water in considerable quantities, operates upon the internal organs somewhat as douches and wet sheets upon the skin; but with this difference, that the viscera are more constant in their reactions, and under the direction of life, are more able to recover their tone than the skin.

The water cure is invaluable for those persons who are the slaves of long habits of ease and indulgence, and whose constitutions are breaking down from the sheer repetition of these imprudent courses. They go to Malvern, and the axe is laid to the root of their tree of evils. Their impossibilities are made possible for them by the rigorous physician, who endures no remonstrances in his house. In an hour they throw down the accumulated baggage of years, and determine to do penance for having carried it so long. Early hours, long walks, long-forgotten beauties of nature, reacquaintance with the crystal springs whose Naiads had been neglected for "old port," sweet sleep hours before midnight, and the sense that they are clean human beings, or on the way to such—all these means carry health to the men who are jaded with business or pleasure, but not yet struck for death. The cold water is the central mortification of the flesh; it has caught them buried in care or luxuriousness, and like the "cold pig" which their schoolfellows once emptied over them in their boyish beds, it makes them start to their feet, and touch the ground of realities once more. We attribute much of the water power to the frigid morality which it inculcates; to the shock which it gives to the dreaming man, and his lazy organs. For it tells him very plainly that there is to be no comfort in bed or board, but that warmth must be moral and come out of work.

But if this be so, then the capacity which the body possesses, to be shocked into its functions, will depend upon the resource of *morale* in the patient, or, in a word, upon the capabilities of his faith. All that I have seen or read of the water cure, strengthens me in this conviction. It appears to me, that humanly speaking, there is a certain amount of life and reaction in the upper parts of every man, which may be drawn upon as occasion requires. How much there is, we have probably but little idea; for each means of calling it down—each syphon that we put into the reservoir—only runs its own kind of life. The quantity of this fluid spirit is therefore practically limited enough, though perchance other means might open a new vein when the old sources dry. But so it is with the water cure; it taps the life and *morale* in its own direction, and obtains wonderful supplies. We think it is not well to exhaust these, or to allow them to flow until their lees mix with the current; for when they are once gone, there is no more to be had from that tube of arts. It must not, therefore, be supposed, that the second and third resorts to the water cure will have the like success with the first; the life which was obedient then, has been partially or perhaps wholly spent. And if the penances are kept up, they become severe macerations; the faculties are not roused but chilled, and the lamp which might have lasted for a quiet while, is besieged by cold to death. In one word, the limits of the water cure are the limits of personal vigor, which, after a certain point, is wasted in the struggle of this treatment for health.

It is necessary to bear this in mind when we are dealing with an agent whose effects are so powerful and immediate. However, our reading on this subject convinces us that the prudences which belong to the water cure are well attended to by its best advocates in this country; and what we have said applies rather to the public, who require to know that this treatment, so often valuable, is limited in its results; and that in diseases proceeding from irregular habits, the first cure is likely to be more satisfactory and innocent than any subsequent one. As moral imbecility grows, from whatever cause, the physical benefits of the water cure will become less and less.

The system of Preissnitz belongs to a group of sciences which include the whole of the four elements of the ancients, and apply them to the healing art. Earth, air, fire and water are the basis of outward hygiene, and are all represented in some sort under the notion of climate. The acclimation of patients resembles the water cure in many respects; only that it produces the desired elevations and depressions of the body by influences which are less visible, though not less striking in their effects. To be bathed in the light and heat of a new sun, and washed with the winds of a fresh sky; to feel the steam of an unwonted surface of earth, and the tension of a different magnetism and electricity to that to which we are accustomed, are important elements in the recovery of health, particularly where moral circumstances also are favorable. The hygienic map of countries, gathered, as it might be, from the physical character of the inhabitants, or from susceptible temperaments on the spot, would be a guide worth having in the direction of patients to localities of specific benefit. In this respect we require something more precise than the guide books which have been written by the climatic physicians.

But medical systems, like the departments of public health (pp. 347—364), follow the organs and faculties of man, and in the nature of things are as numerous as the latter. The intestinal system of drugs or homœopathy, is the basis (p. 382), and belongs to the alimentary tube: hydropathy and its kindred appertain to the general circumstantial system, and specifically to the skin as the organ of tone and continence: we have now to turn our attention to the muscles of medicine, or to that method of cure which only lately has appeared among us, under the designation of KINESIPATHY, or the Swedish Medical Gymnastics, and which already ranks among the most important means of removing chronic symptoms of disease.

For the modern development of artificial analytical exercise, as applied to the treatment of disease, we are indebted to Ling, the Swedish poet, whose system has now been practised with success for more than thirty years in his own country. This Ling was a stern but versatile genius, worthy of the Scandinavian name—worthy of the land of Eddas, Sagas, Gustavuses, the world's best iron, and its Swedenborg. He had read in the ancient lore of his country the record of a mental and bodily prowess of uncommon virtue; the doings of kings, Jarls and vikings in the olden time, when the sea rovers sallied forth with the summer, and astonished the effeminacy of the known world with their strong arms. And the thought occurred to him, that it was possible by knowledge well directed in practice, to combine the muscles of ancient heroism with the civilization of to-day, and in the physical frames of his Swedes, to re-enact the days of Snorro and Hakon Jarl in those of the fourteenth Charles. In short, taking his cue from classic Greece, he sought, by gymnastic exercises, to compensate for the bent backs and dwindled muscles that modern pursuits and common-place existence have produced. He stood in the age like a kind of human Hecla, reminiscent of the valor of a thousand years, and pouring forth a flood of incentives to his race, to emulate the strength of their sires. His verse breathes with a Homeric spirit of combat, with a delight in the good science of the strokes, as well as in the death of the foe. It has the harshness and boldness of a muscular rhyme. His harp was "strung with bears' sinews." But it is not with his gymnastics in general that we can meddle, but only with their medical part: we have touched on the other, because the subject is less known than it deserves to be in England, and our sign post may direct the curious on its way.

It is told of Ling that, when a youth, on one occasion he was weary of life, and like a bad boy he wandered slowly on a biting winter day, as thinly clad as possible, half a Swedish mile into the country, in the hope of catching a chill which would terminate his existence, without his being guilty of the immediate sin of suicide. He, however, only took a common cold in the head, which led him to his first reflections on the human frame, and the means of rendering it hardy.* On another occasion, when suffering from rheumatism in the arm, he instinctively rapped the part with a ruler which he held in his hand, and found that he cured the pain : this natural experiment, it is said, was the occasion of Kinesipathy.

This art consists in applying external motions, passive and active exercise, to the body; and in rendering these so special as to operate on the various inward organs, or on parts of them specifically. Posture, friction, percussion, motion, are all made use of; and already as many as two thousand different movements have been devised for the purpose of operating upon the failing powers within. There are languages of nudges, to remind brain, liver, spleen, and all, of their neglected duties. The effects produced approve the plan, and stamp it as an art and science. It is admonition, contact, exercise, pursued into details, whereby disease is literally *handled.*

Perhaps there is no malady but tends in some way to alter the bearing, posture or general *status* of the body. In acute cases this is plain. We groan, writhe, wriggle, wince, shake, crawl, creep, dance, and so forth, with our agonies and discomforts, showing that disease is a complete posture master and very good serjeant, whose drill is for the purpose of relief and cure. Very small areas of disease have corresponding to them large movements in the system; and if we understood the movements, we could by reaction play upon the parts and particles of the organs. If a special wince or twist arises primarily out of some one place, then by comprehending the twist, and producing it artificially, we get at that place exactly, were it no bigger than a pin's head. Here is precise gunnery—hitting disease with a fine arrow. Again there are instinctive movements of the hands towards afflicted parts of our frames. We rub ourselves with organic pity like dumb animals where the deep flesh is ill. This is nature working for us, and showing us the beginning of a manual science of soothing, traction, nudging, and so forth, the detail of which is *kinesipathy.*

We have been greatly struck with the common sense which dictated the Lingian art, and with the excellent unexpected results

* Atterbom : *Inträdes Tal i Svenska Akademien*, p. 21.

which flow from such simple means. Exercise is often demanded, not so much for the whole frame, as for particular organs. For instance a sluggish liver may refuse to resume its functions under the general stimulus of a walk. The kinesipathist exercises the liver itself: by his jerks and suggestive poking, he commands it to make bile; and sure enough the liver does make it. By a like preciseness of application he cures sluggish bowels. He exerts the physical force of cure with the gentleness of art and science. He strengthens special muscles by adequate ingenious exercises. He cures hot heads and cold feet, by briskly rotating the feet upon the ankles, steadying the limb by grasping its lower part. And so forth. This is evidently the *ultima ratio* of treatment in chronic diseases.

In paralytic cases, where the nervous derangement is only functional, kinesipathy is found to be an effective mode of cure. Its doctrine here, as we read it, commends itself to our acceptance. Where a power has been lost, but its potency is left, it is as though the power had never been developed. A palsied man of this kind has forgotten the art of the use of his limbs, and has to learn it afresh. He is an adult in those parts where his power lies; a baby in the paralyzed tracts. The medical gymnast undertakes to teach the latter, first how to creep, and then how to go. He commences by passive movements—nursing, fumbling, and so stimulating, the helpless large infant limb; and by degrees a little reaction against him is perceived. He then makes more extensive movements, stretching the muscles, and producing further reaction; and finally he commands the resistance of the patient, and then by his superior force slowly overcomes it: in all these processes steadily keeping in view the end, of educating the limb into self-reliance, or as we term it, sense of power (p. 238). Many an old paralytic is cured by these apparently trivial means; the mind and will which had alienated themselves, are coaxed back into his arms and legs.

Like all real agents, kinesipathy is capable of abuse. It appears to be contraindicated in nearly all acute diseases, and we should also say, in those where rest, and not motion, is demanded. But it seems probable, from the obviousness of the method, that neither the gymnast nor the patient would be likely to persevere to any great mischief with the treatment.

The results which have followed this art, are so great as compared with the slightness of the causes set at work, that some have suspected a mesmeric effect from the operator to the patient. It may be so; but at any rate there is a moral cause involved which we think is to be taken into account in all such procedures. The patient feels that something is being done for him; that another human being is active and anxious on his behalf, and does not disdain to toil for his bodily restoration. To many a sick man this is an element of health; so that the bad pun by which I have heard Kinesipathy changed into *kindly sympathy*, conveys a serious truth. I even deem that the dumb hidden organs know the touches of a brother's hand and heart, and are organically comforted by them; for they all have feelings of their own, and spirits; as we have shown in many places in these Chapters (pp. 226—228, 233—236). This points to a defect in mere drug medicining: a physician writes a prescription and leaves it: he has done nothing ostensible to the sufferer, still less to the viscera and vitals of his patient; and the *rapport* between the two persons is very feeble, and by no means of that fraternal warmth which is curative wherever it is truly experienced.*

Ling's system has the merit, a great one in our eyes, of continuing practices that have existed in nearly all nations, from India to Sweden; for rubbing, shampooing, and various forms of gymnastics are almost as widely diffused as language itself. Nor until of late ages has gymnastics disappeared from formal medicine. Kinesipathy†

* The following works give a general notion of the Swedish Medical Gymnastics, though specific movements are difficult to describe, and should be witnessed in order to be comprehended.

Ling, *Gymnastiken's Almanna Grunder.* Upsala, 1834, 1840.

De Betou, *Therapeutic Manipulation.* London, 1846.

Georgii, *Kinesitherapie, ou Traitement des Maladies par le Mouvement, selon la Methode de Ling.* Paris, 1847.

——— *Kinesipathy, or the Cure of Diseases by specific active and passive Movements.* London, 1850.

H. Doherty, *Kinesipathy, or Medical Gymnastics for the Cure of Chronic Disease.* London, 1851.

† Ling's aim was nothing less than the physical education of man corresponding to the mental. This branch is deserving of an attention which it has not

replaces it there, and in such a shape that we are emboldened to hope from what we know of itself and its advocates, that it will never cede its place again.

We now come to another part in the organism of healing, namely, MESMERISM, or what we might term *Anthropopathy*, as it cures by the application of man to man. Of its virtues in cases that have resisted all other means, there cannot be a question: its facts are established not alone by rigid experiment, but by deeds worthy of a mural crown, because they have saved the lives of many citizens. It is for this reason, and also because our organic philosophy, which knows no fear, demands it, that we pass on to Mesmerism, although a storm of hatred rages about it, and every step of its advance is a fight.

It would be interesting in the history of science to canvass the reasons why certain large classes of facts have been rejected from time to time: why, for instance, the Church of Rome felt peculiarly aggrieved that the earth should go round the sun, and not *vice versâ*; why certain moderns dislike to live on a planet which took more than seven days for its creation; why skeptics have a call to blink all evidence for spiritual communications, and afterwards opening their sockets widely, complain of the absence of facts; and lastly, why the medical profession fumes and shivers whenever mesmerism is brought forward. In all these cases, as we deem, it is the instinct of self-preservation that like a skin (p. 282) defends the parties against the reception of the facts. They know instinctively that the limitation and eggshell of their state is in danger, and that if the obnoxious point be admitted, they will have the trouble of building a new house on a larger scale. At present the pill-boxes are arranged in pretty rows; but allow this mesmerism and its consequences, and

received. To our mind, no school ought to be without a physical inspector; if official, so much the better. In the plastic period of youth, physical defects and awkwardnesses may be corrected, which are past relief at a later time. The bodies of boys and girls ought to be developed by inspection, instruction, and emulation, and especially by the universal means of dancing, fencing, and the politer movements. Nor should such trifling matters as biting nails and picking noses be tolerated. Greek and Latin are of less importance to youth than a *corpus sanum* out of which all manifest unseemliness has been weeded.

how they would rattle and dance—what a long period of confusion and elimination they must pass, before any second order as neat as the first could be established! It is the dread of death, that shabbiest of fears, that everlastingly hates truth, because truth leads to death, future states, or integral enlargements, of which there is no end. Such is the motive of this very poor kind of conservatism, though there are as many pretexts as there are ingenious lazy minds who fancy that they have an interest in a well-arranged stagnation of the arts.

As we have said already, mesmerism stands upon its facts, which, in proportion to time, are as numerous and rigorous as those of any other science. But the facts appear so strange, and so little in the order of our knowledge, that they want at least the support which association of principles gives. Those who cure and are cured obtain a grasp of the facts from mere gratitude and service; but the public require also something like a *rationale* in order to steady their minds. Hence, though the facts are the main thing for those intimately concerned, yet a view of the facts is quite necessary for the fixation of mesmerism in the scientific sky.

Nor on the principles of this Book do we find much difficulty in tracing the mesmeric mechanism and its results. If herbs, waters, airs, fire and motion, which are such remote kindred to us (p. 310), will cure our ills, surely man himself, who is comparatively own brother to every man, will go home to disease with a directer relationship of beneficence. If ever it has been good to be under a course of mercury, shall it not be better still to be under a course of humanity? We have seen throughout, on disintegrating man, that he is full of human fluids which no more cease with his surface, than his voice ceases with his lips: that his influence is a combined physical and moral fact whose lengths and durations can hardly be measured. We also find that he subsists in an equilibrium of which his own will is the centre, and that his sphere of radiance, and swoop of powers, depend on the pulsations of his will. According as this heart (pp. 251—287) is little, or large, his world is a nutshell, or an empire. In mesmerism the equilibrium is voluntarily ceded, a human vacuum or obeisance is created, and the radiance of the mesmerizer or active power rushes in to fill the space. A pressure of fresh spiritual air goes right through the frame. This alone is the greatest of alteratives—more than traveling, more than the difference between hot and cold, clear and misty, winter and summer. The organism breathes with another life in all its parts. For the time, every sense, in stomach, spleen, liver, heart and brain and every idea that travels in the fluids (pp. 55, 152—154, 200—224, 351), is changed; self is voluntarily absent; and the functions go on *minus* their old routine of habits. The nerves and arteries which are secreting a morbid growth, a tumor, for example, and which have got to think that they must lay stone to stone of this vice, and build a fresh piece of it every day, suddenly lose their habitude in the interest of a new spirit, change their minds, set the absorbents to work to remove the architecture of evil, and pursue their daily course for their legitimate objects. The better man, in short, is persuasive upon the worser, inside just as outside: it is no senseless *odyle* or fluid that causes the effect, but a human although molecular operation. Fluids, unless they are men also, illustrate nothing, nor will any explanation satisfy, unless it be a rest, and exclude the call for a second explanation.

Mesmerism emphatically gives new or other life to those who need it; and it does this by the mere form and attitude which the agent and patient assume relatively to each other. The human world is full of powers in a state of balance and indifference. Change the posture of anything therein, and the whole has to re-adjust itself to a new balance—a rush of forces takes place, and currents pass to and fro until the equilibrium is recovered. The moral and the physical are both under this statical law. Hence, if you desire to produce forces, you have only to find the neighborhood of the fluids or spirit powers, and by creating a low level on the one hand, you also make relative height, and have a deeper fall for your forceful waters to descend.

These remarks are too brief to be called a theory of mesmeric cure, nor do they touch manifestly upon those parts of mesmerism that are most liable to be discredited. We labor however under want of space to develop the subject as it deserves, and we can now only say, that out of the same law of Correspondence which is the master-principle of science, it is easy to deduce the explanation of the marvels of clairvoyance and transmission of thought, and in short the psychical phenomena of mesmerism. But then under this view, our world and our matter so change, that we are counting upon admissions which few readers can make, but without which we cannot proceed many steps. Let it suffice then to say, that man in the posture of giving up himself, becomes a bodily representative of the powers of unselfishness for the time being; that miracle haunts him, because unselfishness is a miracle; that he enters upon universal sight, second sight, third sight, and more sights than you please, because unselfing is the core of all wide vision. That he dips his body into vigors and cures by the same abnegation, because unselfishness is the tree whose leaves are for the healing of all men. This comes just as the soul comes to the body: the due circumstance or posture of affairs is there, and by the law of correspondence, whereby the equilibrium of creation is maintained, its spirit is "in the midst of it" (pp. 242—244, 294—300). The very theatricism of what is good, true and unselfish, is healing to the scenework of the body.

We have no certain knowledge of the limits of mesmerism as a curative agent, nor of the conditions which should exclude cases from this treatment. In functional disorders of the nervous system it is especially indicated, and, as a number of diseases even seemingly organic spring from this root, it appears that it has a large field of applications here. Hysteria, epilepsy, catalepsy, and those other maladies in which the visceral motions predominate over the rhythmical or rational motions of the lungs, come very markedly under its benefits. But it is not in our power to lay down any rule for the distribution to it of cases generally; and therefore we wait upon experiment, which shows that the utilities to be derived from its employment are very extensive.

Like drugs, cold water, movements, and stimulants, mesmerism is capable of abuse in many ways. Ill-disposed persons may use it to acquire an influence for bad ends. It is however probably more often abused by the patients than by the agents, being resorted to as a kind of opiate to compensate for the want of moral determination. In this case it keeps up a pernicious valetudinarianism that saps the foundations of resolve in the patient, by causing him to rely upon others where there is strength sufficient, were it exerted, in his own organization. A fatal mistake is made whenever we treat with petting and coddling, in our own minds, or in others, states which we ought to discipline with a moral lash; and this mistake, I fear, is often committed by mesmeric patients. They must know that there is no patent outward means that can be a substitute for sanity of will; that sooner or later they must exert themselves, and waken from their delusions; and that every dose of their mesmeric opium over and above what was required, is the vehicle of a weakness which it will cost them a fresh struggle to conquer, whenever the time when they must arise shall come.

We had almost forgotten to place to the credit of mesmerism its introduction of a painless surgery, which is among the most brilliant discoveries of the age. The doctors were totally incredulous of this matter, until ether and chloroform came and did the same thing in a grosser shape. If there were shame in the world they must have felt it, when they found how easy their impossibilities of a fortnight before had become. They doubted the testimony of honest men where mesmerism was concerned; they accepted the same facts when chloroform produced them. It was like them to believe in bottles and to disbelieve in man. But let them pass. This discovery of extinction of pain has no end of results, moral as well as physical. The least it does is to annihilate severe material sufferings; its next fruit is in time to strike out their dread, which is the great body killer. It is plain also to see, that as an idea it is very penetrating, and suggests a painless moral surgery among the ends of man: that all operations in which our dear properties are taken from us, shall be, like the first abstraction of Adam's rib, performed upon us in mercy's "deep sleep."

In quitting mesmerism we notice of it, as we also observed of the Lingian movements, that it gathers up under its banner a number of practices that have existed from time immemorial in most parts of the earth. This is to say, that it communicates with a wide common sense. And although we do not choose to call everything mesmerism which is strange or mysterious looking, yet we recognize a certain family likeness between it and many things which are even venerable. There is however nothing divine in it, and it has no relation to miracles and revelations, excepting that it imitates them afar off, and, like all other things in the world, has a common connection with God.*

Very different from mesmerism, and yet suggested by it, is the process discovered by JAMES BRAID of Manchester, and by him called HYPNOTISM. This is probably but one of a number of arts to which we shall give the generic name of PHRENOPATHY, for it produces its effects principally as actions of mind upon mind.

Being unsatisfied with the pretensions of mesmerism, and skeptical of its truth, Mr. Braid entered the field as a disprover, but soon witnessed phenomena which appeared to him to be real, though susceptible of another explanation than the mesmeric. It struck him that the facts were the result of abstraction or attention carried to excess, and he accordingly tried the experiment of causing patients to stare at any object secured or held above their foreheads, so as to make the stare assume the attitude of intense contemplation. Success attended this mode in many instances, and sleep was induced in a few minutes. Mr. B. usually selects some bright object, as a silver lancet case, held in the mid-line between the eyes, and the patient gazes thereat with fixed stare until the effects are produced —the "double internal squint" upwards being the most potential direction of the eyes for the purpose. Soon the eyes shut, and a state is produced, varying in depth from mere somnolency to double consciousness, or catalepsy.

The preliminary state is that of abstraction, produced by fixed gaze upon some unexciting and empty thing (for poverty of object engenders abstraction), and this abstraction is the logical premise of what follows. Abstraction tends to become more and more abstract, narrower and narrower; it tends to unity, and afterwards to nullity. There then the patient is, at the summit of attention, with no object left—a mere statue of attention—a listening, expectant life; a perfectly undistracted faculty, dreaming of a lessening and lessening mathematical point; the end of his mind sharpened away to nothing.

* *The Zoist*, a quarterly journal, in its ninth year's existence, affords the best history of the progress of mesmerism as a branch of the healing art. We refer our readers to it as an evidence of what mesmerism is good for.

What happens? Any sensation that appeals, is met by this brilliant attention, and receives its diamond glare, being perceived with a force of leisure of which our distracted life affords only the rudiments. External influences are sensated, sympathized with, to an extraordinary degree: harmonious music sways the body into graces the most affecting; discords jar it as though they would tear it limb from limb. Cold and heat are perceived with similar exaltation; so also smells and touches. In short, the whole man appears to be given to each perception. The body trembles like down with the wafts of the atmosphere; the world plays upon it as upon a spiritual instrument finely attuned.

This is the *natural* hypnotic state, but it may be modified artificially. The power of suggestions over the patient is excessive. If you say, What animal is it? the patient will tell you it is a lamb, or a rabbit, or any other. Does he see it? Yes. What animal is it *now?* putting depth and gloom into the tone of *now,* and thereby suggesting a difference. "Oh," with a shudder, "it is a wolf." What color is it? still glooming the phrase. "Black." What color is it *now?* giving the *now* a cheerful air. "Oh, a beautiful blue"—spoken with the utmost delight. And so you lead the subject through any dreams you please, by variation of questions, and of inflections of the voice; and he sees and feels all as real.

Another curious study is the influence of the patient's postures on his mind in this state. Double his fist, and put up his arm, if you dare, for you will have the strength of your ribs rudely tested. Put him on his knees and clasp his hands, and the saints and devotees of the artists will pale before the trueness of his devout actings. Raise his head while in prayer, and his lips pour forth exulting glorifications, as he sees heaven opened, and the Majesty of God raising him to his place; then in a moment depress the head, and he is dust and ashes, an unworthy sinner, with the pit of hell yawning at his feet. Or compress the forehead so as to wrinkle it vertically, and this little attitude of gloom, glooms the whole mind, and thorny-toothed clouds contract in from the very horizon; and what is remarkable, the smallest pinch and wrinkle, such as will lie between your nipping nails, is sufficient nucleus to crystalize the

34

man into that shape, and to make him all foreboding; as again the smallest expansion, in a moment, brings the opposite state, with a full breathing of delight. Raise the head next, and ask, if it be a young lady, whether she or some other is the prettier: and observe the inexpressible hauteur, and the puff sneers let off from the lips, which indicate a conclusion too certain to need utterance; depress the head, and repeat the question; and mark the self-abasement with which she now says, "She is," as hardly worthy to make the comparison. In this state, whatever posture of any passion is induced, the passion comes into it at once, and dramatizes the body accordingly.

Moreover, the patient's mind directed to his own body, does physical marvels. He can do in a manner what he thinks he can. Place a handkerchief on a table, and beg him to try to lift it, observing, however, that you know it to be impossible, and he will groan and sweat over the cambric as though it were the anchor of a man-of-war; on the other hand, tell him that a fifty-six pound weight is a light cork, to be held out at arm's length on his little finger, and he will hold it out with ease. Tell him that a tumor on his body is about to disappear, and his mind will often realize your prophecy. Of the following case Mr. B. himself informed me. A lady who was leaving off nursing from defect of milk, the baby being thirteen months old, was hypnotized by him. He made passes over her right breast to call her attention to it, and in a few moments her gestures showed that she dreamt the baby was sucking. In two minutes, the breast was distended with milk. She was awoke, and on being questioned whether any part of her frame felt differently from the rest, she perceived the state of her bosom, and mentioned it; to which Mr. B. replied that the baby would soon settle that. The infant was nearly choked with the rush of milk. In three days she came back to Mr. B. and complained that he had disfigured her, for she was protuberant on one side. He promised to take the swelling down; hypnotized her; but drew the other side also by the like means; and she nursed her child from an overflowing bosom for twenty-two months; being nine after the hypnotism.

A patient in the full state obeys all motives in the most natural direction. If the arm is placed up, there it will stay: but a waft of air will cause it to fall down. Why? Because it is already up, and the new motive changes the direction. If the arm be down, another waft will raise it. If down, and prevented from moving up, the impression will send it sideways. When the frame is erect, a touch behind in the bend of the knees, will send it into genuflexion, which will at once suggest prayer, as noticed before, &c.

An interesting question arises: In what does hypnotism differ from mesmerism, and from common sleep? Are the two former identical? We think not. We recognize the three as the main trine of sleep, each depending on its first principle. The atom of sleep is diffusion: the mind and body are dissolved in unconsciousness; they go off into nothing through the fine powder of infinite variety, and die of no attention; common sleep is impersonal. The unit of hypnotism is intense attention, abstraction, the personal *ego* pushed to nonentity. The unit of mesmerism is the common state of the patient, caught as he stands, and subjected to the radiant ideas of another person: it is mediate; or both personal and impersonal. Neither sleep nor hypnotism can exist in the presence of a second person, without partaking more or less of mesmerism. The sleep-brain is fluid, the hypnotic brain movable-pointed, and the mesmeric brain elastic. Sleep=influx; hypnotism=efflux; mesmerism=afflux.

Patients can produce the hypnotic effects upon themselves without any second party; although a second will oftentimes strengthen the result by his acts or presence; just as one who stood by, and told you that you were to succeed in a certain work, would nerve your arm with fresh confidence.

We presume it is evident to the reader, what a power Mr. Braid has methodized and called into play for the treatment of diseases. As a curative agent, hypnotism contains two elements—each valuable of its kind: 1. Where it produces trance, it has the benefits of the mesmeric sleep, or furnishes so strong a dose of rest, that many cases are cured by that alone. But 2. The suggestion of ideas of health, tone, duty, hope, which produce dreams influential upon the organization, enables the operator by this means to fulfil the indication of directly ministering to that mind diseased, which always accompanies and aggravates physical disorders. We have a direct proof of the continuation of the mind through the body, in the way in which suggestions directed to the mind respecting the organs, operate upon the latter. By touching the abdomen over the colon, and suggesting the effect, we can, in susceptible persons, produce the results of aperient medicines, and abolish constipation for years. This order of facts has an important bearing upon the origin as well as cure of disease, rendering it probable that a large num- of ills come directly out of the patient's mind; for if alteration of fancy heals, this suggests that fancy first engenders, the complaint. Viewed in this light, many physical changes of structure may be regarded as organic insanities or spells, which only require the right word from the physician to dissipate them; and although, if they went on, they would kill by their virulence, yet they are curable by a simple impression from without. Next to the self-control of the patient, which is the top of the mortal medicines, we may justly reckon this control of the doctor, who makes use of his own health and knowledge to give faith in the moments when it can be received.*

* Mr. Braid's process consists in fixing the eyes, or rather the attention, which itself fixes the eyes. After the patient has gazed steadily for one or two minutes, the lids are closed by the fingers of the operator, should they not previously close by themselves. This is to prevent straining, which might produce headache, or cerebral disturbance. In case the sleep is not produced, Mr. B. still keeps the attention awake by bending the patient's arm and setting it upright, and also by putting out the legs at right angles to the body. The effort to maintain these positions energizes the will, and fixes the patient's mind upon the operation. Where volition is weakened, this is sometimes an excellent means of again bringing it into the organism. In cases of paralysis, patients can frequently move in this state where they have no power in the ordinary one. It is known that the emotions will move limbs that are palsied, although the will has no effect upon them; and that afterwards the will frequently retains the powers thus conferred. In the hypnotic state, the operator can play upon the emotions by a variety of suggestive means; and in this way give power to impotent parts, and hand them over to the will. Mr. Braid's devices for these ends stamp him as a man of inventive genius, and we are surprised that such a piece of combined intellectual and scientific sagacity as hypnotism, has not placed him long ago in the first rank of metropolitan physicians.

The titles of Mr. Braid's works are as follow:—

Neurypnology, or the Rationale of Nervous Sleep. London, 1843.

The Power of the Mind over the Body. London, 1846.

Observations on Trance, or Human Hybernation. London, 1850.

Shall we then say that here we have a stream from the ancients' arts of phrenopathy or *pistopathy*,* and a dawning recognition of the spirituality of disease and of cure? I do not think that loss of faith and the other inward graces is the tap root of bodily sickness; and that fears, apathies, hatreds, and self-seekings are the sowers who go forth to sow poison through our frames. On the other hand, the renewal of faith,† even when impressed from without,

* Πιστις, *faith.*

† With respect to the operation of medicines upon and through the imagination, and to the operation of charms, this belongs to the personal side of the art of healing, and medicine cannot relinquish it without quitting its hold upon some of the principal departments of man. When incantation ceased, and was exploded by newer thoughts, its principle should have been carefully investigated, and the virtue and goodness, if any, which lay at the core, should have been retained. But in acting upon the imagination, one objection may be noticed, which has a large, perhaps the principal share in banishing this kind of hygiene from practice; and yet, which rightly considered, is no objection at all. The patients, so the doctors tell us, no longer believe in imagination-treatment; and because they do not believe, the treatment will have no efficacy. We reply, it makes little matter whether they believe or not. And the reason is clear. For belief is two-fold: first, educational or scientific, but second, fundamental or organic. To illustrate my meaning, I will relate an anecdote which happened to myself with a friend who was a person of strong faculties, and deeply imbued with current modes of philosophizing. We were discussing the reality of ghosts and apparitions, and my friend argued with honest stoutness that they were flimsy illusions, and had no claim on the attention of strong-minded people. But being as candid as he was strong, he presently added: "I don't know how it is, but when I talk of such things, I feel a cold stream down the back." Here was a case in which the same good man was plaintiff in one part of his body, and defendant in another: he denied ghosts with his lips and believed them with his back-bone: and fortified as he was by such arguments negative as could be had, he was still, by witness of the "cold stream," as capable of receiving a dose of fright as if he had been "a sick girl." His body, which is substantial, credited more than his soul: Balaam's ass saw the spirits which Balaam could not see. This is merely an illustration of the existence of a corporeal faith, when the mental faith is deficient; and of the efficacy of the corporeal faith in producing bodily alterations. Thus it is that charms, personal suggestions, mesmerism, minute doses of medicine, &c. &c., act upon the conscious man without the active concurrrence of his reason, nay, often in spite of it. For the body itself knows bodily what is true and influential, whether the mind knows it mentally or not. We have therefore every ground, despite the absence of faith in the higher sense, to persevere in the remedies of faith, which will not cease to cure though we be unfaithful. Thus it is well known that warts can be charmed away where their owner has no faith in the process; and for this reason; that though he has no educational belief in the charm, his imag-

handles the organism as the will handles the muscles, and, if we may use the expression, converts and Christianizes the body, that is to say, heals it. The virtue of hypnotism, where it succeeds, is just this, that for the moment it unweeds the human soil so completely, that whatever faith is impressed, can work and grow. It surely points to human agency all through disease, when we find that monomanias can be given, or removed, in a moment, by the suggestion of another from without; it points to a scientific theory of the influx of ideas from other men, visible and invisible, as an account of the outward supplies of life.

We regard hypnotism as the most intellectual phenomenon which has yet been produced by the *phrenopathists*, as suggestive of a new *personnel* in the bearing of the physician. It seems to show that there is scarcely a case in which the latter should not do something actual besides the administration of drugs; for the sick organism expects to be handled. It also proclaims to us what an artistry should be cultivated by those who practice medicine; what tact should electrify their fingers, what resolve should vertebrate their words, what cordials should drop from their mouths, what airs of reassurance should surround them, and how ease and cheerfulness should radiate from their presence, as they move from bed to bed. To simplify all this to them, they must verily believe that medicine is the daughter of heaven, and that they live to be inspired, and to inspire.

We do not mean that the character of the good physician should be the corollary of any juggle, however useful for a time, or that he

ination, as a body and a structure, is capable of being influenced by it, and of operating its own lordly removals and avaunts upon the warts: for in imagination, what you seem to have, you have; and the body is, among other things, full of imagination. Nevertheless, where the two faiths coincide, the effect will be more rapid; and they are always tending to coincide, either by the conversion of the mind to the catholicity of the body, or of the body to the principles of the mind. A good instance of this is seen in Nicolai's ghost-case, where, in spite of seeing and hearing, Nicolai, being a considerable philosopher, discredited his ghosts, and cured them by the application of leeches: his body became as philosophical as his mind, as much bent upon seeing nothing, and was emptied of its spiritual instincts. In this case Balaam and his beast were both orthodoxly one-sighted, and trudged the common high-road of the bookselling business, each as wise as the other.

should cure ultimately by impressing imaginations upon his flock. We only use these as signs pointing to a truth. What the physician should be, I dare not attempt to sketch. But I see that already he is called out of the ranks as the most humane man of his time. I see that he wants the largest faith, in addition to the largest science; gentleness and sternness also moulded together, as the lamb with the lion. Nor can heroism using all the rest as a resource, be dispensed with, to the very brink of death; for while there is life there is hope. Perhaps without attempting more, we may sum him up in saying that he should be the model of the health of the age.

We use the pronoun "he," leaving it to progress to say, whether our own sex has exclusive rights in the healing art. We cannot settle that question; but at all events we know, that the better half of the health of the world depends upon the partners of our toil. It is clear to us, that in no long time, the various organic systems of medicine, (many of which have still to appear, being *due* to the organs of the frame,) will necessitate the constitution of a tribunal for distributing cases under their proper treatments. Perhaps it may then be found that there is a female side to this as well as to the other arts. Certainly many of the qualities of the physician seem to belong pre-eminently to woman. Instant presence of mind, fine tact, observation quick and subtle, instinctive promptings often surpassing science, might, it seems, sit at the bedside in a female form. Or is it that the nurse-function is to be so far educated and elevated, as at length, without trenching upon professional titles, to touch the heights of the physician's skill? The urn of events must speak for us, for here we are at fault.*

But after all our systems of health, public and private, many in number as they are, though not enough, there is one means remaining, which we should be guilty of much base terror, as well as historical neglect, if we did not dare to bring forth. In all the branches of the New Medicine, we have seen the united principle of faith and works assuming an additional importance, as we have risen from the

* In the meantime our brave friend, Dr. Elizabeth Blackwell, has run the gauntlet of the world's prejudices, and taken perhaps the first medical degree bestowed upon a woman. May success go with her in her attempt to open up a fresh avenue of occupations to her sex!

administration of drugs stage by stage to the phrenopathic art. In the means to which we have alluded, and which is linked with our common faith, this principle becomes all in all. Of course we allude to the healing powers exerted by Christ and His apostles, and by Him bequeathed to the race of man. As we read the Gospels, we see how the Divine Man was also the great Physician; how he went about healing all manner of sickness and all manner of disease among the people; and how as many as touched the hem of his garments were made whole, every one. He also commanded his followers to do the like, and founded cure as the grand evidence of the Christian religion. His proofs of his mission were sound bodies; the deaf hearing; the dumb speaking; lepers cleansed; the dead raised; those who before were blind, now they see. The channel of this was no learned science, but a simple command in His name who has all the power in heaven and on earth. Where is the lineal priesthood of this great restoration? Where are the claimants for this substantial apostolical successorship? Where are the layers on of hands who give man to himself by casting out his devils, and increase the prime wealth of the earth as the sign and seal of the advent of the kingdom of heaven? Where is the clergy to whom sickness makes its last appeal for health, when doctors have pronounced the death words, No hope? We find them among the fishermen of the first century, but not among the prelates of the nineteenth: in mean-clad Peter and Paul, James and John, but not under the lawn of any right reverend bench. Our pontiffs say that the age of miracles is past; but no New Testament ever told them so; Christianity, as we read it, was the institution of miracle as in the order of nature; and if the age of miracles is gone, it is because the age of Christianity is gone. The age of mathematics would be past, if no man cultivated them. On the other hand we aver, by all our honesty to our faith, that for every reason we can perceive, a duty is neglected here which is a main cause of irreligion and skepticism among men. As in the sciences, which are the kings of these late days, let this mode and matter of healing be fairly experimented. It belongs to the priesthood. Let them turn out into the inclemencies of society, and try their adjurations against the storm of physical evil that exasperates the nations to their core.

Let them put on the proofs of the apostolic power. Let them peril all in this great attempt. Let the weak excuse of their age of virtue being past, be exchanged for a godly resolve to bring it back again. If they fail, it will be because they are not Christian, or else because Christianity cannot bide its own proofs. If they succeed, there will be no need of missionaries any more, but mankind will sit in a right mind under them, and bless their privilege, and their Master's name. The *vis medicatrix Christi* will be the physical demonstration of the life of a Christian church.

Under all these means, co-working for good, shall not the body be redeemed, and evil begin to lose the footing that sickness gives it? By heaven's law the sick have claims which the healthy have not, and there is more joy over one man cured, that over ninety and nine who are sound. This is a test of every society—how it speeds, or how it lags, in administering to its sick. They are the weakest parts of our common body, and care and thought turn to them with longings that are the flesh of the physician's heart. And the more that are healed, the more concentrate is the love upon those who suffer still; so that at length the world's whole skill and tenderness shall surround with arts and healing tears the bed of the last sick man.

Yet we are all to die, though in time neither by the sword of war, nor by the violence of disease. The embryo passes without fear into a larger world, which is meant to be kinder than the mother's womb. The man is to be born again, with as little pain of sense and thought, into the next expansion of the spirit. Death is the angel to the irremediable. *O præclarum illum diem!* Let us set our houses in order, make our wills, and take our leave of all things every day: we shall be wanted among our fathers afar off on the morrow.

APPENDIX.

Note to p. 112.

THROUGHOUT this work we have been anxious to register the cases in which the body *expresses* the soul; for we have a conviction that such effects go deep into the nature and theory of our corporeal incarnation. If the outward universe be a manifestation of spirit, it is a fair question whether in the human sphere the realm of *expression* be not the essence of manifestation. And as expression is either self-evidently intelligible, or nothing, so it amounts when fully traced, to a complete exposition of nature in its universal principles. It is but another name for symbolism and correspondency. Hence to our mind the importance of the gestures of the body, in expounding the functions, and the ultimate structure, of the organs. For example, if the liver, as bile maker, be the wise organic anger of the frame, then the whole personality of anger, traced through every part, from the furies of the face, to the violences of the arms, and the stampings of the feet, come to be put back into the liver, and indeed into the bile, which is thus a wrath full of human significance, and thence capable of acting, with due terror-strikings, upon the society of the body. So too the manifestation or gestures of love go into the heart, and show what it and the blood are doing in their auricular privacies, &c., &c., &c. We therefore deem that there is no more important subject in psychology, than this of expression, or organic Words. And as we indicated before (p. 380), it is the plainest symptom of all that gives the clue to the expression, all whose parts follow that bell-wether as its flock.

Note to p. 166.

Trace whatever department we will, we find that ASSOCIATION is the new world of this century, whose symptoms have been long preparing; growing up, like little coral apices, the germinating points of future islands and continents. One early dot of this kind which arose from the ocean of vagueness, was the doctrine of the association of ideas, seized by the strong common sense of Hobbes, and afterwards methodized and mechanized by the genius of Hartley.

That the human mind stimulates itself into action in certain lines of ideas, and that there are trains, regiments and bands of thoughts; that pleasures and pains determine the formation of these intellectual cohorts, is a fact in which there is but little novelty; and yet when it is seized, and taken as a stand-point, it leads by these very regiments, into new kinds of operation. It gives militariness, and march, and in high cases music to the soul, preparing for conquest under strategical principles; for the discipline of thought, whether discovered as nature's, or commanded by man, has the same results upon science, as the drill of armies upon material warfare. We therefore lay it down, that the doctrine of the association of ideas, when set in motion, invades every subject with fresh force, and with the new element of breadth of attack. This is not indeed a department important in practice: we cite it rather from its smallness, as showing that the associative air has permeated even into the minute sphere of individual thought.

One step larger, and we come to the material sciences: and what has been the process here? Every victory has been gained by looking at subjects not in themselves alone, but as connected with their neighbors. How barren each thing is while it stands upon its own individuality; how its properties one by one die down, as we cut it off from the influences of the surrounding natures! The association of single things with each other in a common knowledge, is what brings them under the grasp of a particular science; which is no sooner constituted, than we feel that it too is unfruitful in itself; and that it begins to yearn, like a young maiden, with strange desires for conjunction with some other whole science. The two together are more than twice either; they are three at the very first;

they have a progeny which may rise to any numbers. At the points where they unite, they must of course fit; to fit, they must be of cup and ball construction; in other words, there must be a likeness or likingness between them, and this likeness, when studied, becomes a science of analogies. When the marriages of a dozen sciences have been observed, they all begin to group upon each other, and the thought strikes us that all the sciences fit to each, or that analogy is universal. Everything likes everything. This sends our curiosity abroad, and we come upon the thought of universal organization, and then we find that certain parts of nature, as vegetables, beasts, and men, are shining lights of organization; that flesh, for instance, loves its parts so well, and that they embrace so closely, that nothing can come between them without pain, or destruction. Thus we only get to individuality as the result of a compact of sentient beings whom God has so closely linked in liking, that they cannot be sundered without mischief. The scientific individuality thus obtained, illustrates our common individuality, and we are constrained to think is a true statement of one part of its ground.

From this point our thoughts become organic, and we are sure to make organization or association into a rule of judgment, and a method of discovery; insisting, by faith and science, that the whole world is covertly what the highest things in it are manifestly. We have found that the best and most speaking bodies are disposed in mutual order and helpfulness; we dictate downwards, from the heights of nature, that this is the case with the other parts; and that their imperfection consists in nothing else, than the fact that they do not so openly express the everlasting order. The plants are poor timid things which cannot tell us what they have in them; the beasts have better voices; and mankind are different over again, precisely according as they can display that association and organization which is the lesson of the whole. The best men indeed are still markedly individual; but why? Only because they are the general officers of regiments or armies, and not because they have no world under them. They are dependent on much larger and broader bases than other people. The apex of the pyramid has a free point, but of all the parts, it rests at last the most heavily upon the ground.

Knowledge displays the associative power in its width not less than in its height. Observe the new light that comes simply and solely by putting things together! This act constitutes the strength of modern attainments. A remarkable instance of it is presented in the law of storms, as developed by Mr. Redfield and Colonel Reid. In our incoherent days storms were thought to be gusts more capricious than our tempers are now *thought to be.* They pitched one ship on its beam ends to the south, swooped another to the east, and sucked down a third into a perpendicular grave. At length came a man, who entertained the notion, of putting the bits of a storm together; and he soon found that the pieces fitted; that one side of the cup of windy wrath looked east; others, north, west and south; and that there was a hollow in this, as in other cups. The man found practically that a whole storm was quite different from the conception of separated parts, and that it was made up of associated parts. He began to think that storms "obey regular laws." The world is now beginning to think that all things do the same; and will henceforth look down from that summit which this one hardy man, by strong efforts, gained.

So too, in physical geography—a science consisting purely of the association of our observations of the world into one globe of observations. How differently intelligible does the planet become simply by being all represented, and with its parts in harmony with themselves. And here another blessing of associative over isolated thought may be characterized; the former is *pictorial;* the latter invisible, and made only of wandering points. So soon as you get fine surface knowledge, picture is painted thereupon; the white memory then puts itself down on nature's press, and the soul pulls an impression, which is durable, nay, everlasting, like other print. The imagination is the sharp type of this very memory; and man becomes thereby an everlasting possessor of the world; he takes it with him wherever he goes; and from its plans, in the portfolios of his own nature, he can govern the old estate, let him be roaming among the fixed stars, or safely moored in his final kingdom of the heavens. We may surely deny that the isolated thought has any pretty pictures to show; has anything in it to remember, or can last through successive hours; much less, after death has thrown the lumber away forever. Not so, however, the organic and colored thought which is shelled off by every thing when once we see how much all things love each other.

And is not this to be applied to practice, and to higher thought? Will not man, when put together, be found to be the variegated landscape of a new earth? Will not his chances then first be developed? Will not his problems become parts of a pictorial human life, which in point of reality, and unquestionableness, will take rank with land and sea, with the skies, and the verdurous earth? We dare not answer the question in the negative, until the experiment of association, successful, so far, in every other sphere, has been tried in this crowning matter of human development and happiness. Observe, however, that the putting together of man, is not a mechanical, but a human or religious act.

Note to p. 378.

It seems very evident that disease has as many centres in the human organism, as there are great spheres of its powers. There are diseases which spring from the body, and others that emanate from the mind. In short, as there are hepatic, pulmonary and cerebral complaints, both idiopathic and sympathetic, so there are others of the reason, imagination, senses, equally following this two-fold distribution. It is probable that disease can only be cured by attacking it in its own centres: bodily disease by drugs and bodily means, and spiritual diseases by their appropriate administrations. Thus where a disease from the mind is sympathetically felt in the body considered as the extremity of the mind, phrenopathic means will cure it: but these can only relieve the mental symptoms, not the root of the mischief, if the malady be primary in the body itself.

THE END.

CATALOGUE
OF
VALUABLE BOOKS,
PUBLISHED BY
LIPPINCOTT, GRAMBO & CO.,
NO. 20 NORTH FOURTH STREET, PHILADELPHIA;
CONSISTING OF A LARGE ASSORTMENT OF
Bibles, Prayer-Books, Commentaries, Standard Poets,
MEDICAL, THEOLOGICAL AND MISCELLANEOUS WORKS, ETC.,
PARTICULARLY SUITABLE FOR
PUBLIC AND PRIVATE LIBRARIES.
FOR SALE BY BOOKSELLERS AND COUNTRY MERCHANTS GENERALLY THROUGHOUT THE UNITED STATES.

THE BEST & MOST COMPLETE FAMILY COMMENTARY.

The Comprehensive Commentary on the Holy Bible;
CONTAINING
THE TEXT ACCORDING TO THE AUTHORIZED VERSION,
SCOTT'S MARGINAL REFERENCES; MATTHEW HENRY'S COMMENTARY, CONDENSED, BUT RETAINING EVERY USEFUL THOUGHT; THE PRACTICAL OBSERVATIONS OF REV. THOMAS SCOTT, D. D.;
WITH EXTENSIVE
EXPLANATORY, CRITICAL AND PHILOLOGICAL NOTES,
Selected from Scott, Doddridge, Gill, Adam Clarke, Patrick, Poole, Lowth, Burder, Harmer, Calmet, Rosenmueller, Bloomfield, Stuart, Bush, Dwight, and many other writers on the Scriptures.

The whole designed to be a digest and combination of the advantages of the best Bible Commentaries, and embracing nearly all that is valuable in

HENRY, SCOTT, AND DODDRIDGE.

Conveniently arranged for family and private reading, and, at the same time, particularly adapted to the wants of Sabbath-School Teachers and Bible Classes; with numerous useful tables, and a neatly engraved Family Record.

Edited by Rev. WILLIAM JENKS, D. D.,
PASTOR OF GREEN STREET CHURCH, BOSTON.

Embellished with five portraits, and other elegant engravings, from steel plates; with several maps and many wood-cuts, illustrative of Scripture Manners, Customs, Antiquities, &c. In 6 vols. super-royal 8vo.

Including Supplement, bound in cloth, sheep, calf, &c., varying in

Price from $10 to $15.

The whole forming the most valuable as well as the cheapest Commentary published in the world.

NOTICES AND RECOMMENDATIONS
OF THE
COMPREHENSIVE COMMENTARY.

The Publishers select the following from the testimonials they have received as to the value of the work:

We, the subscribers, having examined the *Comprehensive Commentary*, issued from the press of Messrs. L., G. & Co., and highly approving its character, would cheerfully and confidently recommend it as containing more matter and more advantages than any other with which we are acquainted; and considering the expense incurred, and the excellent manner of its mechanical execution, we believe it to be one of the *cheapest* works ever issued from the press. We hope the publishers will be sustained by a liberal patronage, in their expensive and useful undertaking. We should be pleased to learn that every family in the United States had procured a copy.

B. B. WISNER, D. D., Secretary of Am. Board of Com. for For. Missions.
WM. COGSWELL, D. D., " " Education Society.
JOHN CODMAN, D. D., Pastor of Congregational Church, Dorchester.
Rev. HUBBARD WINSLOW, " " Bowdoin street, Dorchester.
Rev. SEWALL HARDING, Pastor of T. C. Church, Waltham.
Rev. J. H. FAIRCHILD, Pastor of Congregational Church, South Boston.
GARDINER SPRING, D. D., Pastor of Presbyterian Church, New York city.
CYRUS MASON, D. D., " " " " "
THOS. M'AULEY, D. D., " " " " "
JOHN WOODBRIDGE, D. D., " " " " "
THOS. DEWITT, D. D., " Dutch Ref. " " "
E. W. BALDWIN, D. D., " " " " "
Rev. J. M. M'KREBS, " Presbyterian " " "
Rev. ERSKINE MASON, " " " " "
Rev. J. S. SPENCER, " " " Brooklyn.
EZRA STILES ELY, D. D., Stated Clerk of Gen. Assem. of Presbyterian Church.
JOHN M'DOWELL, D. D., Permanent " " " "
JOHN BRECKENRIDGE, Corresponding Secretary of Assembly's Board of Education.
SAMUEL B. WYLIE, D. D., Pastor of the Reformed Presbyterian Church.
N. LORD, D. D., President of Dartmouth College.
JOSHUA BATES, D. D., President of Middlebury College.
H. HUMPHREY, D. D., " Amherst College.
E. D. GRIFFIN, D. D., " Williamstown College.
J. WHEELER, D. D., " University of Vermont, at Burlington.
J. M. MATTHEWS, D. D., " New York City University.
GEORGE E. PIERCE, D. D., " Western Reserve College, Ohio.
Rev. Dr. BROWN, " Jefferson College, Penn.
LEONARD WOODS, D. D., Professor of Theology, Andover Seminary.
THOS. H. SKINNER, D. D., " Sac. Rhet. " "
Rev. RALPH EMERSON, " Eccl. Hist. " "
Rev. JOEL PARKER, Pastor of Presbyterian Church, New Orleans.
JOEL HAWES, D. D., " Congregational Church, Hartford, Conn.
N. S. S. BEAMAN, D. D., " Presbyterian Church, Troy, N. Y.
MARK TUCKER, D. D., " " " " "
Rev. E. N. KIRK, " " " Albany, N. Y.
Rev. E. B. EDWARDS, Editor of Quarterly Observer.
Rev. STEPHEN MASON, Pastor First Congregational Church, Nantucket.
Rev. ORIN FOWLER, " " " " Fall River.
GEORGE W. BETHUNE, D. D., Pastor of the First Reformed Dutch Church, Philada.
Rev. LYMAN BEECHER, D. D., Cincinnati, Ohio.
Rev. C. D. MALLORY, Pastor Baptist Church, Augusta, Ga.
Rev. S. M. NOEL, " " " Frankfort, Ky.

From the Professors at Princeton Theological Seminary.

The Comprehensive Commentary contains the whole of Henry's Exposition in a condensed form, Scott's Practical Observations and Marginal References, and a large number of very valuable philological and critical notes, selected from various authors. The work appears to be executed with judgment, fidelity, and care; and will furnish a rich treasure of scriptural knowledge to the Biblical student, and to the teachers of Sabbath-Schools and Bible Classes.

A. ALEXANDER, D. D.
SAMUEL MILLER, D. D.
CHARLES HODGE, D. D

The Companion to the Bible.

In one super-royal volume.

DESIGNED TO ACCOMPANY

THE FAMILY BIBLE,

OR HENRY'S, SCOTT'S, CLARKE'S, GILL'S, OR OTHER COMMENTARIES:

CONTAINING

1. A new, full, and complete Concordance;

Illustrated with monumental, traditional, and oriental engravings, founded on Butterworth's, with Cruden's definitions; forming, it is believed, on many accounts, a more valuable work than either Butterworth, Cruden, or any other similar book in the language.

The value of a Concordance is now generally understood; and those who have used one, consider it indispensable in connection with the Bible.

2. A Guide to the Reading and Study of the Bible;

being Carpenter's valuable Biblical Companion, lately published in London, containing a complete history of the Bible, and forming a most excellent introduction to its study. It embraces the evidences of Christianity, Jewish antiquities, manners, customs, arts, natural history, &c., of the Bible, with notes and engravings added.

3. Complete Biographies of Henry, by Williams; Scott, by his son; Doddridge, by Orton;

with sketches of the lives and characters, and notices of the works, of the writers on the Scriptures who are quoted in the Commentary, living and dead, American and foreign.

This part of the volume not only affords a large quantity of interesting and useful reading for pious families, but will also be a source of gratification to all those who are in the habit of consulting the Commentary; every one naturally feeling a desire to know some particulars of the lives and characters of those whose opinions he seeks. Appended to this part, will be a

BIBLIOTHECA BIBLICA,

or list of the best works on the Bible, of all kinds, arranged under their appropriate heads.

4. A complete Index of the Matter contained in the Bible Text.

5. A Symbolical Dictionary.

A very comprehensive and valuable Dictionary of Scripture Symbols, (occupying about *fifty-six* closely printed pages,) by Thomas Wemyss, (author of "Biblical Gleanings," &c.) Comprising Daubuz, Lancaster, Hutcheson, &c.

6. The Work contains several other Articles,

Indexes, Tables, &c. &c., and is,

7. Illustrated by a large Plan of Jerusalem,

identifying, as far as tradition, &c., go, the original sites, drawn on the spot by F. Catherwood, of London, architect. Also, two steel engravings of portraits of seven foreign and eight American theological writers, and numerous wood engravings.

The whole forms a desirable and necessary fund of instruction for the use not only of clergymen and Sabbath-school teachers, but also for families. When the great amount of matter it must contain is considered, it will be deemed exceedingly cheap.

"I have examined 'The Companion to the Bible,' and have been surprised to find so much information introduced into a volume of so moderate a size. It contains a library of sacred knowledge and criticism. It will be useful to ministers who own large libraries, and cannot fail to be an invaluable help to every reader of the Bible."

HENRY MORRIS,
Pastor of Congregational Church, Vermont.

The above work can be had in several styles of binding. Price varying from $1 75 to $5 00.

ILLUSTRATIONS OF THE HOLY SCRIPTURES,

In one super-royal volume.

DERIVED PRINCIPALLY FROM THE MANNERS, CUSTOMS, ANTIQUITIES, TRADITIONS, AND FORMS OF SPEECH, RITES, CLIMATE, WORKS OF ART, AND LITERATURE OF THE EASTERN NATIONS:

EMBODYING ALL THAT IS VALUABLE IN THE WORKS OF

ROBERTS, HARMER, BURDER, PAXTON, CHANDLER,

And the most celebrated oriental travellers. Embracing also the subject of the Fulfilment of Prophecy, as exhibited by Keith and others; with descriptions of the present state of countries and places mentioned in the Sacred Writings.

ILLUSTRATED BY NUMEROUS LANDSCAPE ENGRAVINGS,

FROM SKETCHES TAKEN ON THE SPOT.

Edited by Rev. George Bush,

Professor of Hebrew and Oriental Literature in the New York City University.

The importance of this work must be obvious, and, being altogether *illustrative*, without reference to doctrines, or other points in which Christians differ, it is hoped it will meet with favour from all who love the sacred volume, and that it will be sufficiently interesting and attractive to recommend itself, not only to professed Christians of *all* denominations, but also to the general reader. The arrangement of the texts illustrated with the notes, in the order of the chapters and verses of the authorized version of the Bible, will render it convenient for reference to particular passages; while the *copious Index* at the end will at once enable the reader to turn to every subject discussed in the volume.

This volume is not designed to take the place of Commentaries, but is a distinct department of biblical instruction, and may be used as a companion to the Comprehensive or any other Commentary, or the Holy Bible.

THE ENGRAVINGS

in this volume, it is believed, will form no small part of its attractions. No pains have been spared to procure such as should embellish the work, and, at the same time, illustrate the text. Objections that have been made to the pictures commonly introduced into the Bible, as being mere creations of fancy and the imagination, often unlike nature, and frequently conveying false impressions, cannot be urged against the pictorial illustrations of this volume. Here the fine arts are made subservient to utility, the landscape views being, without an exception, *matter-of-fact views of places mentioned in Scripture, as they appear at the present day*; thus in many instances exhibiting, in the most forcible manner, *to the eye*, the strict and *literal* fulfilment of the remarkable prophecies; "the present ruined and desolate condition of the cities of Babylon, Nineveh, Selah, &c., and the countries of Edom and Egypt, are astonishing examples, and so completely exemplify, in the most minute particulars, every thing which was foretold of them in the height of their prosperity, that no better description can now be given of them than a simple quotation from a chapter and verse of the Bible written nearly two or three thousand years ago." The publishers are enabled to select from several collections lately published in London, the proprietor of one of which says that "several distinguished travellers have afforded him the use of nearly *Three Hundred Original Sketches*" of Scripture places, made upon the spot. "The land of Palestine, it is well known, abounds in scenes of the most picturesque beauty. Syria comprehends the snowy heights of Lebanon, and the majestic ruins of Tadmor and Baalbec."

The above work can be had in various styles of binding.

Price from $1 50 to $5 00.

THE ILLUSTRATED CONCORDANCE,

In one volume, royal 8vo.

A new, full, and complete Concordance; illustrated with monumental, traditional, and oriental engravings, founded on Butterworth's, with Cruden's definitions; forming, it is believed, on many accounts, a more valuable work than either Butterworth, Cruden, or any other similar book in the language.

The value of a Concordance is now generally understood; and those who have used one, consider it indispensable in connection with the Bible. Some of the many advantages the Illustrated Concordance has over all the others, are, that it contains near two hundred appropriate engravings: it is printed on fine white paper, with beautiful large type.

Price One Dollar.

LIPPINCOTT'S EDITIONS OF

THE HOLY BIBLE.

SIX DIFFERENT SIZES.

Printed in the best manner, with beautiful type, on the finest sized paper, and bound in the most splendid and substantial styles. Warranted to be correct, and equal to the best English editions, at much less price. To be had with or without plates; the publishers having supplied themselves with over fifty steel engravings, by the first artists.

Baxter's Comprehensive Bible,

Royal quarto, containing the various readings and marginal notes; disquisitions on the genuineness, authenticity, and inspiration of the Holy Scriptures; introductory and concluding remarks to each book; philological and explanatory notes; table of contents, arranged in historical order; a chronological index, and various other matter; forming a suitable book for the study of clergymen, Sabbath-school teachers, and students.

In neat plain binding, from $4 00 to $5 00.—In Turkey morocco, extra, gilt edges, from $8 00 to $12 00.—In do., with splendid plates, $10 00 to $15 00.—In do., bevelled side, gilt clasps and illuminations, $15 00 to $25 00.

The Oxford Quarto Bible,

Without note or comment, universally admitted to be the most beautiful Bible extant.

In neat plain binding, from $4 00 to $5 00.—In Turkey morocco, extra, gilt edges, $8 00 to $12 00.—In do., with steel engravings, $10 00 to $15 00.—In do., clasps, &c., with plates and illuminations, $15 00 to $25 00.—In rich velvet, with gilt ornaments, $25 00 to $50 00.

Crown Octavo Bible,

Printed with large clear type, making a most convenient hand Bible for family use.

In neat plain binding, from 75 cents to $1 50.—In English Turkey morocco, gilt edges, $1 00 to $2 00.—In do., imitation, &c., $1 50 to $3 00.—In do., clasps, &c., $2 50 to $5 00.—In rich velvet, with gilt ornaments, $5 00 to $10 00.

The Sunday-School Teacher's Polyglot Bible, with Maps, &c.,

In neat plain binding, from 60 cents to $1 00.—In imitation gilt edge, $1 00 to $1 50.—In Turkey, super extra, $1 75 to $2 25.—In do. do., with clasps, $2 50 to $3 75.—In velvet, rich gilt ornaments, $3 50 to $8 00.

The Oxford 18mo., or Pew Bible,

In neat plain binding, from 50 cents to $1 00.—In imitation gilt edge, $1 00 to $1 50.—In Turkey super extra, $1 75 to $2 25.—In do. do., with clasps, $2 50 to $3 75.—In velvet, rich gilt ornaments, $3 50 to $8 00.

Agate 32mo. Bible,

Printed with larger type than any other small or pocket edition extant.

In neat plain binding, from 50 cents to $1 00.—In tucks, or pocket-book style, 75 cents to $1 00.—In roan, imitation gilt edge, $1 00 to $1 50.—In Turkey, super extra, $1 00 to $2 00.—In do. do gilt clasps, $2 50 to $3 50.—In velvet, with rich gilt ornaments, $3 00 to $7 00.

32mo. Diamond Pocket Bible;

The neatest, smallest, and cheapest edition of the Bible published.

In neat plain binding, from 30 to 50 cents.—In tucks, or pocket-book style, 60 cents to $1 00.—In roan, imitation gilt edge, 75 cents to $1 25.—In Turkey, super extra, $1 00 to $1 50.—In do. do gilt clasps, $1 50 to $2 00.—In velvet, with richly gilt ornaments, $2 50 to $6 00.

CONSTANTLY ON HAND,

A large assortment of BIBLES, bound in the most splendid and costly styles, with gold and silver ornaments, suitable for presentation; ranging in price from $10 00 to $100 00.

A liberal discount made to Booksellers and Agents by the Publishers.

ENCYCLOPÆDIA OF RELIGIOUS KNOWLEDGE;

OR, DICTIONARY OF THE BIBLE, THEOLOGY, RELIGIOUS BIOGRAPHY, ALL RELIGIONS, ECCLESIASTICAL HISTORY, AND MISSIONS.

Designed as a complete Book of Reference on all Religious Subjects, and Companion to the Bible; forming a cheap and compact Library of Religious Knowledge. Edited by Rev. J. Newton Brown. Illustrated by wood-cuts, maps, and engravings on copper and steel. In one volume, royal 8vo. Price, $4 00.

SPLENDID LIBRARY EDITIONS.

ILLUSTRATED STANDARD POETS.

ELEGANTLY PRINTED, ON FINE PAPER, AND UNIFORM IN SIZE AND STYLE.

The following Editions of Standard British Poets are illustrated with numerous Steel Engravings, and may be had in all varieties of binding.

BYRON'S WORKS.

COMPLETE IN ONE VOLUME, OCTAVO.

INCLUDING ALL HIS SUPPRESSED AND ATTRIBUTED POEMS; WITH SIX BEAUTIFUL ENGRAVINGS.

This edition has been carefully compared with the recent London edition of Mr. Murray, and made complete by the addition of more than fifty pages of poems heretofore unpublished in England. Among these there are a number that have never appeared in any American edition; and the publishers believe they are warranted in saying that this is *the most complete edition of Lord Byron's Poetical Works* ever published in the United States.

The Poetical Works of Mrs. Hemans.

Complete in one volume, octavo; with seven beautiful Engravings.

This is a new and complete edition, with a splendid engraved likeness of Mrs. Hemans, on steel, and contains all the Poems in the last London and American editions. With a Critical Preface by Mr. Thatcher, of Boston.

"As no work in the English language can be commended with more confidence, it will argue bad taste in a female in this country to be without a complete edition of the writings of one who was an honour to her sex and to humanity, and whose productions, from first to last, contain no syllable calculated to call a blush to the cheek of modesty and virtue. There is, moreover, in Mrs. Hemans's poetry, a moral purity and a religious feeling which commend it, in an especial manner, to the discriminating reader. No parent or guardian will be under the necessity of imposing restrictions with regard to the free perusal of every production emanating from this gifted woman. There breathes throughout the whole a most eminent exemption from impropriety of thought or diction; and there is at times a pensiveness of tone, a winning sadness in her more serious compositions, which tells of a soul which has been lifted from the contemplation of terrestrial things, to divine communings with beings of a purer world."

MILTON, YOUNG, GRAY, BEATTIE, AND COLLINS'S POETICAL WORKS.

COMPLETE IN ONE VOLUME, OCTAVO.

WITH SIX BEAUTIFUL ENGRAVINGS.

Cowper and Thomson's Prose and Poetical Works.

COMPLETE IN ONE VOLUME, OCTAVO.

Including two hundred and fifty Letters, and sundry Poems of Cowper, never before published in this country; and of Thomson a new and interesting Memoir, and upwards of twenty new Poems, for the first time printed from his own Manuscripts, taken from a late Edition of the Aldine Poets, now publishing in London.

WITH SEVEN BEAUTIFUL ENGRAVINGS.

The distinguished Professor Silliman, speaking of this edition, observes: "I am as much gratified by the elegance and fine taste of your edition, as by the noble tribute of genius and moral excellence which these delightful authors have left for all future generations; and Cowper, especially, is not less conspicuous as a true Christian, moralist and teacher, than as a poet of great power and exquisite taste."

THE POETICAL WORKS OF ROGERS, CAMPBELL, MONTGOMERY, LAMB, AND KIRKE WHITE.

COMPLETE IN ONE VOLUME, OCTAVO.

WITH SIX BEAUTIFUL ENGRAVINGS.

The beauty, correctness, and convenience of this favourite edition of these standard authors are so well known, that it is scarcely necessary to add a word in its favour. It is only necessary to say, that the publishers have now issued an illustrated edition, which greatly enhances its former value. The engravings are excellent and well selected. It is the best library edition extant.

CRABBE, HEBER, AND POLLOK'S POETICAL WORKS.

COMPLETE IN ONE VOLUME, OCTAVO.

WITH SIX BEAUTIFUL ENGRAVINGS.

A writer in the Boston Traveller holds the following language with reference to these valuable editions:—

"Mr. Editor:—I wish, without any idea of puffing, to say a word or two upon the 'Library of English Poets' that is now published at Philadelphia, by Lippincott, Grambo & Co. It is certainly, taking into consideration the elegant manner in which it is printed, and the reasonable price at which it is afforded to purchasers, the best edition of the modern British Poets that has ever been published in this country. Each volume is an octavo of about 500 pages, double columns, stereotyped, and accompanied with fine engravings and biographical sketches; and most of them are reprinted from Galignani's French edition. As to its value, we need only mention that it contains the entire works of Montgomery, Gray, Beattie, Collins, Byron, Cowper, Thomson, Milton, Young, Rogers, Campbell, Lamb, Hemans, Heber, Kirke White, Crabbe, the Miscellaneous Works of Goldsmith, and other masters of the lyre. The publishers are doing a great service by their publication, and their volumes are almost in as great demand as the fashionable novels of the day; and they deserve to be so: for they are certainly printed in a style superior to that in which we have before had the works of the English Poets."

No library can be considered complete without a copy of the above beautiful and cheap editions of the English Poets; and persons ordering all or any of them, will please say Lippincott, Grambo & Co.'s illustrated editions.

A COMPLETE

Dictionary of Poetical Quotations:

COMPRISING THE MOST EXCELLENT AND APPROPRIATE PASSAGES IN THE OLD BRITISH POETS; WITH CHOICE AND COPIOUS SELECTIONS FROM THE BEST MODERN BRITISH AND AMERICAN POETS.

EDITED BY SARAH JOSEPHA HALE.

As nightingales do upon glow-worms feed,
So poets live upon the living light
Of Nature and of Beauty.

Bailey's Festus.

Beautifully illustrated with Engravings. In one super-royal octavo volume, in various bindings.

The publishers extract, from the many highly complimentary notices of the above valuable and beautiful work, the following:

"We have at last a volume of Poetical Quotations worthy of the name. It contains nearly six hundred octavo pages, carefully and tastefully selected from all the home and foreign authors of celebrity. It is invaluable to a writer, while to the ordinary reader it presents every subject at a glance."—*Godey's Lady's Book.*

"The plan or idea of Mrs. Hale's work is felicitous. It is one for which her fine taste, her orderly habits of mind, and her long occupation with literature, has given her peculiar facilities; and thoroughly has she accomplished her task in the work before us."—*Sartain's Magazine.*

"It is a choice collection of poetical extracts from every English and American author worth perusing, from the days of Chaucer to the present time."—*Washington Union.*

"There is nothing negative about this work; it is *positively* good."—*Evening Bulletin.*

THE DIAMOND EDITION OF BYRON.

THE POETICAL WORKS OF LORD BYRON,

WITH A SKETCH OF HIS LIFE.

COMPLETE IN ONE NEAT DUODECIMO VOLUME, WITH STEEL PLATES.

The type of this edition is so perfect, and it is printed with so much care, on fine white paper, that it can be read with as much ease as most of the larger editions. This work is to be had in plain and superb binding, making a beautiful volume for a gift.

"*The Poetical Works of Lord Byron*, complete in one volume; published by L., G. & Co., Philadelphia. We hazard nothing in saying that, take it altogether, this is the most elegant work ever issued from the American press.

"'In a single volume, not larger than an ordinary duodecimo, the publishers have embraced the whole of Lord Byron's Poems, usually printed in ten or twelve volumes; and, what is more remarkable, have done it with a type so clear and distinct, that, notwithstanding its necessarily small size, it may be read with the utmost facility, even by failing eyes. The book is stereotyped; and never have we seen a finer specimen of that art. Everything about it is perfect—the paper, the printing, the binding, all correspond with each other; and it is embellished with two fine engravings, well worthy the companionship in which they are placed.

"'This will make a beautiful Christmas present.'

"We extract the above from Godey's Lady's Book. The notice itself, we are given to understand, is written by Mrs. Hale.

"We have to add our commendation in favour of this beautiful volume, a copy of which has been sent us by the publishers. The admirers of the noble bard will feel obliged to the enterprise which has prompted the publishers to dare a competition with the numerous editions of his works already in circulation; and we shall be surprised if this convenient travelling edition does not in a great degree supersede the use of the large octavo works, which have little advantage in size and openness of type, and are much inferior in the qualities of portability and lightness." — *Intelligencer.*

THE DIAMOND EDITION OF MOORE.

(CORRESPONDING WITH BYRON.)

THE POETICAL WORKS OF THOMAS MOORE,

COLLECTED BY HIMSELF.

COMPLETE IN ONE VOLUME.

This work is published uniform with Byron, from the last London edition, and is the most complete printed in the country.

THE DIAMOND EDITION OF SHAKSPEARE,

(COMPLETE IN ONE VOLUME,)

INCLUDING A SKETCH OF HIS LIFE.

UNIFORM WITH BYRON AND MOORE.

THE ABOVE WORKS CAN BE HAD IN SEVERAL VARIETIES OF BINDING.

GOLDSMITH'S ANIMATED NATURE.

IN TWO VOLUMES, OCTAVO.

BEAUTIFULLY ILLUSTRATED WITH 385 PLATES.

CONTAINING A HISTORY OF THE EARTH, ANIMALS, BIRDS, AND FISHES; FORMING THE MOST COMPLETE NATURAL HISTORY EVER PUBLISHED.

This is a work that should be in the library of every family, having been written by one of the most talented authors in the English language.

"Goldsmith can never be made obsolete while delicate genius, exquisite feeling, fine invention, the most harmonious metre, and the happiest diction, are at all valued."

BIGLAND'S NATURAL HISTORY

Of Animals, Birds, Fishes, Reptiles, and Insects. Illustrated with numerous and beautiful Engravings. By JOHN BIGLAND, author of a "View of the World," "Letters on Universal History," &c. Complete in 1 vol., 12mo.

THE POWER AND PROGRESS OF THE UNITED STATES.

THE UNITED STATES; Its Power and Progress.

BY GUILLAUME TELL POUSSIN,

LATE MINISTER OF THE REPUBLIC OF FRANCE TO THE UNITED STATES.

FIRST AMERICAN, FROM THE THIRD PARIS EDITION.

TRANSLATED FROM THE FRENCH BY EDMOND L. DU BARRY, M. D.,

SURGEON U. S. NAVY.

In one large octavo volume.

SCHOOLCRAFT'S GREAT NATIONAL WORK ON THE INDIAN TRIBES OF THE UNITED STATES,

WITH BEAUTIFUL AND ACCURATE COLOURED ILLUSTRATIONS.

HISTORICAL AND STATISTICAL INFORMATION

RESPECTING THE

HISTORY, CONDITION AND PROSPECTS

OF THE

Indian Tribes of the United States.

COLLECTED AND PREPARED UNDER THE DIRECTION OF THE BUREAU OF INDIAN AFFAIRS, PER ACT OF MARCH 3, 1847,

BY HENRY R. SCHOOLCRAFT, LL.D.

ILLUSTRATED BY S. EASTMAN, CAPT. U. S. A.

PUBLISHED BY AUTHORITY OF CONGRESS.

THE AMERICAN GARDENER'S CALENDAR,

ADAPTED TO THE CLIMATE AND SEASONS OF THE UNITED STATES.

Containing a complete account of all the work necessary to be done in the Kitchen Garden, Fruit Garden, Orchard, Vineyard, Nursery, Pleasure-Ground, Flower Garden, Green-house, Hot-house, and Forcing Frames, for every month in the year; with ample Practical Directions for performing the same.

Also, general as well as minute instructions for laying out or erecting each and every of the above departments, according to modern taste and the most approved plans; the Ornamental Planting of Pleasure Grounds, in the ancient and modern style; the cultivation of Thorn Quicks, and other plants suitable for Live Hedges, with the best methods of making them, &c. To which are annexed catalogues of Kitchen Garden Plants and Herbs; Aromatic, Pot, and Sweet Herbs; Medicinal Plants, and the most important Grapes, &c., used in rural economy; with the soil best adapted to their cultivation. Together with a copious Index to the body of the work.

BY BERNARD M'MAHON.

Tenth Edition, greatly improved. In one volume, octavo.

THE USEFUL AND THE BEAUTIFUL;

OR, DOMESTIC AND MORAL DUTIES NECESSARY TO SOCIAL HAPPINESS.

BEAUTIFULLY ILLUSTRATED.

16mo. square cloth. Price 50 and 75 cents.

www.ingramcontent.com/pod-product-compliance
Lightning Source LLC
LaVergne TN
LVHW011149110826
845150LV00006B/1198

* 9 7 8 1 4 2 5 5 4 6 6 1 8 *